Study Guide   Volume 1   Chapters 1–21

Study Guide

Volume 1   Chapters 1–21

to accompany **Paul A. Tipler**

Physics

for Scientists and Engineers

Fourth Edition

Gene Mosca
United States Naval Academy

Granvil C. Kyker, Jr.
Rose-Hulman Institute of Technology

Ronald Gautreau
New Jersey Institute of Technology

 W. H. FREEMAN AND COMPANY/WORTH PUBLISHERS

Study Guide, Volume 1, Chapters 1–21
by Gene Mosca, Granvil C. Kyker, Jr., and Ronald Gautreau
to accompany
Tipler: *Physics for Scientists and Engineers*, Fourth Edition

Printed in the United States of America

ISBN: 1-57259-511-6

Printing: 1 2 3 4 5—03 02 01 00 99

Cover image by Max Aguilera-Hellweg.

W. H. Freeman and Company
41 Madison Avenue
New York, New York 10010
http://www.whfreeman.com

Contents

Part I   Mechanics

Part II   Oscillations and Waves

Part III   Thermodynamics

To the Student

This Study Guide was written to help you master Chapters 1 to 21 of Paul Tipler's *Physics for Scientists and Engineers,* Fourth Edition. Each chapter of the Study Guide contains the following sections:

I.   **Key Ideas**   A brief overview of the important concepts presented in the chapter.

II. **Numbers and Key Equations**   A list of the constants, units, and basic equations introduced in the chapter.

III. **Potential Pitfalls**   Warnings about mistakes that are commonly made.

IV. **True and False and Responses**   Statements to test whether or not you understand essential definitions and relations. All the false statements are followed by explanations of why they are false. In addition, many of the true statements are followed by explanations of why they are true.

V.   **Questions and Answers**   Questions that require mostly qualitative reasoning. A complete answer is provided for each question so that you can compare it with your own.

VI. **Problems and Solutions**   With few exceptions, the problems come in pairs; the first of the pair is followed by a detailed solution to help you develop a model, which you can then implement in the second problem. One or more hints accompany each problem. These problems will help you build your understanding of the physical concepts and your ability to apply what you have learned to physical situations.

What Is the Best Way to Study Physics?

Of course there isn't a single answer to that one. It's clear, however, that you should begin early in the course to develop the methods that work best for you. The important thing is to find the system that is most comfortable and effective for you, and then stick to it.

In this course you will be introduced to numerous concepts. It is important that you take the time to be sure you understand each of them. You will have mastered a concept when you fully understand its relationships with other concepts. Some concepts will *seem* to contradict other concepts or even your observations of the physical world. Many of the True and False statements and the Questions in this Study Guide are intended to test your understanding of concepts. If you find that your understanding of an idea is incomplete, don't give up; pursue it until it becomes clear. I recommend that you keep a list of the things that you come across in your studies that you do not understand. Then, when you come to understand an idea, remove it from your list. After you complete your study of each chapter, bring your list to your most important resource, your physics instructor, and ask for assistance. If you go to your instructor with a few well-defined questions, you will very likely be able to remove any remaining items from your list.

Like the example problems presented in the textbook, the problem solutions presented in this Study Guide *start with basic concepts*, not with formulas. I encourage you to follow this practice. Physics is a collection of interrelated basic concepts, not a seemingly infinite list of disconnected, highly specific formulas. Don't try to memorize long lists of specific formulas, and then use these formulas as the starting point for solving problems. Instead, focus on the concepts first and be sure that you understand the ideas before you apply the formulas.

Probably the most rewarding (but challenging) aspect of studying physics is learning how to apply the fundamental concepts to specific problems. At some point you are likely to think, "I understand the theory, but I just can't do the problems." If you can't do the problems, however, you probably don't understand the theory. Until the physical concepts and the mathematical equations become your tools to apply at will to specific physical situations, you haven't really learned them. There are two major aspects involved in learning to solve problems: drill and skill. By *drill* I mean going through a lot of problems that involve the direct application of a particular concept until you start to feel familiar with the way it applies to physical situations. Each chapter of the Tipler textbook contains about thirty-five single-concept problems for you to use as drill. *Do a lot of these!*—at least as many as you need in order to feel comfortable handling them.

At the same time, however, you need to be tackling problems that go beyond the direct application of a single concept. These are the intermediate-level and advanced-level problems in the text. As you develop skill in recognizing which concepts are involved and in applying them to particular situations, you will master the material and become empowered. As you find that you can deal with more complex problems—even some of the advanced-level ones—you will gain confidence and enjoy applying your new skills. The examples in the textbook and the problems in this Study Guide are designed to provide you with a pathway from the single-concept to the intermediate-level and advanced-level problems.

All the problems in the Study Guide are accompanied by suggestions on how to solve them. As mentioned on page vii the problems come in pairs, with a detailed solution for the odd-numbered problem in each pair and an answer for the following, even-numbered, companion problem. I suggest that you start with an odd-numbered problem and study the suggestions and the worked-out solution. Be sure to note how the solution implements the suggestions. Then try the second problem in the pair. When attacking a problem, read the problem statement several times to be sure that you can visualize the physical situation being presented. Then make an illustration of this situation. Now you are ready to solve the problem.

You should budget time to study physics on a regular, preferably daily, basis. Plan your study schedule with your course schedule in mind. One benefit of this approach is that when you study on a regular basis, more information is likely to be transferred to your long-term memory than when you are obliged to cram. Another benefit of studying on a regular basis is that you will get much more from lectures. Because you will have already studied some of the material presented, the lectures will seem more relevant to you. In fact, you should try to familiarize yourself with each chapter before it is covered in class. An effective way to do this is first to read the Key Concepts of that Study Guide chapter. Then thumb through the textbook chapter, reading the headings and examining the illustrations. By orienting yourself to a topic *before* it is covered in class, you will have created a receptive environment for encoding and storing in your memory the material you will be learning.

Another way to enhance your learning is to explain something to a fellow student. It is well known that the best way to learn something is to teach it. That is because in attempting to articulate a concept or procedure, you must first arrange the relevant ideas in a logical sequence. Also, a dialogue with another person may help you to consider things from a different perspective. After you have studied a section of a chapter, discuss the material with another student and see if you can explain what you have learned.

I wish you success in your studies and encourage you to write to me at mosca@nadn.navy.mil if you find errors, or if you have comments or suggestions.

Acknowledgments

I want to thank Granvil C. Kyker, Jr., who authored the first edition of this Study Guide. Much of the material in this edition comes either directly or indirectly from his efforts. Also, I want to thank Paul Tipler for writing a textbook that has been a delight to work with. I am deeply indebted to Valerie Neal and Anne Vinnicombe for their confidence in me, as well as to Yuna Lee and Lee A. Young, who have worked long and hard to make this a better study guide than it could otherwise have been. I am grateful to Edward Adelson, Bruce Liby, Gordon Aubrecht, and to my colleagues at the United States Naval Academy for their counsel on various points, particularly Larry Tankersley for his many useful comments and suggestions. Lastly, I wish to thank my wife Vivian for her patience and support.

June 1998

Gene Mosca
United States Naval Academy

Chapter 1

Systems of Measurement

I. Key Ideas

1-1 Units When we measure any physical quantity, we are comparing it to some agreed-on, standard **unit.** Scientists (and others) use a system of units called the **Système Internationale (SI)** in which the standard units for length, time, and mass are the meter, the second, and the kilogram, respectively.

When the names of units are written out, they always begin with a lowercase letter, *even when* the unit is named for a person. Thus, the unit of temperature named for Lord Kelvin is the kelvin. The abbreviation for a unit also begins with a lowercase letter, *except when* the unit is named for a person. Thus, the abbreviation for the meter is m, whereas the abbreviation for the kelvin is K. The exception is the abbreviation for the liter, which is L. Abbreviations of units are not followed by periods.

All written-out prefixes for powers of 10, such as *mega* for 10^6, begin with lowercase letters. For multiples less than or equal to 10^3, the abbreviations of the prefixes are lowercase letters, such as k for *kilo*. For multiples greater than 10^3, however, the prefix abbreviations are uppercase letters, such as M for *mega*. The prefixes for powers of 10, as well as their abbreviations, are listed in Table 1-1 on page 5 of the text.

1-2 Conversion of Units One unit can be converted to another by multiplying it by a **conversion factor** that equals 1. If we divide each side of the relation

$$60 \text{ s} = 1 \text{ min}$$

by 1 min, we obtain the conversion factor

$$\frac{60 \text{ s}}{1 \text{ min}} = 1$$

To convert 7.4 min to seconds, we multiply 7.4 min by this conversion factor:

$$7.4 \text{ min} \times \frac{60 \text{ s}}{1 \text{ min}} = 444 \text{ s}$$

In this expression, we think of the "min" units as canceling each other. One may draw lines through units that cancel as shown.

1-3 Dimensions of Physical Quantities The **dimensions** of a physical quantity express what kind of quantity it is—whether it is a length, a time, a mass, or whatever. For example, the dimensions of velocity are length per unit time. The corresponding units might be miles per hour or meters per second. Whenever we add or subtract quantities, they must have the same dimensions. Also both sides of an equation must have the same dimensions. Checking that the dimensions are in fact the same is often a useful way of checking for mistakes in setting up equations.

1-4 Scientific Notation Very often in physics we find ourselves dealing with very large or very small numbers. This is much simpler to do if the numbers are written in **scientific notation,** that is, as a number between 1 and 10 multiplied by the appropriate power of 10. An example is 1.67×10^5 for the number 167,000. When numbers are multiplied (or divided), the powers of 10 are added (or subtracted).

1-5 Significant Figures and Order of Magnitude The quantities we deal with in physics are not always pure numbers but instead are often (in principle or in fact) the results of measurements. They are thus known to only limited precision. The number of digits we use to express a quantity is, by implication, an expression of its precision. We therefore use only those digits that are **significant figures,** that is, those digits that have meaning. The term *figure* is synonymous with the term *digit*, and physicists commonly use them interchangeably.

The **order of magnitude** of a value is an estimate of the value to the nearest power of 10.

II. Numbers and Key Equations

Numbers

1 inch (in) = 2.54 centimeters (cm)

1 foot (ft) = 0.3048 meters (m)

1 mile (mi) = 1.61 kilometer (km)

1 liter (L) = 1×10^{-3} m^3

Key Equations

There are no key equations for this chapter.

III. Potential Pitfalls

Without units, expressions of physical quantities do not have meaning. Usually, it is best to use SI units. If you are given problems with quantities in other units, you should usually convert them to SI units before proceeding with the problem. When converting units, remember that all conversion factors must have a magnitude of 1.

This course will require you to solve a large number of problems. For most of these, you will be using a calculator. Do not blindly believe whatever answer your calculator displays. Check your results by estimating the order of magnitude of the calculation on a piece of scratch paper. Do not use all of the digits displayed by the calculator. Write down only the significant digits.

IV. True and False and Responses

True and False

_____1. The length of the meter depends on the duration of a second.

_____2. The SI unit of mass is the gram.

_____3. Two quantities having different dimensions can be multiplied, but they cannot be subtracted.

_____4. All conversion factors are equal to 1.

_____5. The second is defined in terms of the frequency of a certain kind of light emitted by cesium atoms.

_____6. The SI units of force, energy, and other physical quantities are defined in terms of the fundamental units of mass, length, and time.

_____7. Two quantities having the same dimensions must be measured in the same units.

Responses

1. True. According to the current definition of the meter, if the duration of the second doubled, the length of the meter would also double.

2. False. It is the kilogram.

3. True

4. True

5. True

6. True

7. False. For example, time can be measured in hours, minutes, seconds, or other units.

V. Questions and Answers

Questions

1. Is it possible to define a system of units for measuring physical quantities in which one of the fundamental units is not a unit of length?

2. What properties should an object, system, or process have for it to be a useful standard of measurement for a physical quantity such as time or length?

3. A furlong is one-eighth of a mile and a fortnight is two weeks. Of what physical quantity is a furlong per fortnight a unit? What is the SI unit for this quantity? Find the conversion factor between furlongs per fortnight and the corresponding SI unit.

4. Acceleration a has dimensions L/T^2, and those of velocity v are L/T. The velocity of an object that has accelerated uniformly through a distance d is either $v^2 = 2ad$ or $v^2 = (2ad)^2$. Which one must it be?

5. If you use a calculator to divide 3411 by 62.0, you will get something like 55.016 129. (Exactly what you will get depends on your calculator.) Of course, you know that not all these figures are significant. How should you write the answer?

6. If two quantities are to be added, do they have to have the same dimensions? The same units? What if they are to be divided?

Answers

1. Certainly. It could have fundamental units for time and speed. The unit for length could then be defined in terms of these fundamental units.

2. It should be reproducible at a variety of times and places, and it should be precise.

3. The quantity is speed. The SI unit for speed is the meter per second. We determine the conversion factor as follows:

$$\left(\frac{1 \text{ furlong}}{\text{fortnight}}\right)\left(\frac{1 \text{ fortnight}}{14 \text{ days}}\right)\left(\frac{1 \text{ day}}{24 \text{ h}}\right)\left(\frac{1 \text{ h}}{3600 \text{ s}}\right)$$

$$\times \left(\frac{1 \text{ mi}}{8 \text{ furlongs}}\right)\left(\frac{1610 \text{ m}}{1 \text{ mi}}\right) = 1.66 \times 10^{-4} \text{ m/s}$$

The conversion factor is

$$\frac{1.66 \times 10^{-4} \text{ m/s}}{1 \text{ furlong/fortnight}} = 1$$

4. It must be $v^2 = 2ad$ because both sides of the equation have the dimensions L^2/T^2.

5. You are dividing a number with four significant figures by a number with three significant figures. The result should therefore have three significant figures. It should be written 55.0.

6. To be added, the quantities have to have the same dimensions but not the same units. For example, if you add 4 feet and 3 inches you get 4 feet 3 inches. (Normally, you will want to convert the units of one of the quantities to those of the other before adding them.) Quantities can be divided even if they have different dimensions, as when distance is divided by time to get speed.

VI. Problems and Solutions

Problems

1. Write the following in scientific notation without prefixes: (*a*) 22 μm, (*b*) 233.7 Mm, (*c*) 0.4 kK, (*d*) 20.0 pW.

How to Solve It
- Represent each number in scientific notation.

- Replace each prefix with the corresponding power of 10.

- Add the exponents.

2. Write the following in scientific notation without prefixes: (*a*) 330 km, (*b*) 33.7 μm, (*c*) 0.03 K, (*d*) 77.5 GW.

How to Solve It
- Represent each number in scientific notation.

- Replace the prefix with the corresponding power of 10.

- Add the exponents.

3. In the following equation, the distance x is in meters, the time t is in seconds, and the velocity v is in meters per second. What are the dimensions and SI units of the constants A, B, C, D, and E? *Hint:* Both exponents and the arguments of trigonometric functions must be dimensionless. (The argument of cos ϕ is ϕ.)

$$x = Avt + B \sin(Ct) + Dt^{1/2} 3^{x/E}$$

How to Solve It
- The argument of the sine function and the two exponents are dimensionless. Thus, Ct and x/E are dimensionless.

- Each term on the right of the equal sign has the same dimensions as the x on the left of the equal sign. Therefore, the terms Avt, B, and $Dt^{1/2}$ each have dimensions of length. (The sine function is dimensionless.)

4. In the following equation, the distance x is in meters, the time t is in seconds, and the velocity v is in meters per second. What are the dimensions and SI units of the constants A, B, C, D, and E?

$$v = Bt[\sqrt{Ax} + \cos^2(Ct)] - D2^{Et}$$

How to Solve It
- Both the cosine function and its argument are dimensionless. The exponent is also dimensionless. Determine the dimensions of $\sqrt{Ax}$.

- The dimensions of the two terms on the right of the equal sign are the same as those of v on the left of the equal sign.

5. Write the following using prefixes and abbreviations for the SI units: (*a*) 25,000 meters per second, (*b*) 0.004 seconds, (*c*) 4,200,000 watts, (*d*) 0.1 kilograms. For example, 0.0001 meters = 100 μm. (The number preceding the prefix should always be less than 1000 and equal to or greater than 1.)

How to Solve It
- Factor the number by 10^{+3} or 10^{-3} until you have a number less than 1000 and equal to or greater than 1.

- Determine the prefix by looking at Table 1-1 on page 5 of your text.

6. Write the following using prefixes and abbreviations for the SI units: (*a*) 0.000 025 meters per second, (*b*) 0.0445 seconds, (*c*) 0.000 000 032 watts, (*d*) 25,640 kilograms. For example, 0.0001 meters = 100 μm. (The number preceding the prefix should always be less than 1000 and equal to or greater than 1.)

How to Solve It
• Factor the number by 10^{+3} or 10^{-3} until you have a number less than 1000 and equal to or greater than 1.

• Determine the prefix by looking at Table 1-1 on page 5 of your text.

7. A very fast sprinter can run the 100-m dash in slightly under 10 s. This means his average speed is slightly greater than 10 m/s. Using conversion factors, convert 10 m/s to miles per hour.

How to Solve It
• Multiply the speed in meters per second by conversion factors, first to convert meters to miles and then to convert seconds to hours.

8. The speed of light in empty space is 3×10^8 m/s. Use conversion factors to determine the speed of light in feet per nanosecond.

How to Solve It
• Multiply the speed in meters per second by conversion factors, first to convert meters to feet and then to convert seconds to nanoseconds.

Solutions

1. (a) 22 μm = $(2.2 \times 10^1)10^{-6}$ m = $\underline{2.2 \times 10^{-5}\,\text{m}}$
(b) 233.7 Mm = $(2.337 \times 10^2)10^6$ m = $\underline{2.337 \times 10^8\,\text{m}}$
(c) 0.4 kK = $(4 \times 10^{-1})10^3$ K = $\underline{4 \times 10^2\,\text{K}}$
(d) 20.0 pW = $(2.00 \times 10^1)10^{-12}$ W = $\underline{2.00 \times 10^{-11}\,\text{W}}$

2. (a) $\underline{3.3 \times 10^5\,\text{m}}$ (b) $\underline{3.37 \times 10^{-5}\,\text{m}}$
(c) $\underline{3 \times 10^{-2}\,\text{K}}$ (d) $\underline{7.75 \times 10^{10}\,\text{W}}$

3. The terms Avt, B, and $Dt^{1/2}$ each has dimensions of length. This means that A is a dimensionless number (because vt has dimensions of L), B has dimensions of L, and D has dimensions of $L/T^{1/2}$. The argument of the sine function and the exponents are dimensionless, so C has dimensions of $1/T$ and E has dimensions of L. The SI units of B, C, D, and E are m, 1/s, m/s$^{1/2}$, and m, respectively.

4. The dimensions of A are $1/L$, those of B are L/T^2, those of C are $1/T$, those of D are L/T, and those of E are $1/T$. The SI units of A, B, C, D, and E are, respectively, $1/\text{m}$, m/s^2, 1/s, m/s, and 1/s.

5. (a) 25,000 meters per second = 25×10^3 m/s = $\underline{25\,\text{km/s}}$
(b) 0.004 seconds = 4×10^{-3} s = $\underline{4\,\text{ms}}$
(c) 4,200,000 watts = 4.2×10^6 W = $\underline{4.2\,\text{MW}}$
(d) 0.1 kilograms = 100×10^{-3} kg = $\underline{100\,\text{g}}$

6. (a) 25 μm/s   (b) $\underline{44.5\,\text{ms}}$ (c) $\underline{32\,\text{nW}}$
(d) $\underline{25.64\,\text{Mg}}$

7. $10\dfrac{\text{m}}{\text{s}}\left(\dfrac{1\,\text{km}}{10^3\,\text{m}}\right)\left(\dfrac{1\,\text{mi}}{1.61\,\text{km}}\right)\left(\dfrac{60\,\text{s}}{1\,\text{min}}\right)\left(\dfrac{60\,\text{min}}{\text{h}}\right) =$
$\underline{22.4\,\text{mi/h}}$

8. $\underline{1.02\,\text{ft/ns}}$

Chapter 2

Motion in One Dimension

I. Key Ideas

When describing motion in this chapter, we are restricting ourselves to studying the motion of a single particle. The particle concept is central to much of physics because any extended object can often be treated as a collection or "system" of particles. When studying motion, an extended object can be modeled as a particle if variations in its internal structure do not occur, that is, if the object is perfectly rigid. Strictly speaking, if an object is rotating it cannot be modeled as a particle.

Initially we will consider only motion in one dimension.

2-1 Displacement, Velocity, and Speed If we place an x axis along the line of motion, with its origin O at some reference point, we can specify a particle's position by a single number x. The sign of x indicates on which side of the origin O the particle is located. We describe motion of the particle by specifying how its position changes with time. Note that you choose both the position of the origin and which direction is positive (the direction of increasing x).

The **average speed** of a particle is the total distance it has traveled divided by the time taken to travel that distance. For example, suppose a mouse in a 2.00-m-long pipe takes 30 s to run from end A to end B, then back to end A, and then again to end B. The total distance traveled is 6.00 m and the average speed is the total distance divided by the time (0.20 m/s).

When a particle starts at position x_1 and moves to position x_2, the **displacement** x of the particle is defined as the net change in its position, that is

Displacement

$$x = x_2 - x_1$$

The **average velocity** v_{av} of the particle is this displacement divided by the time interval t in which the displacement occurs.

Average velocity

$$v_{av} = \frac{\Delta x}{\Delta t}$$

Like position and displacement, average velocity has a direction in space (in one dimension, a sign) as well as a magnitude. The velocity is positive when the particle moves in the positive x direction, and negative when it moves in the negative x direction.

Instantaneous Velocity What we usually mean when we say "velocity" is instantaneous velocity. The **instantaneous velocity** is the displacement divided by the time interval, in the limit that the time interval—and thus the displacement—approaches zero. In mathematical jargon we'd say, "the velocity of a particle is the time derivative of its position." On a position-versus-time graph, average velocity is the slope of a line connecting the points representing the initial and final positions of the particle. The instantaneous velocity at a point is the slope of the line tangent to the curve at that point.

Instantaneous velocity

$$v = \lim_{\Delta t \to 0} \frac{\Delta x}{\Delta t} = \frac{dx}{dt}$$

The **speed** of a particle is the magnitude (absolute value) of the instantaneous velocity.

Relative Velocity Velocity is always measured relative to a rigid object called a **reference frame.** For example, either the ground or a railroad car can serve as a suitable reference frame. So can the flowing water in a river. "Hold on," you say, "water is not rigid." Under certain circumstances, the water in a smoothly flowing river can be thought of as being rigid. Suppose that dye markers simultaneously released at three arbitrary points in a river are observed to drift downstream in such a manner that the triangle formed by lines connecting the markers maintains its size and shape. We can then think of the water as rigid.

"Wait a minute," you say, "rivers have eddies, and if one of the markers were released behind an eddy the triangle would not maintain its size and shape." Correct. Flowing water that includes a significant eddy cannot be thought of as rigid and cannot be thought of as a reference frame.

Position measurements are always made relative to coordinate systems that are attached to reference frames. For example, suppose you're on a railroad car moving to the right, as shown in Figure 2-1, and are walking toward its front. Also suppose we measure your position using a coordinate axis that is attached to the car, with the origin O at the car's rear, and with the forward direction taken as positive. Your position x is then the distance d between you and the rear of the car. Your velocity v is the rate that x changes over time. If we measure your position with a second coordinate axis along the same line and in the same direction, but with its origin O' at the front of the car, your position x' is the negative of the distance d' between you and the front of the car. Your velocity v' is the rate that x' changes over time.

Figure 2-1

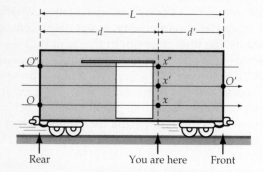

Rear   You are here   Front

Your distance from the rear of the car plus your distance from the front of the car equals the length L of the car. That is, $d + d' = L$. Expressing this relation in terms of the positions x and x' we have $x = d$ and $x' = -d'$, so $d + d' = L$ becomes $x - x' = L$. To find the velocities in these two coordinate systems we differentiate each side of this equation to get $v - v' = 0$, or

$$v = v'$$

where L is constant so that $dL/dt = 0$. That is, the velocities measured using the different coordinate axes

are the same. This is always the case; the velocity of a particle is the same when measured by any two coordinate axes as long as they are parallel and attached to the same reference frame.

"This doesn't make sense!" you exclaim. "Suppose we measure my velocity relative to the train using a third coordinate axis identical with the first but with the positive direction opposite to the direction of the car's motion. My position x'' is then the negative of my distance d from the rear of the train, so $d = d$ becomes $x = -x''$ and differentiating both sides of this equation gives $v = -v''$. In this case it is clear that the velocities are not equal even though both coordinate axes are attached to the train."

You raise a good point. However, the velocities v and v'' describe the same physical motion. If you walk toward the front of the train at 0.44 m/s, then $v = +0.44$ m/s and $v'' = -0.44$ m/s. Both of these values represent the same speed (0.44 m/s) and direction (that of the train's motion), so it is correct to say that the velocity relative to the train is 0.44 m/s toward the front in both cases. The velocities v and v'' have different signs because they refer to coordinate axes with opposite directions taken as positive. To avoid this potentially confusing situation always choose coordinate axes with the same direction taken as positive.

"Fascinating" you say. "To get this stuff correct you sure have to be careful."

Next we discuss the motion of an object that moves relative to two distinct reference frames that move relative to each other. For example, your pal swims directly downstream in a river. Let the shore be one reference frame and the river water be the other. Her velocity relative to the shore v_{ps} equals her velocity relative to the water v_{pw} plus the velocity of the water relative to the shore v_{ws}:

Relative velocity

$$v_{ps} = v_{pw} + v_{ws}$$

Suppose we choose two coordinate axes, one fixed to the shore and the other fixed to the water, and that for each, downstream is the positive direction. If her velocity relative to the water is +2 mi/h and the velocity of the water relative to the shore is +3 mi/h, then her velocity relative to the shore is +5 mi/h.

2-2 Acceleration The relation between acceleration and velocity is precisely the same as the relation between velocity and position. The **average acceleration** is the change in the velocity v divided by the time interval t in which the velocity changes. The **instantaneous acceleration** is the average acceleration in the limit that the time interval—and thus the change in velocity—approaches zero. That is, the acceleration is the time derivative of the velocity.

Average acceleration

$$a_{av} = \frac{\Delta v}{\Delta t}$$

Instantaneous acceleration

$$a = \lim_{\Delta t \to 0} \frac{\Delta v}{\Delta t} = \frac{dv}{dt}$$

2-3 Motion With Constant Acceleration The simplest form of accelerated motion is motion in which the acceleration remains constant. When the acceleration is constant, the velocity changes linearly with time. Only then does the average velocity equal the numerical average of the initial velocity v_1 and final velocity v_2 so that $v_{av} = \frac{1}{2}(v_1 + v_2)$. When we combine these observations with the fact that the displacement is the average velocity times the time, we get the following equations:

Constant acceleration equations

$$v_{av} = \tfrac{1}{2}(v_1 + v_2)$$
$$v = v_0 + at$$
$$x = x_0 + v_0 t + \tfrac{1}{2}at^2$$
$$v^2 = v_0^2 + 2a\,\Delta x$$

2-4 Integration Integration is the inverse of differentiation. Therefore we can express the kinematic relations as

Integral form of the kinematic relations

$$\frac{dx}{dt} = v \quad \Leftrightarrow \quad \Delta x = \int_{t_1}^{t_2} v\,dt$$

$$\frac{dv}{dt} = a \quad \Leftrightarrow \quad \Delta v = \int_{t_1}^{t_2} a\,dt$$

It follows that the area under the velocity-versus-time curve is the change in position (the displacement) of the particle, and the area under the acceleration-versus-time curve is the change in velocity.

Average value always refers to the time average. The time average of a function u is given by

$$u_{av} = \frac{1}{t_2 - t_1} \int_{t_1}^{t_2} u(t)dt$$

It follows that the average value of the velocity v is then given by

$$v_{av} = \frac{1}{t_2 - t_1} \int_{t_1}^{t_2} v(t)dt$$

However, average velocity is defined as $v_{av} = \Delta x/\Delta t$ (Section 2-1). By substituting Δx for $\int_{t_1}^{t_2} v\,dt$ (see the first integral form in this section) the integral expression for average velocity reduces to

$$v_{av} = \frac{1}{t_2 - t_1} \Delta x = \frac{\Delta x}{\Delta t}$$

the definition of average velocity.

II. Numbers and Key Equations

Numbers

Acceleration due to gravity

$$g = 9.81 \text{ m/s}^2 = 32.2 \text{ ft/s}^2$$

Key Equations

Displacement

$$x = x_2 - x_1$$

Average velocity

$$v_{av} = \frac{\Delta x}{\Delta t}$$

Instantaneous velocity

$$v = \lim_{\Delta t \to 0} \frac{\Delta x}{\Delta t} = \frac{dx}{dt}$$

Relative velocity

$$v_{ps} = v_{pw} + v_{ws}$$

Average acceleration

$$a_{av} = \frac{\Delta v}{\Delta t}$$

Instantaneous acceleration

$$a = \lim_{\Delta t \to 0} \frac{\Delta v}{\Delta t} = \frac{dv}{dt}$$

Constant acceleration equations

$$v_{av} = \tfrac{1}{2}(v_1 + v_2)$$

$$v = v_0 + at$$

$$x = x_0 + v_0 t + \tfrac{1}{2}at^2$$

$$v^2 = v_0^2 + 2a\,\Delta x$$

Integral form of the kinematic relations

$$\Delta x = \int_{t_1}^{t_2} v\,dt$$

$$\Delta v = \int_{t_1}^{t_2} a\,dt$$

III. Potential Pitfalls

In physics we model nature by representing physical properties with algebraic variables. We use t for the time of an event, for example. Don't confuse the value of a quantity with a change in, or increment of, that quantity. For instance, 10:30 A.M. on Tuesday is a specific point (instant) in time, but 10.5 hours is a time interval. Similarly, x is the position of a particle at some instant, but the change in position during a time interval, its displacement, is $x = x_2 - x_1$.

When doing kinematics problems, pay careful attention to signs. Signs indicate direction in space. Once you choose the positive direction, all signs must be consistent with that choice. When doing a free-fall problem with the positive direction chosen as upward, the object's acceleration is $-g$, because the object is accelerating downward. (Don't ever write $g = -9.81$ m/s^2. g itself denotes the magnitude of the free-fall acceleration—not its direction—and thus is never negative.)

Speed is the magnitude of the velocity, but average speed is not necessarily the magnitude of the average velocity. Average speed is defined as the total distance traveled divided by the time interval, whereas average velocity is defined as the net displacement divided by the time interval. Velocity and acceleration can be either positive or negative. In everyday usage we say "deceleration" for the rate of decrease in speed. However, when a particle is slowing down, the acceleration can be either positive or negative, depending on the choice for the positive direction. When a particle is slowing down, the acceleration and the velocity are always opposite in direction. In one-dimensional motion, opposite in direction goes hand-in-hand with opposite in sign.

The equations of motion with constant acceleration are used throughout the text. We recommend that you spend a few minutes now and commit them to memory. Don't forget that they only apply to situations where the acceleration is constant. These equations are developed with reference to the initial position x_0 and initial velocity v_0. In working a problem, it is often convenient to choose the origin at the initial position so that x_0 equals zero.

IV. True and False and Responses

True and False

_____ 1. An object can be treated as a particle if both its rotational motion and its internal structure do not vary.

_____ 2. If the average speed of a particle over a certain 3-s interval is 1.5 m/s, the average velocity over the same 3-s interval must be either +1.5 m/s or −1.5 m/s.

_____ 3. The average speed of a particle is never negative.

_____ 4. If a car is driven for 1 h at an average velocity of +60 km/h, the distance it travels in that same 1-h interval is 60 km.

_____ 5. An instant is not a time interval, but a point in time like 10:23 A.M.

_____ 6. A particle with a position that is given by $x(t) = At + B$ (A and B are constants) is moving with a constant velocity.

_____ 7. A particle with a position that is given by $x = Ct^2$ (C is a constant) has an acceleration given by $4C$.

_____ 8. Over a certain time interval the velocity of a certain particle changes from zero to v_f. It necessarily follows that its average velocity over the same interval is $v_f/2$.

_____ 9. If at a certain instant the acceleration of a particle is positive, its velocity must also be positive.

_____ 10. During a certain time interval, the area under a graph of the velocity of a particle versus time is the total distance the particle has traveled during that interval.

_____ 11. Relative to a track, John runs with a velocity of +8 mi/h due north and Marie runs with a velocity of −8 mi/h due south. It follows that John and Marie run with the same speed in the same direction relative to the track.

_____ 12. If the velocity of train A relative to train B is +80 km/h due north, then it follows that the velocity of train B relative to train A is −80 km/h due north (+80 km/h due south).

Responses

1. True.

2. False. The average velocity depends only on the initial and final positions and not on the distance traveled.

3. True.

4. True.

5. True. In technical usage, the terms instant, instantaneously, moment, and momentarily refer to a point in time and not a finite time interval.

6. True. Differentiating will show that the velocity is A.

7. False. The acceleration is $2C$ (Differentiate x twice.)

8. False. The average velocity necessarily equals $v_f/2$ only when the acceleration is constant. If the acceleration varies, the average velocity may or may not equal $v_f/2$.

9. Not necessarily. If a particle is moving in the negative direction and is slowing down, its velocity is negative while its acceleration is positive.

10. False. The area under a velocity-versus-time curve equals displacement, not distance.

11. True.

12. True.

V. Questions and Answers

Questions

1. Figure 2-2 shows the position of a particle versus time. The following questions refer to the intervals between the times shown. (*a*) During which interval(s) does the velocity remain zero? (*b*) During which interval(s) does the velocity remain constant?

Figure 2-2

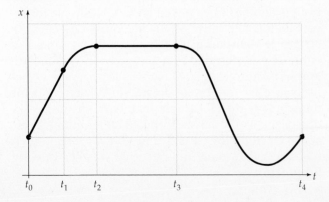

(*c*) During which interval(s) is the average velocity zero? (*d*) During which interval(s) does the acceleration remain negative?

2. Starting from rest, a world-class sprinter can run 100 m in 10 s, but he cannot run 30 m in 3 s. Neither can he run 400 m in 40 s. Why not?

3. At some instant a car's velocity is 15 m/s; 1 s later it is 11 m/s. If the car's acceleration is constant, what is its average velocity in this one-second interval? How far does the car go during that interval? Can you answer these questions if you don't know whether or not the acceleration is constant?

4. An elevator is moving upward at a speed of v_e when a bolt falls off its undercarriage. Does the bolt immediately descend, or does it continue to ascend until its velocity becomes zero?

5. A bolt falls off its undercarriage when an elevator is (*a*) moving upward at v_0, (*b*) moving downward at speed v_0, (*c*) at rest. Assuming the elevator is at the same height in all three cases and disregarding the effects of air resistance, in which case(s) does the bolt reach the bottom of the shaft at the greatest speed?

6. Is the relation between velocity and position identical to the relation between acceleration and velocity?

Answers

1. (*a*) The velocity of the particle equals the slope of the tangent to the curve. For $t_2 < t < t_3$ the slope is zero, so for that interval $v = 0$; (*b*) The slope is constant for $t_0 < t < t_1$ and again for $t_2 < t < t_3$; (*c*) The average velocity for an interval equals the slope of the straight line that connects the points at opposite ends of the interval. The slope of the line connecting the points at t_0 and t_4 is zero, as is that connecting the points at t_2 and t_3. It follows that the average velocity is zero for the intervals $t_0 < t < t_4$ and $t_2 < t < t_3$; (*d*) The acceleration is negative when the slope of the tangent to the curve decreases as t increases, which it does throughout the interval from $t_1 < t < t_2$. *Remark:* The acceleration is also negative during the first third of the interval $t_3 < t < t_4$ because the slope of the tangent decreases (becomes more negative) as t increases. After the inflection point, about one third the way through the interval, the acceleration becomes positive.

2. It takes a sprinter several seconds to reach top speed. Thus in the first three seconds he is traveling slower than 10 m/s most of the time. He can't average 10 m/s over 40 s because a sprinter is unable to sustain top speed for that length of time.

3. If the acceleration is constant, $v_{av} = (v_1 + v_2)/2 = 13$ m/s and $\Delta x = v_{av}\,\Delta t = 13$ m. If you do not know whether or not the acceleration is constant, you cannot determine these quantities.

4. The bolt continues to ascend until its velocity reaches zero.

5. The three cases are (a) the initial velocity is $+v_0$, (b) the initial velocity is $-v_0$, and (c) the initial velocity is zero, where we have chosen upward as the positive direction. Consider the formula $v^2 = v_0^2 + 2a\,(x - x_0)$, where a and $x - x_0$ are the same and v_0 is different for each case. By examining this equation you can see that the speed v is least for case (c) where the initial velocity equals zero, and is the same for cases (a) and (b) where the initial velocity squared is the same.

6. Yes.

VI. Problems and Solutions

Problems

1. A car travels 40 km along a straight road at a speed of 86 km/h and then goes 40 km farther at a speed of 50 km/h. What is the car's average velocity for the entire trip?

How to Solve It
• Average velocity is the displacement divided by the time interval.

• The total displacement is the sum of the two displacements, and the total time interval is the sum of the two time intervals.

• In *each* of the two intervals the car was moving at constant speed.

2. You're bicycling across central Oklahoma on a perfectly straight road. You started at 8:30 A.M. and you've covered 21 miles by 11:15 A.M. when your chain breaks. You haven't any spare parts with you, so you have to walk the bike back to the last town at a speed of 2.6 mi/h. If you get there at 1:30 P.M., what was (a) your displacement, (b) your average velocity, and (c) your average speed for the whole trip?

How to Solve It
• Displacement is *net* change in position, never mind what happened along the way. What's the difference between your initial and your final positions? Don't forget that you changed directions!

• Average velocity is the displacement divided by the time interval.

• The average speed is the *total distance gone,* without regard to direction, divided by the time interval.

3. Starting at a time we'll call $t = 0$, a man walks east from his office to McDonald's, has lunch, and then walks to his bank. His trip is shown on the graph of position versus time in Figure 2-3. (a) What was his average velocity from his office to the bank? (b) How much time did he spend at lunch? (c) How far is it from McDonald's to the bank? (d) At what point in the trip was he walking fastest?

Figure 2-3

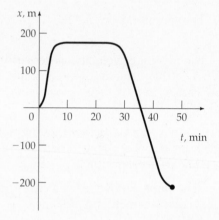

How to Solve It
• On an x-versus-t graph, the average velocity between any two points is the slope of a straight line drawn between those points.

• Where is McDonald's? What's happening on the graph while the man is having his lunch?

• The instantaneous velocity at a point is the slope of a line drawn tangent to the curve at that point.

4. A woman drives west from Nashville toward Memphis on I-40. (Assume that the highway is straight.) At Bucksnort, Tennessee she develops car trouble and has to be towed back to Nashville for repairs before she can continue her journey. She finally arrives at Memphis at 1700 hours, military time. The entire trip is described on the position-versus-time graph in Figure 2-4. (a) What was her average velocity before her car broke down? (b) What was it for the whole trip? (c) Assume that, without the breakdown, she would have continued at her initial (average) velocity. How much time did the breakdown cost her? (d) In what time interval was her velocity the greatest?

Figure 2-4

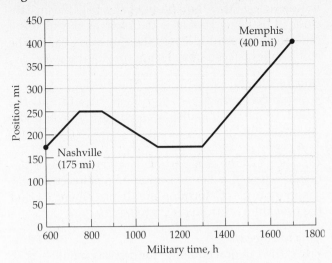

Table 2-1   Data for Problem 6

t, s	v, m/s
0	0
1	12.3
2	17.7
3	21.5
4	24.6
5	27.3
6	29.9

(b) Draw a graph of these data and determine the car's instantaneous acceleration at $t = 2$ s.

How to Solve It
• Average acceleration is the change in (instantaneous) velocity over a time interval divided by the duration of the interval.

• On a graph of velocity versus time, the acceleration at a particular time is the slope of a line drawn tangent to the curve at that point.

7. A world-class sprinter can run the 100-m sprint in 10.0 s. Assume that he starts from rest and runs with uniform acceleration for the first 50 m and thereafter runs at his top (constant) speed. (a) What is his acceleration in the first part of the run? (b) Assuming his acceleration and top speed are the same as in part (a), how long would it take him to run 200 m?

How to Solve It
• There's nothing subtle about the physics here, but you need to give a little thought to the algebra!

• There are two time intervals—from 0 to some time t_1, during which the sprinter is accelerating, and from t_1 to $t = 10$ s.

• The final velocity attained in the first part of the sprint equals the velocity during the second part. Determine the time t_1 for the first part using the relation $\Delta x = v_{av}t$ and the fact that the distances traveled in the first and second parts are equal. Remember, $t_2 = 10$ s $- t_1$.

• Once you've solved for t_1, you can calculate his acceleration in the first 50 m.

8. A toy rocket starting from rest moves straight upward with an acceleration three times that of gravity until it is 300 m above the ground. At that point its engine quits. How long after it was fired does the rocket hit the ground? Neglect air resistance.

How to Solve It
• First write an equation for the *powered* part of the flight. How long does this part take?

How to Solve It
• Average velocity is the displacement divided by the time interval.

• The (instantaneous) velocity is greatest where the slope of a line drawn tangent to the curve is steepest.

• No, I didn't make up Bucksnort. It's there.

5. A ball rolls down a hill, starting from rest at the top of the hill at time $t = 0$ and picking up speed as it goes. Suppose that its velocity is given by the equation

$$v = (0.06 \text{ m/s}^3)t^2$$

(a) What is the ball's average acceleration over the interval from $t = 0$ to $t = 2$ s? (b) What is its average acceleration from $t = 2$ to $t = 4$ s? (c) Is the ball moving with constant acceleration?

How to Solve It
• Average acceleration is the change in (instantaneous) velocity over a time interval divided by the duration of the interval.

• The equation stated in the problem gives you the (instantaneous) velocity at any time t.

• If the ball's acceleration were constant, what would you expect to get for the average acceleration in the two different time intervals?

6. Table 2-1 shows acceleration data for an automobile (a 2010 Whizbang) from a standing start. From these data, (a) calculate the average acceleration of the car over the interval from $t = 1$ s to $t = 3$s.

• When the engine quits, what is the rocket's upward speed?

• You know the rocket's height at the moment the engine quits. Find out how long (starting from this point) it's in the air.

9. A man standing on a cliff 50 m above the level of the ground below throws a stone straight up in the air. The stone falls back past the edge of the cliff and strikes the ground 5 s after it was thrown. With what initial velocity did the man throw the stone?

How to Solve It
• The stone is in free fall, so you know how to write an equation of motion for it.

• Notice that you don't need to solve for any intermediate quantities such as how high the stone went.

• If you write an equation for the stone's position at time $t = 5$ s, you know everything in the equation except the initial velocity.

10. The speed of sound in air is about 340 m/s. If you drop a stone into a well and hear the splash 3.80 s later, how deep is the well?

How to Solve It
• Use the free-fall equations to relate the time the stone took to fall to the depth of the well.

• Don't forget the time the sound takes to get back to you.

• These two times add up to 3.80 s.

• This results in a quadratic equation that can be solved for the depth of the well using the quadratic formula.

11. A ball thrown straight up is caught by the thrower. Use the constant-acceleration equations to show that (neglecting air resistance) for a ball thrown straight up, the rise time equals the fall time.

How to Solve It
• The constant-acceleration equations for the ball are $v = v_0 - gt$, $y = y_0 + v_0 t - \frac{1}{2}gt^2$, and $v^2 = v_0^2 - 2g(y - y_0)$.

• Use the equation $v = v_0 - gt$ to find the rise time t_{rise}. What is the velocity v when t equals the rise time?

• Now use $y = y_0 + v_0 t - \frac{1}{2}gt^2$ to find the total flight time t_{tft}. What is the height y when t equals t_{tft}? To find the fall time, subtract the rise time from the total flight time.

12. For the same thrown ball in Problem 11, use the constant-acceleration equations to show that (neglecting air resistance) the speed of the ball the instant prior to its being caught is the same as the speed of the ball the instant it leaves the thrower's hand. The ball is caught at the same height that it is thrown.

How to Solve It
• Use the equation $v^2 = v_0^2 - 2g(y - y_0)$ to find the speed. What is the displacement $y - y_0$ when the ball is caught?

13. The velocity of a particle is given by the equation $v = 3t^2 + 4$, where v is in meters per second and t is in seconds. Determine the displacement during the time interval from $t = 2$ s to $t = 5$ s.

How to Solve It
• Use the integral equation for displacement

$$\Delta x = \int_{t_1}^{t_2} v \, dt$$

14. The acceleration of a particle is given by the equation $a = 4t - 5$, where a is in meters per second squared and t is in seconds. The velocity at $t = 0$ is -6 m/s. Determine the velocity at $t = 4$ s.

How to Solve It
• Use the integral equation for the change in velocity

$$\Delta v = \int_{t_1}^{t_2} a \, dt$$

15. A car is traveling at 18 m/s along a straight, level road. Suddenly, the driver sees a truck ahead of her in her lane, moving in the same direction at 8 m/s. At the instant she hits the brakes the truck is 28 m ahead of the car. If we assume her acceleration is constant, what must it be if she is to avoid a collision?

How to Solve It
• If we take the forward direction as positive, the car's acceleration is negative since it is slowing down. We want to calculate the acceleration at which the car barely touches the truck. For any acceleration more negative than this, the collision will be avoided completely.

• Write an equation expressing velocity of the car in terms of the time and the car's yet-to-be-determined acceleration.

• If the collision is to be *extremely* gentle, the velocity of the car will equal the velocity of the truck at the time of impact t_i. Solve for the time t_i at which the velocity of the car is equal to the velocity of the truck. The acceleration of the car will appear in your result; that's okay.

• Write an equation that gives the position x_c of the car as a function of time. Write another equation that gives the position x_t of the truck as a function of time. Remember to include a term for the initial position of the truck.

• At the time of impact, the positions of the vehicles are equal. Equate these positions, substitute in the expression for the time of impact, and solve for the acceleration.

16. Al sees Bob drive past his house at 20 m/s and realizes he must catch Bob to get the day's physics assignment. Al jumps in his car and starts it—this takes 15 s—and takes off after Bob. Suppose Al accelerates at a *uniform* rate of 1.4 m/s², and Bob continues driving at a constant speed of 20 m/s. *Where* and *when* does Al catch up to Bob?

How to Solve It
• Write an equation for Al's motion that gives his position x as a function of time. Then write an equation for Bob's motion.

• At the particular time t when Al catches up with Bob, the two equations will give the same value of x!

• Now you know *when* Al catches up to Bob; use either equation to calculate *where* this happens.

17. You are at O'Hare airport (near Chicago) where there are 400-yd-long moving sidewalks. You and your twin are standing at the entrance of a moving sidewalk and Gate 17 (your gate) is 300 yards in the same direction. Suppose both of you have a walking speed of 3 mi/h and you simultaneously start walking to the gate. However, you walk directly to the gate on the floor, not using the sidewalk, while your twin walks on the sidewalk and then backtracks to the gate. You both arrive at the gate at the same time. How fast is the sidewalk moving?

How to Solve It
• There are three paths: yourself walking to the gate, your twin walking on the sidewalk, and your twin walking back from the end of the sidewalk to the gate. Define times and speeds (relative to the gate) for each.

• Write down an equation of the form distance = speed × time for each of the three paths. Use three equations, and the fact that both you and your twin arrive at the gate simultaneously, to solve for the speed of the sidewalk.

18. You swim 1.00 mi/h in still water. Suppose you jump off a pier and swim against a 0.50 mi/h current to a second pier 0.25 miles upstream. Upon reaching the second pier you immediately turn around and

swim downstream to the first pier. (*a*) How long does the round trip take? (*b*) What total distance do you swim relative to the shore? (*c*) relative to the water?

How to Solve It
• Use the velocities of the swimmer relative to the shore to find the time for the trip up and the time for the trip back.

• The speed relative to the water times the time equals the total distance traveled relative to the water.

Solutions

1. In the first leg of the trip the car covers 40 km at a speed of 86 km/h in time

$$\Delta t_1 = \frac{\Delta x}{v_1} = \frac{40 \text{ km}}{86 \text{ km/h}} = 0.465 \text{ h}$$

The second leg takes

$$\Delta t_2 = \frac{\Delta x}{v_2} = \frac{40 \text{ km}}{50 \text{ km/h}} = 0.80 \text{ h}$$

The car has gone 80 km in a total of 0.465 h + 0.80 h = 1.265 h, so

$$v_{av} = \frac{\Delta x}{\Delta t} = \frac{80 \text{ km}}{1.265 \text{ h}} = \underline{63.2 \text{ km/h}}$$

2. (*a*) $\underline{15.2 \text{ mi}}$ (*b*) $\underline{3.04 \text{ mi/h}}$ (*c*) $\underline{5.36 \text{ mi/h}}$

3. (*a*) The man got to the bank at $x = -200$ m (see Figure 2-2), 46 min after he started. His average velocity was therefore

$$v_{av} = \frac{\Delta x}{\Delta t} = \frac{-200 \text{ m}}{46 \text{ min}} \times \frac{1 \text{ min}}{60 \text{ s}} = \underline{-0.072 \text{ m/s}}$$

(*b*) He is at rest (presumably at McDonald's) from $t = 6$ min to $t = 26$ min, so his lunch must have taken $\underline{20 \text{ min}}$.
(*c*) McDonald's is at $x = 170$ m, and the distance from there to the bank is

$$|\Delta x| = |x_2 - x_1|$$
$$= |(-200 \text{ m}) - (170 \text{ m})| = \underline{370 \text{ m}}$$

(*d*) His velocity at any point is the slope of the x-versus-t curve at that point, so the greatest speed corresponds to the steepest slope. It looks like this occurs at $\underline{\text{between } t = 1 \text{ min and } t = 3 \text{ min}}$; presumably he was in a hurry to get his lunch.

4. (*a*) $\underline{50 \text{ mi/h}}$ (*b*) $\underline{20.5 \text{ mi/h}}$ (*c*) $\underline{6.5 \text{ h}}$
(*d*) from $\underline{1300 \text{ to } 1700 \text{ hours}}$

5. (a) To get the velocity at time $t = 2$ s, we plug into the equation given:

$$v_{2\,s} = (0.06\text{ m/s}^3)(2\text{ s})^2 = 0.24\text{ m/s}$$

Since the ball starts at rest, $v_0 = 0$. The average acceleration over an interval is the change in velocity divided by the time interval, so

$$a_{0-2\,s} = \frac{\Delta v}{\Delta t} = \frac{0.24\text{ m/s} - 0}{2\text{ s}} = 0.12\text{ m/s}^2$$

(b) In the same way, we find $v_{4\,s} = 0.96$ m/s and $a_{2-4\,s} = 0.36$ m/s^2.

(c) Since the acceleration is not the same in the two time intervals, clearly it is not constant.

6. (a) 4.6 m/s^2 (b) You should get about 4.0 m/s^2.

7. (a) The sprint takes place in two parts, with $t_1 + t_2 = t = 10$ s, with the runner traveling 50 m in each part. In the first part the runner, starting from rest, runs with constant acceleration, so his average velocity is $v_1/2$, where v_1 is the final velocity. Because the distance traveled in each part is the same,

$$\frac{v_1}{2} t_1 = v_1(t - t_1)$$

Dividing through this equation by v_1 and solving for t_1 we obtain

$$t_1 = \tfrac{3}{2}t = \tfrac{3}{2}(10\text{ s}) = 6.67\text{ s}$$

The acceleration is related to the distance by the formula

$$d_1 = \tfrac{1}{2}at_1^2$$

so

$$a = \frac{2d_1}{t_1^2} = \frac{2(50\text{ m})}{(6.67\text{ s})^2} = 2.25\text{ m/s}^2$$

(b) The velocity of the runner at the end of the first part of the sprint is

$$v_1 = at_1 = (2.25\text{ m/s}^2)(6.67\text{ s}) = 15.0\text{ m/s}$$

This velocity is sustained for the last 150 m of the sprint. Thus, the time for this last 100 m is

$$\frac{100\text{ m}}{15.0\text{ m/s}} = 6.67\text{ s}$$

The first 100 m took 10 s, so the total time for the 200 m sprint is 16.7 s. Nobody runs the 200 m that fast, of course, which suggests that our model of the sprints is an oversimplified one.

8. The rocket is in the air for 33.7 s.

9. Let's take the ground below the cliff to be height $y = 0$ and the values of height to be positive upward. At $t = 0$, the stone is at $y_0 = 50$ m, and its acceleration $a = -g = -9.81$ m/s^2. At $t = 5$ s, the stone is at $y = 0$. Notice that the stone's acceleration is negative because we chose to treat upward as positive. Thus in the equation

$$y = y_0 + v_0 t + \tfrac{1}{2}at^2$$

we know everything except v_0, so we can solve for that:

$$0 = (50\text{ m}) + v_0(5\text{ s}) + \tfrac{1}{2}(-9.81\text{ m/s}^2)(5\text{ s})^2$$

which gives

$$v_0(5\text{ s}) = 72.6\text{ m}$$

or

$$v_0 = 14.5\text{ m/s}$$

10. The well is 64.0 m deep.

11. To find the rise time t_{rise} we use the equation $v = v_0 - gt$, setting $v = 0$ and solving for the time.

$$0 = v_0 - gt_{\text{rise}}$$

$$t_{\text{rise}} = v_0/g$$

The fall time equals the total flight time minus the rise time. We solve for the total flight time t_{tft} using the equation $y = y_0 + v_0 t - \tfrac{1}{2}gt^2$, setting $y = y_0$ and solving for the time.

$$y_0 = y_0 + v_0 t_{\text{tft}} - \tfrac{1}{2}gt_{\text{tft}}^2$$

$$0 = v_0 t_{\text{tft}} - \tfrac{1}{2}gt_{\text{tft}}^2$$

$$0 = v_0 - \tfrac{1}{2}gt_{\text{tft}} \quad (t_{\text{tft}} \neq 0)$$

$$t_{\text{tft}} = 2v_0/g$$

The fall time equals the total flight time minus the rise time.

$$t_{\text{fall}} = t_{\text{tft}} - t_{\text{rise}}$$

$$= \frac{2v_0}{g} - \frac{v_0}{g} = \frac{v_0}{g}$$

The rise and fall time are both equal to v_0/g.

12. When it is caught, $y = y_0$, so $v^2 = v_0^2$ and $v = \pm v_0$.

13. $\Delta x = \displaystyle\int_{t_1}^{t_2} v\, dt = \int_2^5 (3t^2 + 4)\, dt$

$= \left(3\dfrac{t^3}{3} + 4t\right)\Big|_2^5 = 145 - 16 = \underline{129\text{ m}}$

14. $\underline{6\text{ m/s}}$

15. Let the instant that the driver of the car hits the brakes be $t = 0$. Afterward, the velocity v_c of the car is

$$v_c = v_0 + at$$

where v_0 is the initial velocity of the car and a is the car's acceleration. If the collision is to be a gentle kiss, the velocity of the car and the velocity v_t of the truck must be equal at impact. This allows us to find an expression for the time of impact t_i:

$$v_c = v_t$$
$$v_0 + at_i = v_t$$

so

$$t_i = -\frac{v_0 - v_t}{a} \tag{1}$$

The positions of the car and truck are

$$x_c = v_0 t + \tfrac{1}{2}at^2$$

and

$$x_t = x_0 + v_t t$$

where x_0 is the initial position of the truck, 28 m ahead of the car. The positions of the car and truck are equal when $t = t_i$. Equating the positions at this time gives

$$x_t = x_c$$
$$x_0 + v_t t_i = v_0 t_i + \tfrac{1}{2}at_i^2$$

or

$$x_0 = (v_0 - v_t)t_i + \tfrac{1}{2}at_i^2$$

Substituting in the expression for t_i that we obtained in Equation 1 gives

$$x_0 = (v_0 - v_t)\left(-\frac{v_0 - v_t}{a}\right) + \frac{1}{2}a\left(\frac{v_0 - v_t}{a}\right)^2$$

$$x_0 = -\frac{(v_0 - v_t)^2}{2a}$$

Solving this for a gives

$$a = -\frac{(v_0 - v_t)^2}{2x_0} = -\frac{[(18\text{ m/s}) - (8\text{ m/s})]^2}{2(28\text{ m})}$$

$$= \underline{-1.79\text{ m/s}^2}$$

Thus, the acceleration of the car when the car and truck just touch is $a = -1.79\text{ m/s}^2$. For any negative acceleration greater than this, the collision will be avoided.

16. Al catches Bob at $t = \underline{54.5\text{ s}}$ and $x = \underline{1090\text{ m}}$.

17. Let L ($= 400$ yd) be the length of the moving sidewalk. The distance you walk to the gate is $0.75L$ ($= 300$ yd) so the equation, distance equals the walking speed v times the time t, is

$$0.75L = vt \tag{1}$$

The velocity of your twin—while on the moving sidewalk—relative to the gate, v_{tg}, equals the velocity of your twin relative to the sidewalk, v_{ts}, plus the velocity of the sidewalk relative to the gate, v_{sg}. That is, $v_{tg} = v + v_{sg}$. Since v_{ts} equals the walking speed v ($= 3$ mi/h), it follows that

$$L = v_{tg}t_1 = (v + v_{sg})t_1 \tag{2}$$

where t_1 is the time he is on the moving sidewalk.

Let t_2 be the time for your twin to walk the distance $0.25L$ from the sidewalk exit back to the gate. Then,

$$0.25L = vt_2 \tag{3}$$

You both arrive at the gate at the same time. Thus

$$t = t_1 + t_2 \tag{4}$$

Solving Equation (1) for t, Equation (2) for t_1, and Equation (3) for t_2 gives

$$t = \frac{0.75L}{v}, \quad t_1 = \frac{L}{v + v_{sg}}, \quad \text{and } t_2 = \frac{0.25L}{v}$$

Substituting these expressions into Equation (4) we obtain

$$\frac{0.75L}{v} = \frac{L}{v + v_{sg}} + \frac{0.25L}{v}$$

Dividing through by L and reordering gives

$$\frac{1}{v + v_{sg}} = \frac{0.75}{v} - \frac{0.25}{v} = \frac{0.5}{v}$$

Taking the reciprocal of both sides we have

$$v + v_{sg} = \frac{v}{0.5} = 2v$$

Solving for the v_{sg} we obtain

$$v_{sg} = 2v - v = v = \underline{3.0 \text{ mi/h}}$$

Remark 1: Another way to solve this problem is to realize that at the instant your twin exits the moving sidewalk both you and your twin must be the same distance $(0.25L)$ from the gate. At this instant, relative to the gate you will have moved a distance of $0.50L$ while your twin has traveled a distance L. It follows that while on the sidewalk your twin's speed, relative to the gate, is twice the walking speed. Thus the sidewalk's speed is equal to the walking speed (3.0 mi/h).

Remark 2: To save you from doing extra work, it is a good idea to simplify solutions with variables as much as possible before inserting actual numbers. In this case, you might have initially thought that solving the problem would require converting yards to miles or vice versa. However, by working through the solution with variables, you find that only the ratios of distances are needed; in the final expression for sidewalk speed, none of the actual distances are required and conversion from yards to miles (or vice versa) is not necessary.

18. (*a*) $\underline{40 \text{ min}}$ (*b*) $\underline{0.50 \text{ mi}}$ (*c*) $\underline{0.67 \text{ mi}}$

Remark: You swim at 1.00 mi/h relative to the water for the entire 40 min. Thus, relative to the water you swim a distance of

$$(1.00 \text{ mi/h})(40 \text{ min})[(1 \text{ h})/(60 \text{ min})] = 0.67 \text{ mi.}$$

Chapter 3

Motion in Two and Three Directions

I. Key Ideas

The motion of a particle can be described in terms of its position, velocity, and acceleration, all of which are vectors. (Often the motion of an object can be represented by the motion of a single point, and then the object can be treated as a particle.) **Vectors** are quantities that have both magnitude and direction. For example, consider a train with a velocity of 80 mi/h toward the northwest. The speed of the train—how fast it's going—is the magnitude of the velocity vector, while northwest—the direction it's going—is the direction of the velocity vector. The tug of a rope pulling a sled is an example of another vector quantity—force.

Scalars are quantities that have magnitude but no direction, like the volume of water in a swimming pool or the temperature of a cup of coffee.

In printed materials vectors are frequently represented by boldfaced symbols, for example, **A** and **B**. Both here and in the text they will be represented by placing an arrow over an italicized, boldfaced symbol; for example, $\vec{A}$ and $\vec{B}$.

3-1 The Displacement Vector

The **displacement vector** of a moving particle is the quantity that is graphically represented by an arrow with a tail that starts at the initial position of the particle and a head that ends at its final position. The length and direction of this arrow are the magnitude and direction of the displacement vector. As illustrated in Figure 3-1, the displacement vector depends only on the initial and final positions of the particle, not on the path taken by the particle as it moves from the initial to the final position.

3-2 General Properties of Vectors

All vectors combine by the same rules. If you understand the rules by which displacements combine, you'll understand the rules by which all vectors combine. Figure 3-2a shows a particle that moves from point 1 to point 2 and then from point 2 to point 3. The displacement $\vec{A}$ (from point 1 to point 2) followed by displacement $\vec{B}$ (from point 2 to point 3) is equivalent to displacement $\vec{C}$ (from point 1 to point 3). In the mathematics of vectors, this combining of $\vec{A}$ and $\vec{B}$, which results in $\vec{C}$, is called **vector addition**. Figure 3-2b illustrates the vector addition of $\vec{A}$ and $\vec{B}$ to get $\vec{C}$; that is $\vec{A} + \vec{B} = \vec{C}$. The prescription to follow, in order to add $\vec{B}$ to $\vec{A}$ to get $\vec{C}$, is to draw the arrow representing $\vec{A}$, then draw the arrow representing $\vec{B}$ with its tail at the head of the arrow representing $\vec{A}$. Then draw the arrow representing $\vec{C}$ from the tail of $\vec{A}$ to the head of $\vec{B}$. The arrows, of course, must be drawn to some scale, and the direction of the arrows is the same as the direction of the actual displacements.

Vector subtraction We are all familiar with the subtraction of ordinary numbers. In the equation $a - b = d$, d is the number that must be added to b to

Figure 3-1

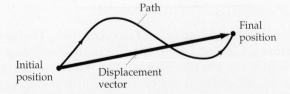

Figure 3-2

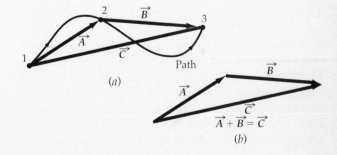

17

Figure 3-3

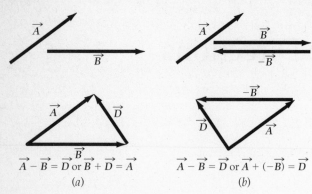

$$\overrightarrow{A} - \overrightarrow{B} = \overrightarrow{D} \text{ or } \overrightarrow{B} + \overrightarrow{D} = \overrightarrow{A} \qquad \overrightarrow{A} - \overrightarrow{B} = \overrightarrow{D} \text{ or } \overrightarrow{A} + (-\overrightarrow{B}) = \overrightarrow{D}$$
$$(a) \qquad\qquad\qquad (b)$$

get a; that is, $b + d = a$. Analogously, in the vector equation $\overrightarrow{A} - \overrightarrow{B} = \overrightarrow{D}$, $\overrightarrow{D}$ is the vector that must be added to $\overrightarrow{B}$ to get $\overrightarrow{A}$; that is $\overrightarrow{B} + \overrightarrow{D} = \overrightarrow{A}$, as shown in Figure 3-3a.

The vector $-\overrightarrow{B}$ is equal in magnitude and opposite in direction to the vector $\overrightarrow{B}$. Figure 3-3b illustrates that subtracting $\overrightarrow{B}$ from $\overrightarrow{A}$ is equivalent to adding $-\overrightarrow{B}$ to $\overrightarrow{A}$. Note that vector $\overrightarrow{A}$ has the same direction and length in both Figure 3-3a and 3-3b— as do vectors $\overrightarrow{B}$ and $\overrightarrow{D}$.

Addition of Vectors by Components When a particle undergoes a displacement $\overrightarrow{A}$, the x component of the displacement A_x is the number of units the particle travels in the $+x$ direction, as illustrated in Figure 3-4. If the particle moves in the $-x$ direction, A_x is a negative number.

Figure 3-4

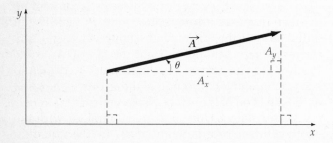

Component-vector relations

$$A_x = A \cos \theta \qquad A = \sqrt{A_x^2 + A_y^2}$$
$$A_y = A \sin \theta \qquad \tan \theta = A_y / A_x$$

Components of vectors are not themselves vectors, but are ordinary signed (real) numbers that can be added, subtracted, multiplied, and divided like ordinary numbers. When adding vectors, the x component of the sum equals the sum of the x components of the individual vectors. The corresponding relation is valid for the y and z components. Thus

if $\overrightarrow{C} = \overrightarrow{A} + \overrightarrow{B}$, then $C_x = A_x + B_x$, $C_y = A_y + B_y$, and $C_z = A_z + B_z$.

For an application using vector components we look to the American game of football. In this game, the yards gained during each play equals the forward component of the displacement $\overrightarrow{D}$ of the football during the play (see Figure 3-5); that is, yards gained = $D \cos \theta$. The forward component of the football's displacement is negative if the play ends with the ball behind the line of scrimmage (an imaginary line across the field through the position of the ball at the start of each play). In a series of plays, the total yards gained is the sum of the forward components of the ball's displacements for each play.

Figure 3-5

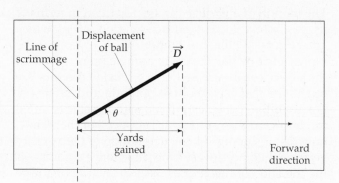

Unit Vectors and Multiplication of Vectors by Scalars **Unit vectors** are vectors that are used to indicate directions, so they are dimensionless with a magnitude of unity. (A unit vector can be formed by dividing a vector by its magnitude.) The unit vectors $\hat{i}$, $\hat{j}$, and $\hat{k}$ are commonly used to indicate the positive x, y, and z directions, respectively. The vector $A_x \hat{i}$ is the product of the component A_x and the unit vector $\hat{i}$. This vector is parallel to the x axis (or antiparallel if A_x is negative) and has a magnitude $|A_x|$. A general vector $\overrightarrow{A}$ can thus be written as the sum of three vectors, each of which is parallel to a coordinate axis. That is,

$$\overrightarrow{A} = A_x \hat{i} + A_y \hat{j} + A_z \hat{k}$$

If a displacement of 3 m to the east is denoted $\overrightarrow{A}$, a sequence of two such displacements is denoted $2\overrightarrow{A}$; that is, $\overrightarrow{A} + \overrightarrow{A} = 2\overrightarrow{A}$. The generalization of this process is that the product of a number s (for scalar) and a vector $\overrightarrow{A}$ is a vector whose magnitude is $|sA|$ and whose direction is the same as that of $\overrightarrow{A}$ if s is positive and opposite to $\overrightarrow{A}$ if s is negative.

3-3   *Position, Velocity, and Acceleration* Position vectors are vectors from the origin of a coordinate system to the location of a particle. As shown in Figure 3-6, the position vector $\overrightarrow{r}$ of a particle has com-

Figure 3-6

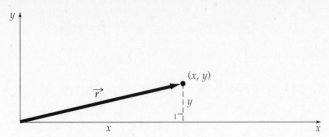

ponents x and y, which are the coordinates of the location of the particle.

The **instantaneous-velocity vector** is the time rate of change of the position vector. Its direction is the direction of motion of the particle. Speed is the magnitude of the velocity. The equations relating velocity, position, and time are

Velocity

$$\vec{v}_{av} = \frac{\Delta \vec{r}}{\Delta t} \qquad \vec{v} = \lim_{\Delta t \to 0} \frac{\Delta \vec{r}}{\Delta t} = \frac{d\vec{r}}{dt}$$

or

$$v_{x,av} = \frac{\Delta x}{\Delta t} \qquad v_x = \lim_{\Delta t \to 0} \frac{\Delta x}{\Delta t} = \frac{dx}{dt}$$

$$v_{y,av} = \frac{\Delta y}{\Delta t} \qquad v_y = \lim_{\Delta t \to 0} \frac{\Delta y}{\Delta t} = \frac{dy}{dt}$$

$$v_{z,av} = \frac{\Delta z}{\Delta t} \qquad v_z = \lim_{\Delta t \to 0} \frac{\Delta z}{\Delta t} = \frac{dz}{dt}$$

Relative Velocity A person is standing on a railroad car that is moving relative to the ground. Consider the position of the person relative to the moving car $\vec{r}_{pc}$, the position of the person relative to the ground $\vec{r}_{pg}$, and the position of the car relative to the ground $\vec{r}_{cg}$. These relative position vectors are illustrated in Figure 3-7, where g, p, and c represent points fixed to the ground, person, and car, respectively.

Figure 3-7

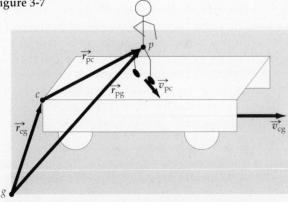

The relation between these position vectors is $\vec{r}_{pg} = \vec{r}_{pc} + \vec{r}_{cg}$. Because points p and c are moving, a short time later they will be at slightly different locations. The changes in their position are related by the equation $\Delta \vec{r}_{pg} = \Delta \vec{r}_{pc} + \Delta \vec{r}_{cg}$. By dividing each term of this last equation by the time interval t, and identifying the terms (in the limit that the time interval t approaches zero) as relative velocity vectors (see Figure 3-8), we get

Relative velocity equation

$$\vec{v}_{pg} = \vec{v}_{pc} + \vec{v}_{cg}$$

Figure 3-8

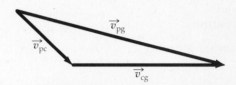

Relative velocity problems can often be solved by using only this equation. However, it is often helpful to recognize that the velocity of the person relative to the ground and the velocity of the ground relative to the person are equal in magnitude and opposite in direction. That is $\vec{v}_{gp} = -\vec{v}_{pg}$.

The Acceleration Vector The **instantaneous-acceleration vector** is the time rate of change of the velocity vector. Only when the magnitude and the direction of the velocity are both fixed is the acceleration zero. The equations relating acceleration, velocity, and time are

Acceleration

$$\vec{a}_{av} = \frac{\Delta \vec{v}}{\Delta t} \qquad \vec{a} = \lim_{\Delta t \to 0} \frac{\Delta \vec{v}}{\Delta t} = \frac{d\vec{v}}{dt}$$

3-4 Projectile Motion Projectiles are objects moving under the influence of only gravity, such as the massive sphere that is thrown by a shot-putter (neglecting air resistance). Near the earth's surface the horizontal component of a projectile's acceleration is zero and the vertical component of its acceleration is a constant $g = 9.81$ m/s$^2 \pm 0.03$ m/s^2 in the downward direction. On the earth's surface g varies slightly with location, but is usually specified as 9.81 m/s^2. The projectile motion equations are

Projectile motion relations

$$v_{0x} = v_0 \cos \theta \qquad v_{0y} = v_0 \sin \theta$$

$$v_x = v_{0x} \qquad v_y = v_{0y} - gt$$

$$\Delta x = v_{0x} t \qquad \Delta y = v_{0y} t - \tfrac{1}{2} g t^2$$

where the positive y direction is upward, the x axis is horizontal, and θ is the angle of the initial velocity vector above the horizontal.

Many projectile problems involving both horizontal x and vertical y motion can be successfully attacked by solving a y equation for time and then solving an x equation at that value of time.

II. Numbers and Key Equations

Numbers

Acceleration due to gravity

$$\vec{g} = -9.81 \text{ m/s}^2 \hat{j} \qquad g = |\vec{g}| = 9.81 \text{ m/s}^2$$

Key Equations

Component-vector relations

$$A_x = A \cos \theta \qquad A = \sqrt{A_x^2 + A_y^2}$$

$$A_y = A \sin \theta \qquad \tan \theta = A_y / A_x$$

Velocity

$$\vec{v}_{av} = \frac{\Delta \vec{r}}{\Delta t} \qquad \vec{v} = \lim_{\Delta t \to 0} \frac{\Delta \vec{r}}{\Delta t} = \frac{d\vec{r}}{dt}$$

or

$$v_{x, av} = \frac{\Delta x}{\Delta t} \qquad v_x = \lim_{\Delta t \to 0} \frac{\Delta x}{\Delta t} = \frac{dx}{dt}$$

$$v_{y, av} = \frac{\Delta y}{\Delta t} \qquad v_y = \lim_{\Delta t \to 0} \frac{\Delta y}{\Delta t} = \frac{dy}{dt}$$

$$v_{z, av} = \frac{\Delta z}{\Delta t} \qquad v_z = \lim_{\Delta t \to 0} \frac{\Delta z}{\Delta t} = \frac{dz}{dt}$$

Acceleration

$$\vec{a}_{av} = \frac{\Delta \vec{v}}{\Delta t} \qquad \vec{a} = \lim_{\Delta t \to 0} \frac{\Delta \vec{v}}{\Delta t} = \frac{d\vec{v}}{dt}$$

or

$$a_{x, av} = \frac{\Delta v_x}{\Delta t} \qquad a_x = \lim_{\Delta t \to 0} \frac{\Delta v_x}{\Delta t} = \frac{dv_x}{dt}$$

$$a_{y, av} = \frac{\Delta v_y}{\Delta t} \qquad a_y = \lim_{\Delta t \to 0} \frac{\Delta v_y}{\Delta t} = \frac{dv_y}{dt}$$

$$a_{z, av} = \frac{\Delta v_z}{\Delta t} \qquad a_z = \lim_{\Delta t \to 0} \frac{\Delta v_z}{\Delta t} = \frac{dv_z}{dt}$$

Relative velocity equations

$$\vec{v}_{AC} = \vec{v}_{AB} + \vec{v}_{BC} \qquad \vec{v}_{AB} = -\vec{v}_{BA}$$

Constant acceleration equations

$$\Delta \vec{r} = \vec{v}_0 \Delta t + \tfrac{1}{2} \vec{a} (\Delta t)^2 \qquad \vec{v} = \vec{v}_0 + \vec{a} \Delta t$$

or

$$\Delta x = v_{0x} \Delta t + \tfrac{1}{2} a_x (\Delta t)^2 \qquad v_x = v_{0x} + a_x \Delta t$$

$$\Delta y = v_{0y} \Delta t + \tfrac{1}{2} a_y (\Delta t)^2 \qquad v_y = v_{0y} + a_y \Delta t$$

where

$$v_{0x} = v_0 \cos \theta \qquad v_{0y} = v_0 \sin \theta$$

and θ is the angle the initial velocity vector makes with the positive x axis.

For projectile motion use the constant acceleration equations with $\vec{a} = \vec{g}$ and the $+y$ direction vertically upward. That is, with

Projectile motion relations

$$a_x = 0 \qquad \text{and} \qquad a_y = -g$$

III. Potential Pitfalls

Don't forget to use appropriate notation to indicate vectors and unit vectors.

When you add vectors, remember that you can't simply add their magnitudes. The equation $\vec{C} = \vec{A} + \vec{B}$ does not mean that $C = A + B$. This is true only if $\vec{A}$ and $\vec{B}$ are in the same exact direction.

The displacement between two points is defined as the straight-line magnitude and direction between them "as the crow flies." Its magnitude is not necessarily equal to the distance actually traveled by the particle between the points.

It can be challenging to arrive at the equations for relative velocities. To be successful at this, it is best to consistently use the subscript notation described below. With this notation, $\vec{v}_{AB}$ is the velocity of A relative to B, while $\vec{v}_{BA}$ is the velocity of B relative to A. The relation between these velocities is expressed by the equation $\vec{v}_{AB} = -\vec{v}_{BA}$. (This equation formally expresses the observation that if A is moving north at 30 m/s relative to B, then B must be moving south at 30 m/s relative to A.)

Consider reference frames A, B, and C that are in motion relative to each other. An equation for the relative velocities of these frames is $\vec{v}_{AC} = \vec{v}_{AB} + \vec{v}_{BC}$. Note that the left subscript on both sides of this equation is A, the right subscript on both sides is C, and the adjacent interior subscripts are both Bs. When an additional moving frame D is considered, an equation relating the velocities is $\vec{v}_{AD} = \vec{v}_{AB} + \vec{v}_{BC} + \vec{v}_{CD}$. There are two subscript rules for these equations:

Rule 1. The left exterior subscripts of both sides of the equation must be identical and the right exterior subscripts must be identical. In the equation $\vec{v}_{AD} = \vec{v}_{AB} + \vec{v}_{BC} + \vec{v}_{CD}$, A and D are the exterior subscripts for both sides of the equation.

Rule 2. On either side of the equation, adjacent subscripts of adjacent relative velocity terms must be identical. That is, for the expression $\vec{v}_{AB} + \vec{v}_{BC} + \vec{v}_{CD}$, B is the common adjacent subscript for the first and second terms whereas C is the common one for the second and third terms.

The projectile motion equations apply only when an object is in free fall, that is, when the only thing affecting its motion is gravity. If air resistance affects its motion these equations do not apply.

Consider a projectile thrown straight upwards. Don't think that when it reaches the highest point in its trajectory both its velocity and its acceleration are zero. Its velocity, which is momentarily zero at its peak, is changing which means its acceleration is not zero. In fact, its acceleration is the same at the top as it is on the way up and on the way down—9.81 m/s², directed downward.

Consider a projectile thrown both upwards and to the side, like a shot put. Don't think that when such a projectile is at the top of its arc its velocity is zero. Only the y component of its velocity is zero. (The x component of its velocity is constant throughout the motion.)

For motion in one dimension, constant speed means that the acceleration is zero. This is not true for motion along a curved path. As long as the direction of the velocity vector is changing, even if the speed is constant, the acceleration vector cannot be zero.

IV. True and False and Responses

True and False

_____ 1. Three dimensions are necessary to describe the general motion of a particle.

_____ 2. The displacement of a particle is the change in its position.

_____ 3. A component of a vector is itself a vector.

_____ 4. All vector quantities have both a magnitude and a direction, and all follow the same rules for vector addition that consecutive displacements follow.

_____ 5. The magnitude of a vector quantity is a dimensionless number.

_____ 6. Both velocity and time must be expressed as vector quantities.

_____ 7. Vectors may be added by adding corresponding components.

_____ 8. In Figure 3-9, $\vec{A} + \vec{C} = \vec{B} + \vec{D} + \vec{E}$.

_____ 9. In Figure 3-9, $\vec{A} + \vec{B} + \vec{C} = \vec{D} - \vec{E}$.

Figure 3-9

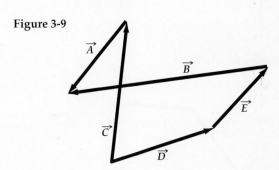

_____ 10. For any two vector quantities $\vec{P}$ and $\vec{Q}$, $\vec{P} - \vec{Q}$ is a vector having the same magnitude as, but a direction opposite to, $\vec{P} + \vec{Q}$.

_____ 11. When a particle is in motion, the difference between its position vector at time t_1 and its position vector at time t_2 is its displacement during the time interval $t_2 - t_1$.

_____ 12. The instantaneous-velocity vector is always in the direction of motion.

_____ 13. If a particle moves in a straight line, its position and velocity vectors are parallel.

_____ 14. The magnitude of the acceleration vector is equal to the rate at which the _speed_ of the particle changes with respect to time.

_____ 15. A projectile is a body in motion that cannot be treated as a particle.

_____ 16. The time it takes for a bullet fired horizontally to reach the ground is the same as if it were dropped from rest from the same height.

Responses

1. True

2. True. This is the definition of displacement in fact.

3. False. A component of a vector is the projection of the vector along a given direction. It is a scalar quantity. Thus, if $\vec{A} = (6\hat{i} + 8\hat{j})$ ft, its x and y components are 6 ft and 8 ft, respectively.

4. True

5. False. It has no direction but it may well have dimensions. Thus, if $\vec{A} = (6\hat{i} + 8\hat{j})$ ft, the magnitude $|A| = 10$ ft.

6. False. Time has no direction in space.

7. True

8. True

9. False

10. False. This describes not $\vec{P} - \vec{Q}$ but $-\vec{P} - \vec{Q}$. (See also Question 2.)

11. True

12. True

13. False. See Figure 3-10.

Figure 3-10

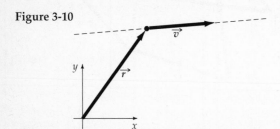

14. False. For instance, consider something moving along a curved path at constant speed. So long as the direction of the velocity vector is changing, the acceleration cannot be zero.

15. False. In fact, we have been assuming that projectiles are particles.

16. True. This is because the initial *vertical* component of the velocity is zero in either case, so the vertical motions are the same.

V. Questions and Answers

Questions

1. You throw a baseball from the outfield to a friend at home plate. In general, the distance the ball travels is not equal to the magnitude of its displacement vector. Which of the two is larger?

2. Two displacement vectors $\vec{S_1}$ and $\vec{S_2}$ add to give a resultant of zero. What can we say about the two displacements?

3. Is it possible to drive your car around a curve without accelerating?

4. Suppose you have "sighted in" a rifle so that it will hit whatever the sights are aimed directly at on a target 150 m away over level ground. When shooting at a target 90 m away, should you aim above, below, or directly at the target?

5. Describe briefly what kind of motion a particle is undergoing when (a) the position vector changes in magnitude but not in direction; (b) the velocity vector changes in magnitude but not in direction; (c) the position vector changes in direction but not in magnitude; and (d) the velocity vector changes in direction, but not in magnitude.

Answers

1. The magnitude of the displacement is the straight-line distance between the two points, so the actual distance cannot be less than the magnitude of the displacement. They would be equal only if the path of the ball were a straight line. Since the ball must rise and fall some in flight, the distance it travels must be greater than the magnitude of the displacement.

2. The equation $\vec{S_1} + \vec{S_2} = 0$ means that $\vec{S_1} = -\vec{S_2}$, so the displacements are equal and opposite. If these are the two legs of a trip, then the second leg is a return to the starting point.

3. No. The question points out the difference between everyday usage and the scientific meaning of the term "accelerating." Scientifically speaking, acceleration occurs whenever the magnitude or direction of the velocity vector is changing. Since your velocity is at least changing in direction, you cannot drive around a curve without accelerating.

4. When you "sight in" a rifle at a particular distance, you are adjusting the sights to compensate for the distance by which the bullet falls below its initial line of flight. The axis of the barrel thus points above the sight line. When the target is closer than the distance for which the rifle was sighted, the bullet will fall a shorter distance. If you aim at the target, the bullet will be high when it gets there, so you should aim *below* the target to compensate.

5. (a) In this case, the particle is moving in a straight line toward or away from the origin. (b) The particle is moving in a straight line (since the direction of v isn't changing) with variable speed. (c) This would mean that the moving particle stays at the same distance from the origin, so it must be traveling in a circle around the origin. (d) This indicates that whatever the moving particle is doing, it's doing it at constant speed.

VI. Problems and Solutions

Problems

1. At 3:00 P.M. you pass mile marker 160 as you are driving due south on Interstate 77. At mile marker

138, you turn off on U.S. 54. At 3:40 P.M. you have gone 14 mi due southwest on U.S. 54. What was your average velocity (magnitude and direction) over this 40-min interval?

How to Solve It
• Where is your car at the end of the 40 min?

• What was the net displacement of your car in this interval?

• Average velocity is this displacement divided by the time interval. Pay attention to units when you do the arithmetic.

2. A car is traveling along a mountain road. At a certain instant it is moving due west across level ground at a speed of 18 m/s; 2 s later it is moving north down a steep hill at an angle of 15° below the horizontal at 10 m/s. Calculate the car's average acceleration over this time interval.

How to Solve It
• Calculate the *change* in the car's velocity over the 2-s time interval.

• The change in velocity divided by the time interval is the definition of average acceleration.

3. A pilot starts a trip from St. Louis to Memphis, which is 385 km due south. She flies due south at her maximum airspeed of 90 m/s but fails to correct for a crosswind of 15 m/s directed due east. (*a*) What is her speed relative to the ground? (*b*) At the time she expects to reach Memphis, where will she actually be? (*c*) How much longer will it take her to get to Memphis?

How to Solve It
• The airplane's velocity with respect to the air is 90 m/s due south. Add this vectorially to the wind velocity to get the airplane's ground velocity.

• How long did the pilot expect it to take to get to Memphis?

• At the time she thought she would get to Memphis, how far away is she in fact?

4. The current at a certain point where the Wabash River is 80 m wide flows at 0.4 m/s. A swimmer sets out for a point directly across the river. (*a*) If the swimmer's maximum speed is 0.75 m/s relative to still water, in what direction should he swim to go directly to his goal? (*b*) How long will it take him to get there?

How to Solve It
• The swimmer's velocity relative to the river bank is the vector sum of the velocity of the swimmer relative to the water plus the velocity of the water relative to the bank.

• Use this to find the direction in which he should swim.

• When you know this, you can calculate how far he has to go through the water and therefore how long it will take him.

5. A ball rolls off a tabletop 0.9 m above the floor and lands on the floor 2.6 m away from a point that is directly under the edge of the table. At what speed did it roll off?

How to Solve It
• Remember that the horizontal and vertical components of the ball's motion can be treated independently.

• How long did it take the ball to fall 0.9 m?

• If it moved horizontally 2.6 m in this time, what was its horizontal speed?

6. A diver leaps from a springboard 6 m above the water with a velocity of 6 m/s at an angle 30° from the vertical. How far from a point directly beneath the end of the board, and at what speed, does she hit the water?

How to Solve It
• Remember that the horizontal and vertical components of her motion can be treated independently.

• What is the vertical component of her initial velocity? Given this, find how long it takes her to hit the water.

• How far does she move horizontally in this time?

7. A basketball player takes a shot when he is standing 24 ft from the 10-ft high basket as shown in Figure 3-11. If he releases the ball at a point 6 ft from the floor and at an angle of 40° above the horizontal, with what speed must he throw the ball for it to go through the hoop?

Figure 3-11

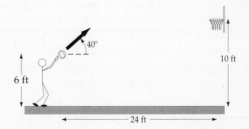

How to Solve It
• In terms of the unknown speed v_0, how much time does it take the ball to reach the basket 24 ft away horizontally?

• Write the equation for the vertical component of the ball's motion in terms of this time. When it reaches the basket, you know that the ball is 4 ft above the point from which it was thrown.

• This equation can be solved for v_0.

8. General Lee's artillery is taking aim on Union troops on the cliffs across the river, 300 m away. (See Figure 3-12.) The muzzle velocity of the cannonballs is 67 m/s. In order to hit the enemy, the cannonballs have to be fired at an angle of elevation of 31°. How high are the cliffs?

Figure 3-12

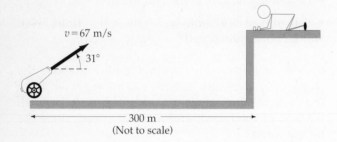

$v = 67$ m/s

31°

300 m
(Not to scale)

How to Solve It
• First, calculate the time a cannonball takes to reach the other side of the river, 300 m away.

• At this time, what is the altitude of the projectile above the point from which it was fired?

• Notice that in both Problems 7 and 8, although you aren't asked for the time, it comes in as an intermediate calculated quantity. If you like algebra, you can eliminate the time between the x and y equations. The result is an equation for y in terms of x, which you can apply directly to this sort of problem.

9. In a game of American football, a quarterback throws a pass from his 20-yard line with an initial velocity of 22 m/s at an angle of 40° above the horizontal. The receiver starts running downfield from the 30-yard line 1.0 s before the pass is thrown. With what speed must he run in order to catch the pass? (Assume that the pass is caught at the same height above the ground at which it was thrown. If you need to convert units, 1 m = 1.09 yd.)

How to Solve It
• In this problem you need to know how long the ball is in the air and how long the receiver has to run for the pass.

• Calculate where the pass lands and, from this, how far the receiver has to go.

• The distance he has to run, divided by the time he has to do it in, gives you his velocity. Is the answer reasonable?

10. In a baseball game, a batter hits a fly ball directly toward the center fielder. The bat strikes the ball at a point 1.1 m directly above home plate, and the ball leaves the bat at a speed of 29.5 m/s at an angle of 35° above the horizontal. The center fielder is standing 116 m from home plate. If the center fielder starts running at the instant the ball is hit and if he catches the ball 1.1 m above the ground, how fast must he run to catch the ball? Assume that air resistance can be neglected.

How to Solve It
• From the vertical motion determine the ball's time of flight.

• Calculate where the ball will be caught. How far does the fielder have to run to reach it?

• Now calculate how fast the fielder has to run to make the catch.

Solutions

1. (See Figure 3-13.) Before turning, you went 22 mi south (from mile marker 160 to mile marker 138). Call this displacement $\vec{s}_1$. After turning, you went 14 mi southwest. Call this displacement $\vec{s}_2$. If we call east the x direction and north the y direction, then in components

$$s_{1x} = 0 \quad \text{and} \quad s_{1y} = -22 \text{ mi}$$
$$s_{2x} = s_{2y} = -(14 \text{ mi}) \cos 45° = -9.90 \text{ mi}$$

Figure 3-13

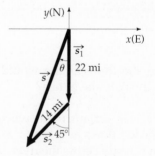

y(N)

x(E)

$\vec{s}_1$

θ 22 mi

$\vec{s}$

14 mi

45°
$\vec{s}_2$

The total displacement is $\vec{s} = \vec{s}_1 + \vec{s}_2$ and its components are

$$s_x = s_{1x} + s_{2x} = 0 + (-9.90 \text{ mi}) = -9.90 \text{ mi}$$
$$s_y = s_{1y} + s_{2y} = (-22 \text{ mi}) + (-9.90 \text{ mi})$$
$$= -31.9 \text{ mi}$$

So

$$s = \sqrt{s_x^2 + s_y^2} = \sqrt{(-9.90)^2 + (-31.9)^2}$$

$$= 33.4 \text{ miles}$$

Its direction, the angle θ, is given by

$$\tan \theta = \frac{s_x}{s_y} = \frac{-9.90 \text{ mi}}{-31.9 \text{ mi}} = 0.310$$

$$\theta = 17°$$

So your net displacement $\vec{s}$ was 33.4 mi directed 17° west of south. Since $\Delta \vec{r} = \vec{s}$, the average velocity is

$$\vec{v}_{av} = \frac{\Delta \vec{r}}{\Delta t} = \left(\frac{33.4 \text{ mi}}{40 \text{ min}}\right)\left(\frac{60 \text{ min}}{1 \text{ h}}\right)$$

$$= \underline{50.1 \text{ mi/h directed } 17° \text{ west of south}}$$

2. $\vec{a}_{av} = \underline{10.3 \text{ m/s}^2}$ directed $\underline{28° \text{ north of east and}}$ $\underline{\text{about } 7° \text{ below the horizontal.}}$

3. (*a*) See Figure 3-14. Let $\vec{v}_{AG}$ be the wind velocity, that is, the velocity of the air relative to the ground. Here $\vec{v}_{AG}$ is 15 m/s due east. The pilot's velocity relative to the air $\vec{v}_{PA}$ is 90 m/s due south. Her velocity relative to the ground $\vec{v}_{PG}$ is the vector sum of this and the air's velocity relative to the ground $\vec{v}_{AG}$:

$$\vec{v}_{PG} = \vec{v}_{PA} + \vec{v}_{AG}$$

$$v_{PG} = \sqrt{v_{PA}^2 + v_{AG}^2}$$

$$= \sqrt{90^2 + 15^2} = \underline{91.2 \text{ m/s}}$$

and

$$\tan \theta = \frac{15}{90}$$

so $\theta = 9.5°$ east of due south.

Figure 3-14

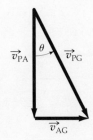

(*b*) Neglecting the crosswind, she expected the trip to take $(385 \text{ km})/(90 \text{ m/s}) = 4280 \text{ s} = 1.19 \text{ h}$. Her ac-

tual ground velocity, however, is 90 m/s south and 15 m/s east, so in this time she has moved (15 m/s) $(4280 \text{ s}) = 64{,}200 \text{ m} = \underline{64.2 \text{ km eastward}}$ of her intended flight path. Thus, she still has to go 64.2 km west to get to Memphis.

(*c*) To finish her trip she has to fly directly upwind. Her ground velocity is thus $90 \text{ m/s} - 15 \text{ m/s} = 75 \text{ m/s}$, and it takes

$$\left(\frac{64.2 \text{ km}}{75 \text{ m/s}}\right)\left(\frac{1000 \text{ m}}{1 \text{ km}}\right) = \underline{856 \text{ s}}$$

or a little under 15 minutes to get there.

4. He must swim $\underline{32.2°}$ upstream from straight across; it takes him $\underline{126 \text{ s}}$.

5. Since the ball is falling freely with an initial vertical velocity of zero, we can use the equation $y = y_0 + v_{0y}t - \frac{1}{2}gt^2$ to find the time it is in the air:

$$0 = (0.9 \text{ m}) + 0 - \tfrac{1}{2}(9.81 \text{ m/s}^2)t^2$$

so

$$t = 0.428 \text{ s}$$

In this time, the ball moved 2.6 m horizontally at constant speed, so

$$v_x = \frac{x}{t} = \frac{2.6 \text{ m}}{0.428 \text{ s}} = \underline{6.07 \text{ m/s}}$$

6. $\underline{5.27 \text{ m}}$; $\underline{12.4 \text{ m/s}}$

7. Let the horizontal component of the ball's initial velocity be

$$v_x = v_{0x} = v_0 \cos \theta = v_0 \cos 40° = 0.766 v_0$$

Then the time the ball is in the air is

$$t = \frac{\Delta x}{v_x} = \frac{24 \text{ ft}}{0.766 v_0} = \frac{31.3 \text{ ft}}{v_0}$$

In this time, it has moved vertically

$$y = v_{0y}t - \tfrac{1}{2}gt^2$$

where

$$v_{0y} = v_0 \sin \theta = v_0 \sin 40° = 0.643 v_0$$

Using $g = 32.2$ ft/s^2, we obtain

$$4 \text{ ft} = 0.643 v_0 \left(\frac{31.3 \text{ ft}}{v_0}\right) - \frac{1}{2}\left(32.2 \text{ ft/s}^2\right)\left(\frac{31.3 \text{ ft}}{v_0}\right)^2$$

$$16.1 \text{ ft} = \frac{15{,}800 \text{ ft}^3/\text{s}^2}{v_0^2}$$

$$v_0 = \underline{31.3 \text{ ft/s}}$$

8. <u>46.4 m</u>

9. If we call the level at which the ball is thrown and caught $y = 0$, then we can use the equation $\Delta y = v_{0y} t - \frac{1}{2}g t^2 = 0$ to find the time it is in the air. Rearranging this equation yields

$$t = \frac{2 v_{0y}}{g} = \frac{2 v_0 \sin \theta}{g} = \frac{(2)(22 \text{ m/s}) \sin 40°}{9.81 \text{ m/s}^2}$$

$$= 2.88 \text{ s}$$

In this time, the ball goes a horizontal distance

$$x = v_x t = v_0 (\cos \theta) t$$

$$= (22 \text{ m/s})(\cos 40°)(2.88 \text{ s})$$

$$= 48.5 \text{ m}\left(\frac{1.09 \text{ yd}}{\text{m}}\right) = 53.1 \text{ yd}$$

The receiver must therefore cover 43.1 yd in (2.88 s + 1 s) = 3.88 s and so would have to run at

$$v_x = \frac{x}{t} = \frac{43.1 \text{ yd}}{3.88 \text{ s}} = \underline{11.1 \text{ yd/s}}$$

This may be possible, but it's an awfully fast speed!

10. <u>9.46 m/s</u>

Chapter 4

Newton's Laws

I. Key Ideas

Newton's laws of motion pertain only to particles. Often the motion of an object can be represented by the motion of a single point, and then the object can be treated as a particle. Newton's laws are as follows:

Law 1. An object continues in its initial state of rest or motion with uniform velocity unless it is acted on by an unbalanced, or net external, force. The net force acting on an object is the vector sum of all the forces acting on it:

$$\vec{F}_{net} = \Sigma \vec{F}$$

Law 2. The acceleration of an object is inversely proportional to its mass and directly proportional to the net external force acting on it. That is,

$$\vec{a} = \frac{\vec{F}_{net}}{m} \quad \text{or} \quad \vec{F}_{net} = m\,\vec{a}$$

Law 3. Forces always occur in pairs. If a force $\vec{F}_{AB}$ is exerted by object A on object B, a force $\vec{F}_{BA}$ is exerted by object B on object A. These forces are equal in magnitude and oppositely directed. That is,

$$\vec{F}_{AB} = -\vec{F}_{BA} \quad \text{or} \quad \vec{F}_{AB} + \vec{F}_{BA} = 0$$

4-1 Newton's First Law: The Law of Inertia When the motion of an object is observed and measured, it must be observed and measured from a specific reference frame. According to Newton's first law, an object moves with constant velocity when there are no forces acting on it. Reference frames where this occurs are called **inertial reference frames.** Newton's three laws require all observations and measurements be taken in inertial reference frames.

4-2 Force, Mass, and Newton's Second Law Forces are pushes or pulls, like the tug of a rope pulling a sled. **Contact forces** are forces exerted on objects by other objects via physical contact (touching). **Action-at-a-distance forces,** like the pull of gravity, are forces that do not require that the objects physically touch.

The SI unit of force is the newton (N), and the SI unit of mass is the kilogram (kg). The kilogram is, by definition, the mass of the standard body—a platinum cylinder kept at the International Bureau of Weights and Measures in Sèvres, France. A net force of **one newton,** acting on the standard body, results in an acceleration of 1 m/s². Forces are defined by the acceleration they produce on the standard body.

A convenient agent for exerting forces on objects is a spring. It takes a force (a pull) to stretch a spring. To maintain the spring in a stretched condition requires that the force be maintained. The greater the force, the greater the extension of the spring. Consider a particular spring scale attached to a 1-kg body on a frictionless horizontal surface, as shown in Figure 4-1. When the spring scale is pulled to the right, the spring inside it stretches. The stretched spring pulls on the string, causing the block to accelerate. The greater the extension of the spring, the greater the acceleration. By noting the extension needed to produce a particular acceleration, and by using Newton's second law ($\vec{F}_{net} = m\vec{a}$), we can calibrate the spring scale. An acceleration of 1 m/s² means that the spring is exerting a force of 1 N; an acceleration of 2 m/s² means that the spring is exerting a force of 2 N; and so on.

Figure 4-1

Using our 1-kg block, we can calibrate other spring scales in the same way. For common springs, the force F_x exerted is proportional to the extension Δx (for small extensions). This means that for any two extensions Δx_1 and Δx_2, with forces F_{x1} and F_{x2} respectively, the ratios of the force to extension are equal. That is, $\dfrac{F_{x1}}{\Delta x_1} = \dfrac{F_{x2}}{\Delta x_2} = k$, where k is a constant that is different for different springs. This is known as **Hooke's law.**

Hooke's law

$$F_x = -k\,\Delta x$$

We can use our calibrated springs to exert known forces on other bodies and measure their acceleration. If equal net forces act on two objects with different masses, the objects will undergo different accelerations. Newton's second law tells us, that if the net forces are equal, the products of the corresponding masses and accelerations will also be equal. Therefore, measuring the accelerations when the net forces are equal allows the masses of objects to be compared:

Mass comparison with equal forces

$$m_1 a_1 = m_2 a_2 \quad \text{or} \quad \frac{m_1}{m_2} = \frac{a_2}{a_1}$$

Do two 1-N forces simultaneously acting in the same direction on a body produce the same acceleration as would a single 2-N force? To investigate this we can use our calibrated spring scales to simultaneously exert two or more forces on a single body. By measuring the acceleration we can determine how forces behave when acting in combination. In doing this we would find that two 1-N forces simultaneously acting in the same direction do indeed produce the same acceleration as would a single 2-N force.

4-3   *The Force Due to Gravity: Weight*   The weight $\vec{w}$ of an object is the gravitational force exerted on it. To determine the weight of an object, we measure the acceleration of an object when the only force acting on it is the gravitational force (the weight). When objects near the surface of the earth fall under the influence of *only* gravity, that is when air resistance is negligible, they all fall with the same acceleration $\vec{g}$, which is 9.8 m/s^2, downward. For these objects, the net force acting on them is their weight $\vec{w}$. It follows that for such falling objects, Newton's second law reduces to the relation between the object's weight and the object's mass, that is

Weight

$$\vec{w} = m\vec{g}$$

The weight of an object does not depend on whether or not the object is falling.

When you stand on a scale "weighing yourself," you are not directly measuring the gravitational force acting on you. Instead, you are measuring the magnitude of the force exerted by the scale upward on your feet. This is the force that counters your weight. The force of the scale prevents you from being accelerated by the weight force. When objects are "weighed" in this manner, the quantity being measured is not the true weight but the **apparent weight.** When you are standing still, your acceleration is zero and so your apparent weight equals your actual weight.

There are instances when the apparent weight measured by a scale does not equal the true weight. For example, if an astronaut stands on a scale to weigh himself while in orbit around the earth, the scale would read zero even though his true weight is not zero (the earth still exerts a gravitational force on him). In this case, the astronaut in orbit accelerates downward under the influence of his weight. The magnitude of the force exerted by the scale upward on his feet is zero because the Shuttle and scale are also accelerating downward at the same rate.

4-4   *Newton's Third Law*   Forces always come in pairs because one body cannot exert a force on another without the other exerting a force on it. These are referred to as an **action–reaction force pair.** The two forces that make up an action–reaction pair are always equal in magnitude and oppositely directed, and they always act on different objects. If you stand at rest and want to run toward the ocean, Newton's second law says that you cannot start moving toward the water unless a net external force pushes you in that direction. To start moving you must cause your surroundings to push you in the direction you want to go. You do this by pushing your surrounding in a direction *opposite* to the direction you wish to go. To start moving toward the water, push backward on the sand under your feet. That the sand will then push you toward the water follows from Newton's third law.

4-5   *Forces in Nature*   All observable natural forces can be described in terms of four basic interactions that occur between elementary particles:

1. The gravitational force
2. The electromagnetic force
3. The strong nuclear force
4. The weak nuclear force

For the next few chapters, we will be working almost exclusively with gravitational and electromagnetic forces. The **contact forces** between touching objects are electromagnetic in nature, and almost all **action-at-a-distance** forces considered in the next few chapters are gravitational in nature. Until we get to Chapter 22 where electric forces are introduced, the only action-at-a-distance force we will consider in any detail is the gravitational force.

When two *solid* objects touch, like a box sliding down a ramp, the contacting surfaces deform slightly and exert forces on each other—forces that are distributed over the contacting surfaces. The contact force acting on a single surface is actually the vector sum of the distributed contact forces that act on that surface. It is often useful to think of this contact force as the sum of two distinct forces, one normal to and one parallel to the surface. (The word normal means perpendicular.) These forces are referred to as the **normal force** and the **friction force,** respectively. Even when an object sits at rest on a surface, deformations are present. These allow the surface to exert a force to counteract the gravitational force on the object.

If we pull on a *string* it stretches slightly and pulls back with an equal but opposite force. Under tension some strings, like monofilament fishing line, stretch a fair amount while other strings, like cotton thread, stretch almost imperceptibly. In physics textbook problems you should assume that the stretching of any string is negligible (as is its mass), unless it is stated otherwise. The strings in physics textbooks act much more like cotton thread than mono-filament fishing line. A cotton string acts much like a very stiff spring, one with a very large force constant.

4-6 *Problem Solving* A typical physics problem describes a physical situation—such as a child swinging on a swing—and asks related questions. For example: If the speed of the child is 5.0 m/s at the bottom of her arc, what is the maximum height the child will reach? Solving such problems requires you to apply the concepts of physics to the physical situation, to generate mathematical relations, and to calculate the requested numbers. The problems presented here and in your textbook are exemplars; that is, they show a method that is used by problem-solving experts. When you master the methodology presented in the worked-out examples, you'll be able to solve problems for a wide variety of physical situations.

To be successful in solving physics problems, study the techniques used in the worked-out example problems. A good way to test your understanding of a specific solution is to take a sheet of paper, and—without looking at the worked-out solution—reproduce it. If you get stuck and need to refer to the presented solution, do so. But then take a fresh sheet of paper, start from the beginning, and reproduce the entire solution. This may seem tedious at first, but it does pay off.

This is not to suggest that you reproduce solutions by rote memorization, but that you reproduce them by drawing on your understanding of the relations involved. By reproducing a solution in its entirety, you'll verify for yourself that you have mastered a particular example problem. As you repeat this process with other examples, you'll build your very own personal base of physics knowledge, a base of knowledge relating occurrences in the world around you—the physical universe—and the concepts of physics. The more complete the knowledge base that you build, the more success you will have in physics.

The recommended method of attack for solving problems using Newton's Laws is as follows:

Physics Problem-Solving Guidelines

1. Draw a neat pictorial diagram. List known and unknown quantities. You may need to return to this step after steps 2 and 3 for further clarification of your visualization of the situation.

2. Determine the direction of the acceleration of each particle. For an object moving in a straight line the acceleration vector will be parallel to the line.

3. Isolate the object (particle) of interest, and draw a free-body diagram showing each external force that acts on the object. Draw the force vectors to scale so that the vector sum is in the direction of the acceleration vector. If there is more than one object of interest in the problem, draw a separate free-body diagram for each. Each free-body diagram should include coordinate axes with the positive direction for each axis clearly indicated. Guidelines for determining optimal coordinate axes directions are presented both below and in Chapter 5.

4. Solve the resulting equations for the unknowns using whatever additional information is available. The unknowns may include the masses, the components of the acceleration, or the components of some of the forces.

5. Finally, inspect your results carefully. Check to see if they correspond to reasonable expectations. It is usually best to put off substituting numerical values until an algebraic expression for the desired quantity is obtained. Then your work can often be checked by evaluating the expression for specific limiting values of pertinent quantities. This is an excellent way to check your work for errors.

You can get the additional information referred to in Rule 4 by applying Newton's third law $\vec{F}_{\text{by A on B}} = -\vec{F}_{\text{by B on A}}$ and by exploiting conditions that constrain the motion. For example, when an object moves along a track, it is constrained to follow the track.

Each free-body diagram should include a set of coordinate axes. Choosing the coordinate axis directions is often an important part of the problem-solving process. In principle, physics problems can be solved using coordinate axes in any direction. However, a well-selected set of coordinate axis directions goes a long way toward reducing the mathematical complexity of a solution.

To choose optimal coordinate axis directions use the following two guidelines:

Guidelines for Coordinate Axis Directions

1. Determine the direction of the acceleration and then select a coordinate system with one axis, say the x axis, parallel to the acceleration vector. *Note:* This rule is frequently followed in the worked-out examples in this and the next chapter.

2. If the acceleration is zero, then select coordinate axis directions that maximize the number of force vectors parallel to an axis.

In a string under tension the molecules of the string pull on each other, keeping the string taut. It is the strength of this pull that is called the **tension.** For most general physics problems the mass of the string is assumed to be negligible.

4-7   *Problems with Two or More Objects*   In many mechanics problems, two or more objects that move are touching each other or are connected by a string. Such problems can be solved by treating each object separately (including the string). Draw a free-body diagram for each object, and then apply Newton's second law ($\Sigma \vec{F} = m\vec{a}$) to each. The resulting equations are then solved simultaneously for the unknown forces or accelerations. Two additional types of relations, one kinematic and one dynamic, are needed to solve multiple-object problems effectively. Kinematics is the study of motion exclusive of the influences of mass and force whereas dynamics is the study of the relation between motion and the forces affecting motion. (1) If two objects are connected by a string (or rope) that remains taut, the motions of the two objects will be related kinematically. For example, if I am dangling from the end of a rope, and you are holding onto the other end on top of a glacier and are being dragged toward the edge of the crevasse—then our motions will be identical, except that yours will be horizontal while mine

will be vertical. That is, we will move equal distances in equal times so at any instant we will have the same speed as well as the same rate of change of speed. It follows that the magnitudes of our acceleration vectors will be equal. (2) If one object is touching another object, like a rope pulling a wagon, the force exerted by the rope on the wagon affects the motion of the wagon, whereas the force exerted by the wagon on the rope affects the motion of the rope. These two forces are equal but opposite—a dynamic relation (Newton's third law).

II. Numbers and Key Equations

Numbers

Acceleration due to gravity

$$\vec{g} = 9.81 \text{ m/s}^2, \text{downward} \qquad g = |\vec{g}| = 9.81 \text{ m/s}^2$$

newton

$$1 \text{ N} = 1 \text{ kg·m/s}^2$$

Key Equations

Newton's second law

$$\vec{a} = \frac{\vec{F}_{\text{net}}}{m} \qquad \text{or} \qquad \vec{F}_{\text{net}} = m\vec{a}$$

Newton's third law

$$\vec{F}_{AB} = -\vec{F}_{BA} \qquad \text{or} \qquad \vec{F}_{AB} + \vec{F}_{BA} = 0$$

Mass comparison with equal forces

$$m_1 a_1 = m_2 a_2 \qquad \text{or} \qquad \frac{m_1}{m_2} = \frac{a_2}{a_1}$$

Hooke's law

$$F_x = -k\,\Delta x$$

Weight

$$\vec{w} = m\vec{g}$$

III. Potential Pitfalls

Don't think that one object exerts a force on another object if they are not physically touching. Except for action-at-a-distance forces, like gravity, this does not occur. For all forces, including action-at-a-distance forces, one object cannot *give* another object a force.

Notice that the action–reaction force pairs referred to by Newton's third law never act on the same body. Thus they never cancel when the net force acting on a body is calculated.

Always choose a coordinate system that is fixed to an inertial reference frame. Newton's laws must be applied in inertial reference frames. For most of the problems considered in this chapter, the earth is considered an inertial reference frame and all coordinate systems will be fixed relative to the earth.

Be sure to put only those forces on a free-body diagram for which you can identify an external physical source. Acceleration is not a force, and neither is the product of mass and acceleration ($m\vec{a}$). Any force on your free-body diagram should be identified as either an action-at-a-distance force or a contact force. With rare exceptions, all action-at-a-distance forces considered during the first semester of a general physics course are gravitational forces. In addition to action-at-a-distance forces, the diagram should contain at least one force vector for each thing that touches the object.

Only those forces that act on a body go on that body's free-body diagram. Forces the body exerts on other things go on their free-body diagrams, not on its own. The net force acting on a body is the name for the vector sum of all the forces that act on the body.

Be aware of the positive direction of the coordinate axes used in each problem. The sign of the x component of a vector depends on the choice of the positive x direction (the direction of increasing x).

IV. True and False and Responses

True and False

_____ 1. If a particle moves with constant velocity, no forces can be acting on it.

_____ 2. Force is a vector quantity.

_____ 3. The mass of an object may be determined in terms of the acceleration produced by a known force.

_____ 4. If a body moves at constant speed, the net force acting on it must be zero.

_____ 5. The weight of a body is the force that the earth's gravity exerts on it.

_____ 6. The weight of an astronaut in an orbiting satellite is zero.

_____ 7. When all the forces that act on a particle and the mass of the particle are known, Newton's laws provide a complete description of the particle's motion.

_____ 8. An object rests on a tabletop. The upward force by the table surface on the object and the downward gravitational force exerted by the earth on the object form an action–reaction pair and thus must be equal in magnitude and oppositely directed.

_____ 9. That the tension in a taut string connecting two objects is the same throughout its length follows from Newton's third law (action equals reaction) alone.

_____ 10. Newton's second law of motion ($\Sigma\vec{F} = m\vec{a}$) is valid only in inertial reference frames.

_____ 11. The forces that bind atoms together into molecules are electromagnetic in origin.

_____ 12. Contact forces between macroscopic objects are electromagnetic in origin.

_____ 13. Nuclear forces are effective only over very short distances.

_____ 14. Hooke's law states that the force exerted by a compressed or extended spring is directly proportional to the compression or extension of that spring.

_____ 15. When a pony pulls a stationary cart and sets it into motion, the force with which the pony pulls the cart forward exceeds the force with which the cart pulls back on the pony.

_____ 16. A large, fully-loaded eighteen wheeler (large truck) runs head on into a Geo Metro (small car). During the crash the magnitude of the force exerted by the truck on the car is the same as that exerted by the car on the truck.

_____ 17. A large, fully-loaded eighteen wheeler (large truck) runs head on into a Geo Metro (small car). During the crash the magnitude of the car's acceleration is greater than that of the truck's acceleration.

_____ 18. A soccer ball flies through the air after being kicked. During the flight the forces acting on it include air drag, the force of gravity, and the force of the kick.

Responses

1. False. The vector sum of all the forces acting on it must be zero.

2. True

3. True

4. False. For the acceleration—and thus the net force—to be zero, the body must move at a constant velocity (at a constant speed in a fixed direction).

5. True

6. False. The astronaut's apparent weight is zero.

The true weight—the gravitational force exerted on the astronaut by the earth—is the force that keeps the astronaut moving in a curved path around the earth.

7. False. Knowledge of all the forces and the mass specifies only the acceleration. Knowledge of both the initial position and the initial velocity is also needed to completely specify the motion.

8. False. These forces both act on the same object, so they can't be an action–reaction pair. Together they constitute the net force acting on the object. Because the acceleration is zero, it follows from Newton's second law ($\vec{F}_{net} = m\vec{a}$) that the net force is zero. If the net force due to two forces is zero, the forces must be equal in magnitude and oppositely directed.

9. False. The tensions at two separate points in a taut string are not an action–reaction pair. The net force on the segment of string between the two points equals the product of its mass and acceleration. If the two tensions are the only forces acting on the segment, they will be equal in magnitude and oppositely directed only if the string's mass is negligible.

10. True

11. True

12. True

13. True

14. True

15. False. The forces are an action–reaction pair, so they are always equal in magnitude and oppositely directed. The net force on the pony is forward because the forward force of the ground on the pony's feet exceeds the backwards force of the cart on the pony. Similarly, the net force on the cart is forward. The forward force of the pony on the cart exceeds any backwards forces tending to slow the cart.

16. True

17. True

18. False. The force of the kick acts on the ball only while it is in contact with the foot. While it is flying through the air following the kick only air drag and the force of gravity act on the ball. Contact forces act only while the objects are in contact.

V. Questions and Answers

Questions

1. Suppose that only a single force of known direction acts on a body. From this information can you tell in what direction the body is moving?

2. When you jump into the air, you have (for a short time) an upward acceleration. What external agent is exerting the upward force on you? What is the reaction force to the upward force exerted by this agent?

3. A car is being driven up a long straight hill at constant speed. What forces act on it and what is the net force?

4. Suppose a force F stretches a spring a distance Δx and another force F' stretches it by $2\,\Delta x$. How can you tell if F' is in fact equal to $2F$?

5. Why are you thrown forward when a car in which you are riding stops abruptly?

6. Why are you thrown outward (to the right) when a car in which you are riding makes a sharp turn to the left?

7. An object of mass m is being weighed in an elevator that has an upward directed acceleration $\vec{a}$. What is the result if the weighing is done using (*a*) a spring scale and (*b*) a balance?

Answers

1. Knowing the direction of the force will give you only the direction of the acceleration, not the direction of the velocity.

2. The floor exerts an upward contact force on the soles of your feet. The reaction force is the downward push of the soles of your feet on the floor.

3. The forces acting on the car are the gravitational force of the earth, the contact force of the road, and the contact force of the air (air resistance). Because the magnitude and direction of the velocity are constant, the acceleration must be zero. Because the acceleration is zero, it follows from Newton's second law ($\vec{F}_{net} = m\vec{a}$) that the net force is also zero.

4. To tell whether or not $F' = 2F$, place a block on a frictionless table top and accelerate it with the spring extension first at Δx and then at $2\,\Delta x$. For each of these spring extensions, measure the acceleration of the block. If the second acceleration is twice the first, then $F' = 2F$.

5. You are not thrown forward. You and the car seat are initially moving forward at the same speed. When the brakes are applied, the seat slows down but inertia keeps you moving forward.

6. You are not thrown outward. You and the car seat are initially moving forward at the same speed. When the car makes a sharp turn—say to the left—your seat moves with it, and you are pulled to the left literally by the seat of your pants. The inertia of your head and upper torso cause them to lag behind, and it "feels" as if you are being thrown to the right.

7. (*a*) The spring scale reading is proportional to the apparent weight of the mass *m*. When a spring scale is used, the spring must exert an additional upward force to cause the mass to accelerate upward. To do this the spring must stretch more; thus the reading on the spring scale increases. (*b*) A balance compares the mass *m* with that of an object of known mass. When a balance is used, the arm of the balance supporting the mass must exert an additional upward force to cause it to accelerate upward. However, the other arm of the balance must also exert an additional upward force on the counterweight—also of mass *m*—to give it the required acceleration $\vec{a}$. The balance will remain balanced regardless of the acceleration and give the same reading as it would if there were no acceleration.

VI. Problems and Solutions

Problems

1. A 2-kg block slides to the right at constant velocity along a frictionless horizontal tabletop, as shown in Figure 4-2. Determine the magnitude and direction of all forces acting on the block.

Figure 4-2

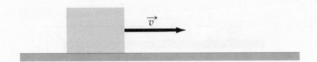

How to Solve It
• Draw a free-body diagram of the block. There is only one object touching the block, the tabletop. Thus there are only two forces acting on the block, the gravitational force of the earth (the weight) and the contact force of the tabletop. If the direction of a force is not known, draw the vector representing that force in an arbitrary direction.

• Apply Newton's second law to the block. Write out the equation for Newton's second law ($\vec{F}_{net} = m\vec{a}$) in vector form to the block. Select a coordinate system and write out the equivalent component equations.

• Express the weight as the product of the mass and the free-fall acceleration $\vec{g}$ and specify as much as possible about the acceleration components. Solve the component equations to determine the magnitude and direction of each force.

2. A 2-kg block on a horizontal, frictionless table is pulled to the right by a string exerting a constant force as shown in Figure 4-3. In 5 s the speed of the block increases from zero to 10 m/s. Determine the magnitude and direction of all forces acting on the block.

Figure 4-3

How to Solve It
• Draw a free-body diagram. There are two objects touching the block, the tabletop and the string. Thus there are three forces acting on it, the gravitational force due to the earth (the weight), the contact force of the tabletop, and the contact force of the string. Use a result from the previous problem—that the contact force exerted by a frictionless surface is normal to the contacting surfaces—to give you the direction of the force exerted on the block by the tabletop.

• Apply Newton's second law to the block. Write out the equation for Newton's second law ($\vec{F}_{net} = m\vec{a}$) in vector form. Select a coordinate system and write out the equivalent component equations.

• Express the weight $\vec{w}$ as the product of the mass *m* and the free-fall acceleration $\vec{g}$, and specify as much as possible about the acceleration components. In particular, specify that the vertical component of the acceleration is zero and determine the horizontal acceleration component from the given information. Solve the component equations to determine the magnitude and direction of each force.

3. As is shown in Figure 4-4, a 25-kg traffic light is suspended from two light strands of wire. Determine the tension in each strand.

Figure 4-4

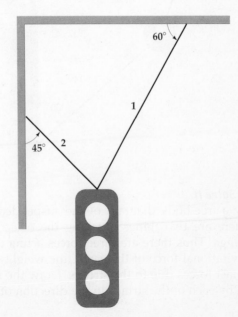

How to Solve It
- Draw a free-body diagram of the traffic light. There are two objects touching the traffic light: the two strands. Thus there are three forces acting on it, the gravitational force of the earth (the weight) and the contact forces of the two strands. Draw the force exerted by each of the strands in the direction of that strand. The weight is straight down, so draw the weight vector straight down.

- Apply Newton's second law to the light. Write out the equation for Newton's second law ($\vec{F}_{net} = m\vec{a}$) in vector form. Select a coordinate system and write out the equivalent component equations.

- Express the weight $\vec{w}$ as the product of the mass m and the free fall acceleration $\vec{g}$ and specify as much as possible about the acceleration components. Use trigonometry to relate the components to the magnitudes and angles and substitute these expressions into the component equations.

- Solve the component equations for the tensions in the strands.

4. A 10-kg object is supported by a string that is attached to the ceiling. The object is pulled to the side by a second string connected to the object, as shown in Figure 4-5. Determine the tensions in the two strings.

Figure 4-5

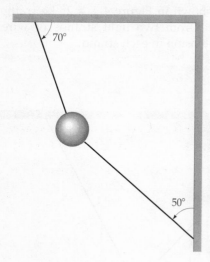

How to Solve It
- Draw a free-body diagram of the suspended object. There are two things touching the object, the two strings. Thus there are three forces acting on it, the gravitational force of the earth (the weight) and the contact forces due to the strings. Draw the force exerted by each of the strings in the direction of that

string. The weight is straight down, so draw the weight vector straight down.

- Apply Newton's second law to the object. Write out the equation for Newton's second law ($\vec{F}_{net} = m\vec{a}$) in vector form. Select a coordinate system and write out the equivalent component equations.

- Express the weight $\vec{w}$ as the product of the mass m and the free-fall acceleration $\vec{g}$ and specify as much as possible about the acceleration components. Use trigonometry to relate the components to the magnitudes and angles and substitute these expressions into the component equations.

- Solve the component equations for the tensions in the strings.

5. Two blocks, connected by a light string as shown in Figure 4-6, are being pulled across a frictionless horizontal tabletop by a second light string. Block A has twice the mass of block B, the blocks are gaining speed as they move toward the right, and the strings remain taut at all times. Find the ratio T_1/T_2 of the two tensions.

Figure 4-6

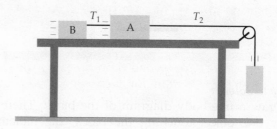

How to Solve It
- Draw a separate free-body diagram for each block. There are two objects touching block B, the tabletop and a string, and three touching block A, the tabletop and the two strings. Thus, in addition to a weight force there are two forces acting on block B and three on block A. The tabletop is frictionless so the forces exerted by it on the blocks are normal to its surface.

- Apply Newton's second law to block A. Write out the equation for Newton's second law ($\vec{F}_{net} = m\vec{a}$) in vector form for the block. Select a coordinate system and write out the equivalent component equations. Repeat this process for block B.

- Assume that the string remains taut but does not stretch and relate the acceleration of block A to that of block B.

- Solve for the ratio of the tensions.

6. Two blocks, connected by a light string as shown in Figure 4-7, are sliding toward the left across a frictionless horizontal tabletop. A second light string attached to block B pulls it to the right and as a result the blocks lose speed. Block A has two times the mass of block B and the strings remain taut at all times. Find the ratio T_1/T_2 of the two tensions.

Figure 4-7

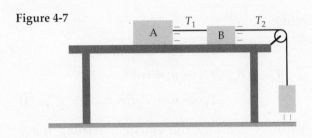

How to Solve It
- Draw a separate free-body diagram for each block. There are two objects touching block A, the tabletop and a string, and three touching block B, the tabletop and the two strings. Thus, in addition to a weight force there are two forces acting on block A and three on block B. The contact force exerted by a frictionless surface is normal to the contacting surfaces.

- Apply Newton's second law to the block A. Write out the equation for Newton's second law ($\vec{F}_{net} = m\vec{a}$) in vector form for the block. Select a coordinate system and write out the equivalent component equations. Repeat this process for block B.

- Assume that the string remains taut but does not stretch and relate the acceleration of block A to that of block B.

- Solve for the ratio of the tensions.

Solutions

1. There are two forces acting on the block, the weight $\vec{w}$ and the contact force $\vec{F}$ exerted by the table on the block. We know the weight acts downward, so we draw the vector $\vec{w}$ pointed downward on our free-body diagram in Figure 4-8. We are not sure of the direction of the contact force, so we draw it in an arbitrary direction. On this diagram we also draw a coordinate system with the y axis parallel to $\vec{w}$, and we draw the components F_x and F_y. *Note:* The components of each vector depend upon the coordinate axis directions but not on their locations. Therefore, it doesn't matter where the coordinate-system origin is placed.

Figure 4-8

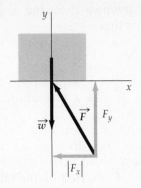

By applying Newton's second law to the block, we obtain the vector equation

$$\vec{w} + \vec{F} = m\vec{a}$$

This vector equation is a concise representation of the two component equations

$$w_x + F_x = ma_x$$
$$w_y + F_y = ma_y$$

We will first consider the equation relating x components. Because the problem states that the velocity is constant, we know that acceleration $\vec{a}$ must be zero. This means, of course, that both a_x and a_y equal zero. Because $\vec{w}$ is perpendicular to the x axis, we know that w_x is equal to zero. Therefore,

$$w_x + F_x = ma_x$$
$$0 + F_x = m(0)$$
$$F_x = 0$$

We now consider the equation relating the y components. Because $\vec{w}$ is in the negative y direction, we know that w_y is equal to $-w$ (w is the magnitude of $\vec{w}$). Therefore,

$$w_y + F_y = ma_y$$
$$(-w) + F_y = m(0)$$
$$F_y = w$$

The weight of the block can be expressed in terms of the block's mass by the relation $\vec{w} = m\vec{g}$. Thus the weight $\vec{w}$ has a magnitude of

$$w = mg$$
$$= (2\ kg)(9.8\ m/s^2)$$
$$= \underline{19.6\ N}$$

and is directed downward (in the negative y direction). This means that

$$F_y = +19.6 \text{ N}$$

We have shown that F_x equals zero and F_y equals $+19.6$ N. Thus $\vec{F}$ has a magnitude of 19.6 N and is directed upward.

Note: We think of a surface as being frictionless when the contact force does not act to oppose sliding motion. The analysis presented here shows that the contact force $\vec{F}$ exerted by the table is directed upward, at a right angle (or normal) to the contacting surfaces.

2. The weight is 19.62 N downward; the force exerted by the table is 19.62 N upward; the acceleration is 2 m/s²; and the force exerted by the string is 4 N toward the right.

3. Figure 4-9 is a free-body diagram of the traffic light. Acting on it are the gravitational force $\vec{w}$, the contact force $\vec{T}_1$ exerted by strand 1, and the contact force $\vec{T}_2$ exerted by strand 2. The directions of $\vec{T}_1$ and $\vec{T}_2$ are along their respective strands. On this diagram we also draw a coordinate system with the y axis parallel to $\vec{w}$.

Figure 4-9

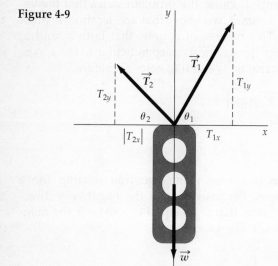

By applying Newton's second law to the light, we obtain the vector equation

$$\vec{w} + \vec{T}_1 + \vec{T}_2 = m\vec{a}$$

This vector equation is a concise representation of the two component equations

$$w_x + T_{1x} + T_{2x} = ma_x$$
$$w_y + T_{1y} + T_{2y} = ma_y$$

We will first consider the equation relating x components. Because the light remains stationary, we know the acceleration $\vec{a}$ is zero. This means, of course, that both a_x and a_y equal zero. Because $\vec{w}$ is perpendicular to the x axis, we know that w_x is equal to zero. By using trigonometry, we obtain the relations

$$T_{1x} = +T_1 \cos \theta_1$$
$$T_{2x} = +T_2 \cos \theta_2$$

Therefore,

$$w_x + T_{1x} + T_{2x} = ma_x$$
$$0 + T_1 \cos \theta_1 - T_2 \cos \theta_2 = m(0)$$
$$T_1 \cos \theta_1 = T_2 \cos \theta_2 \qquad (1)$$

We now consider the equation relating the y components. Because the weight $\vec{w}$ is in the negative y direction, we know that w_y is equal to $-mg$. By using trigonometry, we obtain the relations

$$T_{1y} = +T_1 \sin \theta_1$$
$$T_{2y} = +T_2 \sin \theta_2$$

Therefore,

$$w_y + T_{1y} + T_{2y} = ma_y$$
$$-mg + T_1 \sin \theta_1 + T_2 \sin \theta_2 = m(0) = 0$$
$$T_1 \sin \theta_1 + T_2 \sin \theta_2 = mg \qquad (2)$$

We now simultaneously solve the two equations for the tensions. If we solve the first equation (1) for T_2 and substitute into Equation (2), we have

$$T_1 \sin \theta_1 + \left(T_1 \frac{\cos \theta_1}{\cos \theta_2} \right) \sin \theta_2 = mg$$

By solving this equation for T_1 we obtain

$$T_1 = mg \frac{\cos \theta_2}{\sin \theta_1 \cos \theta_2 + \cos \theta_1 \sin \theta_2}$$

$$= (50 \text{ kg})(9.8 \text{ m/s}^2)$$

$$\times \frac{\cos 45°}{\sin 60° \cos 45° + \cos 60° \sin 45°}$$

$$= 359 \text{ N}$$

We now solve Equation (1) for T_2. This gives

$$T_2 = T_1 \frac{\cos \theta_1}{\cos \theta_2}$$

$$= (359 \text{ N}) \frac{\cos 60°}{\cos 45°} = 254 \text{ N}$$

4. The tension in the string attached to the ceiling is 150 N, and the tension in the other string is 67 N. Can you explain why the tension in the string attached to the ceiling exceeds the weight of the 10-kg object?

5. First, we draw the free-body diagram of block A and apply Newton's second law ($\vec{F} = m\vec{a}$) to it. There are four forces acting on it: the weight force $\vec{w}_A$ exerted by the Earth, the normal force $\vec{F}_{nA}$ exerted by the table, and the two tension forces $\vec{T}_{1A}$ and $\vec{T}_{2A}$. The block is gaining speed as it moves in a straight line, so we know the acceleration vector is in the same direction as the velocity vector. Consequently, on the diagram (Figure 4-10) we draw the x axis parallel with the acceleration vector.

Figure 4-10

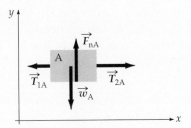

Applying Newton's second law in vector form to block A gives:

$$\vec{T}_{2A} + \vec{T}_{1A} + \vec{F}_{nA} + \vec{w}_A = m_A \vec{a}_A$$

Writing this in component form for the x direction we have

$$T_{2Ax} + T_{1Ax} + F_{nAx} + w_{Ax} = m_A a_{Ax}$$

Expressing the components in terms of symbols for the magnitudes we obtain

$$T_2 - T_1 + 0 + 0 = m_A a_A \qquad (1)$$

where T_1 and T_2 are the tensions in strings 1 and 2, respectively.

Next, we draw a free-body diagram of block B and apply Newton's second law ($\vec{F} = m\vec{a}$) to it. There are four forces acting on it: the weight force $\vec{w}_B$ exerted by planet Earth, the normal force $\vec{F}_{nB}$ exerted by the table, and the tension force $\vec{T}_{1B}$. The block is gaining speed as it moves in a straight line, so we know the acceleration vector is in the same direction as the velocity vector. Consequently, on the diagram (Figure 4-11) we draw the x axis parallel with the acceleration vector.

Figure 4-11

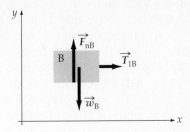

Applying Newton's second law in vector form to block B gives:

$$\vec{T}_{1B} + \vec{F}_{nB} + \vec{w}_B = m_B \vec{a}_B$$

Writing this in component for the x direction we have

$$T_{1Bx} + F_{nBx} + w_{Bx} = m_B a_{Bx}$$

Expressing the components in terms of symbols for the magnitudes we obtain

$$T_1 + 0 + 0 = m_B a_B \qquad (2)$$

Next we relate the accelerations of the blocks. As long as the string does not slacken nor stretch, the speed of block A and the speed of block B will remain equal. This means the two speeds must change at the same rate so the magnitudes of the blocks' accelerations also remain equal. That is:

$$a_A = a_B = a$$

Dividing Equation (1) by Equation (2) gives

$$\frac{T_2}{T_1} - 1 = \frac{m_A a}{m_B a} \qquad \text{or}$$

$$\frac{T_2}{T_1} = 1 + \frac{m_A}{m_B} = 1 + \frac{2}{1} = 3 \qquad (3)$$

We are looking for the ratio T_1/T_2. Solving Equation (3) for this ratio we obtain

$$\frac{T_1}{T_2} = \frac{1}{3}$$

Remark 1: In applying Newton's second law to the blocks only the x component equations were utilized. The y component equations are valid but do not contribute to solving for the tensions as long as friction is negligible.

Remark 2: Tension T_1 is the only force giving block B acceleration a. Since block A has twice the mass, to give block A the same acceleration requires twice the net force. Block A has a force T_1 holding it back. For the net force on block A to equal $2T_1$ in the forward direction, it follows that T_2 must equal $3T_1$.

Remark 3: Applying Newton's second law to block A resulted in an equation where each term had a subscript A. Likewise, in repeating this process for block B each term had a subscript B. This subscripting was done so that the symbol for each force or acceleration term is unique. It is advisable to follow this practice in order to avoid mix-ups.

Remark 4: In Equation (1), T_1 represents the tension in the right end of string 1 at block A, whereas in Equation (2) it represents the tension in its left end. The tensions in these two ends are only approximately equal. By drawing a free-body diagram (see Figure 4-12) of string 1 and by applying Newton's second law, we now show that the tensions at opposite ends are equal only in the limit that the string's mass approaches zero.

Figure 4-12

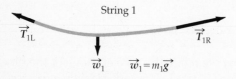

Applying Newton's second law to the string gives:

$$\vec{T}_{1R} + \vec{T}_{1L} + m_1\vec{g} = m_1\vec{a}_1$$

where m_1 is the mass of string 1. Taking the limit of both sides as $m_1 \to 0$ we obtain

$$\vec{T}_{1R} + \vec{T}_{1L} = 0$$

which means the tension forces acting on the opposite ends of the string are equal in magnitude and oppositely directed. This is so even though the force of gravity pulls downward on the string and even though it is being accelerated to the right.

The gravity force is proportional to the string's mass, so a negligible mass results in a negligible gravity force and, as a result, a negligible sag.

Since the mass of the string being accelerated is negligible, to cause it to accelerate to the right requires that the tension force towards the right exceeds the tension force towards the left by only a negligible amount.

6. $\dfrac{T_1}{T_2} = \dfrac{2}{3}$

Chapter 5

Applications of Newton's Laws

I. Key Ideas

Recommendations for problem solving relevant to this chapter are discussed at some length in Chapter 4. These recommendations should be reread and learned well. It is strongly recommended that you develop the habit of approaching a problem by first making a drawing of the situation. Then, for each object, draw a free-body diagram illustrating the forces acting on it.

When working with objects that are in contact, like a box sliding down a ramp, the contacting surfaces deform slightly and exert forces on each other—forces are distributed over the contacting surfaces. The contact force acting on a single surface is actually the vector sum of the distributed contact forces that act throughout that surface. It is often useful to think of this contact force as the sum of two distinct contact forces, one normal to and one parallel to the surface. (The word normal means perpendicular.) These two forces are referred to as the **normal force** and the **friction force** respectively.

5-1 Friction The friction force is called the force of *static friction* f_s when the surfaces in contact *do not* slide on each other. Static frictional forces oppose any tendencies for the contacting surfaces to slide. For a given pair of contacting surfaces, the magnitude of the static frictional force f_s ranges anywhere from zero to a maximum value $f_{s, max}$ that is proportional to the normal force. This empirically established relation is expressed as either $f_s \leq \mu_s F_n$ or

Static friction

$$f_{s, max} = \mu_s F_n$$

where μ_s is the **coefficient of static friction.**

When skidding occurs, that is, when surfaces in contact slide across each other, the frictional force is called the force of *kinetic friction* f_k. The force of kinetic friction is given by the relation

Kinetic friction

$$f_k = \mu_k F_n$$

where μ_k is the **coefficient of kinetic friction.** The coefficients of static and kinetic friction are small for slippery surfaces and larger for surfaces that do not easily slide across one another. Values for μ_s and μ_k for a sampling of materials are listed in Table 5-1 in Chapter 5 of the text. The two coefficients depend upon the nature of the two surfaces in contact.

Guidelines for Solving Problems Involving Friction

1. Select a coordinate system with the x axis in the direction of motion and the y axis in the normal direction. With this choice the normal force F_n appears only in the y component equation and the frictional force f_k or f_s appears only in the x component equation.

2. With the above choice of axes first solve for F_n using the y component equation and next solve for f_k using the relation $f_k = \mu_k F_n$. Then substitute this result for f_k into the x component equation and solve for the desired unknown. (If the friction is static and a maximum then substitute $f_{s, max}$ for f_k and μ_s for μ_k.)

5-2 Circular Motion The direction of the acceleration vector is always in the direction of the change in velocity vector $\Delta \vec{v}$ during a sufficiently short time interval. The acceleration of a particle moving along a circular path has both a tangential component and a centripetal (center-seeking) component. The tangential component equals the rate of change of speed, while the centripetal component equals the speed squared divided by the radius of curvature of the path.

Tangential acceleration

$$a_t = \frac{dv}{dt}$$

Centripetal acceleration

$$a_c = \frac{v^2}{r}$$

For each of these acceleration components there is a component of the net force that is respectively tangential to the path or directed toward the path's center of curvature (the center of the circle of radius r).

When a particle moves along a curved path, at any small segment of the path there is a specific circle whose arc comes closest to coinciding with the path. Thus an arbitrary curved path can be represented as a succession of small circular arcs.

Guideline for Solving Problems
Involving Circular Motion

Select a coordinate system with the x axis in the direction of motion and the y axis in the centripetal direction. With this choice $a_y = v^2/r$ and $a_x = dv/dt$, where v is the speed.

5-3   Drag Forces   When an object moves through a fluid such as air or water, the fluid exerts a **drag force** that opposes the motion of the object through the fluid. Unlike the force of kinetic friction, this force increases with the speed of the object. This drag force increases with the first power of the speed for slow-moving objects and with the square of the speed for faster-moving objects. The magnitude of the drag force is related to the speed by the expression

Fluid drag

$$F_{drag} = bv^n$$

Here b depends on the size, shape, and roughness of the object along with the density and the viscosity of the fluid; and n varies from 1 to 2 as the speed increases and the flow becomes more turbulent. (For low speeds, where turbulence is negligible, n equals 1, and for higher speeds, where turbulence dominates, n equals 2.) When an object is falling straight down through a fluid, the drag force increases as the speed of the object increases until it reaches terminal speed v_t. At terminal speed the acceleration is zero so the drag force equals the object's weight. It follows that

Terminal speed

$$v_t = \left(\frac{mg}{b}\right)^{1/n}$$

For a sky diver, terminal speed is about 60 m/s (134 mi/h).

5-4   Numerical Methods   In many circumstances it is possible to integrate the equation obtained by applying Newton's second law ($\vec{F}_{net} = m\vec{a}$) to an object to determine its position in terms of a simple algebraic function of time. However, in some circumstances this is either very inconvenient or actually impossible to do. In such situations we resort to numerical methods. Numerical methods require knowledge of the dependence of the acceleration on position, velocity, and time.

The basic idea of Euler's method of numerical integration is to divide the time interval into a large number of short intervals and to then use the equations for motion with constant acceleration to solve for the changes in velocity and position during the first short time interval. These new values of velocity and position are then used to update the value of the acceleration, which is used to calculate the changes in the velocity and position during the second short time interval, and so forth.

II. Numbers and Key Equations

Numbers

There are no new numbers for this chapter.

Key Equations

Static friction

$$f_{s,max} = \mu_s F_n \qquad \text{or} \qquad f_s \leq \mu_s F_n$$

Kinetic friction

$$f_k = \mu_k F_n$$

Tangential acceleration

$$a_t = \frac{dv}{dt}$$

Centripetal acceleration

$$a_c = \frac{v^2}{r}$$

Fluid drag

$$F_{\text{drag}} = bv^n$$

Terminal speed

$$v_t = \left(\frac{mg}{b}\right)^{1/n}$$

III. Potential Pitfalls

The value of the force of static friction f_s is not always equal to $\mu_s F_n$. The maximum value that f_s can have is $\mu_s F_n$. Remember that while static friction is always directed so as to oppose relative motion between the contacting surfaces, static friction does not necessarily oppose motion. For example, when you start walking on a horizontal surface, like the floor of a classroom, it is the force of static friction exerted by the surface of the floor on the soles of your shoes that pushes you in the direction that you start to move. This force, which is in the forward direction, opposes relative motion between your foot and the floor by preventing your foot from sliding backward.

Do not treat the net force as a separate force. The net force is just a name for the vector sum of all the forces acting on an object. The centripetal force is not a separate force either. It is the name given to the centripetal component of the net force. For motion along a general curved path, the centripetal direction is the direction toward the center of curvature of the path; thus for a circular path it is toward the center of the circle.

When two surfaces come into contact, both a normal force and a frictional force act on each surface. (In reality, the normal and the frictional forces are not distinct forces but are the normal and tangential components of a single contact force.)

You should not assume that the normal force on an object is equal to its weight. It is equal to the weight only for certain specific situations. The normal force is best determined by applying Newton's second law ($\vec{F}_{\text{net}} = m\vec{a}$) in component form to the object. In doing this choose one coordinate axis to be parallel with the normal force and the other parallel with the frictional force. For an inert object sliding along a flat surface the normal component of the acceleration is zero.

When you use numerical methods to do integration, do not set the time intervals too small or the round-off errors will be too large. The trick is to set them small enough to determine the value of the final velocity or position to the desired accuracy, but not so small that the round-off errors become excessive. One way to tell if you have set the time intervals small enough is to reduce them by a factor of four and then recalculate the results. If the results do not appreciably change, the size of the time interval is probably satisfactory.

IV. True and False and Responses

True and False

_____ 1. The force of static friction is the force that must be exerted to start two surfaces sliding against each other.

_____ 2. It takes more force to keep a given pair of surfaces sliding against one another than it does to get them started.

_____ 3. The speed of an object falling through a fluid under the influence of gravity approaches a limiting value.

_____ 4. The coefficient of kinetic friction is directly proportional to the normal force.

_____ 5. The kinetic frictional force is directly proportional to the normal force so long as the kinetic coefficient of friction remains constant.

_____ 6. The static frictional force is directly proportional to the normal force so long as the static coefficient of friction remains constant.

_____ 7. For a particle in circular motion the average acceleration vector for a time interval is always in the same direction as the change in velocity vector for that time interval.

_____ 8. For a particle in circular motion at constant speed the average acceleration vector for a time interval is in the same direction as the change in velocity vector for that time interval.

_____ 9. For a particle in circular motion at constant speed the instantaneous acceleration vector is directed radially outward.

_____10. For a particle in circular motion at constant speed the instantaneous acceleration vector equals zero.

Responses

1. False. The force of static friction is the component of the contact force that opposes the sliding of the surfaces against each other. Another force creates the tendency to slide.

2. False. It takes more force to start the two surfaces sliding against each other than it does to keep them sliding. Anyone who has had to slide a heavy object like a piece of furniture or a large exercise mat should know this.

3. True

4. False. While the force of kinetic friction is directly proportional to the normal force, the coefficient of kinetic friction only depends upon the nature of the surfaces.

5. True

6. False. It is only the maximum static frictional force that is proportional to the normal force.

7. True

8. True

9. False. It is directed radially inward, in the direction of the change in velocity vector for a sufficiently short time interval.

10. False. The velocity vector is changing, not in magnitude but in direction, so the acceleration is not zero.

V. Questions and Answers

Questions

1. Why are curved roads banked?

2. Do the coefficients of either static or kinetic friction ever exceed (the number) one?

3. When you are standing still and then start walking, what outside force acts upon you to cause your acceleration?

4. When a car travels around an unbanked curve, friction between the tires and the road provides the needed centripetal force. Is this friction kinetic or static?

5. You are driving a car with rear-wheel drive on an icy road. The car starts to skid sideways off the road so you turn the front wheels toward the direction of the skid. Why?

6. Certain racing cars have an airfoil over the top of the car. The airfoil deflects air upward when the car turns corners. What advantage is gained by deflecting the air upward?

Answers

1. Curved roads are banked so the normal force of the road on the tires has a centripetal component. That way friction alone does not have to provide the necessary centripetal force, and there is less danger of skidding off the curve.

2. Yes, they sometimes exceed one. There is nothing special about a coefficient of friction of one. The coefficient of static friction between a dragster (a car

used in drag races) tire and the track is typically about three.

3. The floor applies a frictional force to the soles of your shoes in response to your shoes pushing the floor.

4. Static. If it were kinetic, the tires would leave skid marks.

5. You turn into the skid to reestablish static friction between the front tires and the road. Static friction can occur only when the wheels are rolling in the direction of the car's motion.

6. According to Newton's third law, when the foil deflects the air upward the air will push the foil downward. This increases the normal force of the car on the road, which is desirable because the maximum frictional force is proportional to the normal force, and it is this frictional force that provides the necessary centripetal acceleration.

VI. Problems and Solutions

Problems

1. A 2-kg block on a horizontal table is pulled to the right by a constant force of 16 N, as shown in Figure 5-1. The acceleration of the block is 3 m/s². Determine the coefficient of kinetic friction between the block and the tabletop.

Figure 5-1

How to Solve It

• Draw a free-body diagram of the block. You should represent the contact force exerted by the tabletop as a normal force $\vec{F}_n$ and a frictional force $\vec{f}_k$ directed to oppose the motion of the block. Select a coordinate system with the y axis in the normal direction and the x axis in the direction of motion.

• Apply Newton's second law to the block. Write out the equation for Newton's second law ($\Sigma \vec{F} = m\vec{a}$) in vector form to the block. Write out the equivalent component equations.

• The block is moving to the right with an acceleration of 3 m/s². Therefore $a_y = 0$ and $a_x = 3$ m/s². Solve the y-component equation for F_n, solve for f_k using the relation $f_k = \mu_k F_n$, and then substitute this result into the x component equation and solve for μ_k.

2. A 2.0-kg block on a horizontal table is pulled by a constant force of 16 N at an angle of 20° above the horizontal, as shown in Figure 5-2. The kinetic coefficient of friction is 0.51. Determine the acceleration of the block.

Figure 5-2

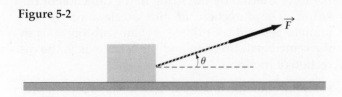

How to Solve It
• Draw a free-body diagram of the block. You should represent the contact force exerted by the tabletop as a normal force $\vec{F}_n$ and a frictional force $\vec{f}_k$ directed to oppose the motion of the block. Select a coordinate system with the y axis in the normal direction and the x axis in the direction of motion.

• Apply Newton's second law to the block. Write out the equation for Newton's second law ($\Sigma \vec{F} = m\vec{a}$) in vector form. Write out the equivalent component equations. On the free-body diagram sketch in the x and y components of the force $\vec{F}$. Use trigonometry to express these components in terms of the magnitude of $\vec{F}$ and the angle θ.

• The block is moving along a flat surface. Therefore the normal acceleration component is zero. Solve the y component equation for F_n and then solve for f_k using $f_k = \mu_k F_n$. Substitute this result into the x component equation and solve for the acceleration.

3. A pendulum consists of a one-inch-diameter steel ball attached to one end of a string, with the other end attached to the ceiling. The ball is pulled to the side and released from rest. It then swings back and forth along the arc of a vertical circle. The instant that the bob is at its highest point, as shown in Figure 5-3, the ball's acceleration vector is best represented by which of the arrows shown?

Figure 5-3

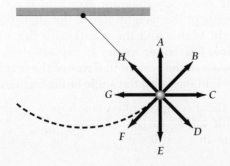

How to Solve It
• Sketch the velocity vector $\vec{v}_i$ shortly before the ball is in the position shown and the vector $\vec{v}_f$ shortly after it is in the position shown.

• Use the rules for vector addition ($\vec{v}_i + \Delta\vec{v} = \vec{v}_f$) and the vectors from the previous hint to draw the change in velocity vector $\Delta\vec{v}$. The acceleration vector is in the direction of the change in velocity vector.

• Check your answer. Draw a free-body diagram of the ball and show that the direction of the net force is compatible with the direction of the acceleration vector obtained from the previous step.

4. A pendulum consists of a one-inch diameter steel ball attached to one end of a string, with the other end attached to the ceiling. The ball is pulled to the side and released from rest. It then swings back and forth along the arc of a vertical circle. The instant that the bob is at the position shown in Figure 5-4 (on the way up and losing speed) the ball's acceleration vector is best represented by which of the arrows shown?

Figure 5-4

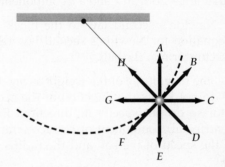

How to Solve It
• Sketch the velocity vector $\vec{v}_i$ shortly before the ball is in the position shown and the vector $\vec{v}_f$ shortly after it is in the position shown.

• Using the rules for vector addition ($\vec{v}_i + \Delta\vec{v} = \vec{v}_f$) and the vectors from the previous hint draw the change in velocity vector $\Delta\vec{v}$. The acceleration vector is in the direction of the change in velocity vector.

• Check your answer. Draw a free-body diagram of the ball and show that the direction of the net force is compatible with the direction of the acceleration vector obtained from the previous step.

5. A 0.1-kg steel ball is suspended from the ceiling by a light string 0.4 m long as shown in Figure 5-5. A physics instructor moves the ball to the side and then flicks it so the ball passes through the lowest point in the arc with a speed of 0.8 m/s. Determine the tension in the string when the ball is passing through the lowest point in its arc. Neglect air drag.

Figure 5-5

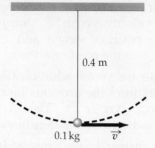

How to Solve It

• Draw a free-body diagram of the ball. There is only one object touching the ball, the string. Thus there are only two forces acting on it, the gravitational force of the earth (the weight) and the contact force of the string. Select a coordinate system with the y axis in the centripetal direction and the x axis in the direction of motion.

• Imagine the path taken by the ball as it swings on the end of the string. It is a circular arc centered at the point where the string is attached to the ceiling. The direction of the acceleration is not known. Therefore assume that it has both an x and a y component.

• Apply Newton's second law to the block. Write out the equation for Newton's second law ($\Sigma \vec{F} = m\vec{a}$) in vector form to the ball.

• Express the magnitude of the weight as mg. In this case, we do not know the direction of the acceleration vector so do not specify its direction. Express the y (centripetal) component of the acceleration in terms of the speed of the ball and the radius of the circular arc.

• Solve for the tension in the string.

6. A boy is swinging a 0.1-kg yo-yo in a vertical circle of radius 0.4 m by a light string, as shown in Figure 5-6. When the string is horizontal and the yo-yo is moving straight upward, the speed of the ball is 2 m/s. At that instant, determine the tension in the string and the rate of change of the yo-yo's speed.

Figure 5-6

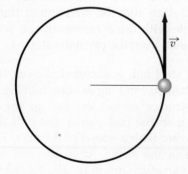

How to Solve It

• Draw a free-body diagram of the yo-yo. There is only one object touching the yo-yo, the string. Thus there are only two forces acting on it, the gravitational force of the earth (the weight) and the contact force of the string. Strings can only pull, thus draw the force exerted by the string in the direction of the string. The direction of the acceleration is not known. Select a coordinate system with one axis in the centripetal direction and the other axis in the direction of motion.

• Apply Newton's second law to the yo-yo. Write out the equation for Newton's second law ($\Sigma \vec{F} = m\vec{a}$) in vector form.

• Express the weight as the product mg. We do not know the direction of the acceleration vector. Express the horizontal (centripetal) component of the acceleration in terms of the speed of the ball and the radius of the circle.

• Solve the component equations for the tension in the string and for the vertical (tangential) component of the acceleration. The rate of change in speed equals the tangential component of the acceleration.

7. A car is moving at 33 m/s down a hill that makes an angle of 20° with the horizontal. If the coefficient of static friction between the tires and the road is 0.58, what is the shortest distance in which the car can stop?

How to Solve It

• Draw a sketch of the car and the incline. The incline makes an angle of 20° with the horizontal. Also draw a free-body diagram of the car. You should represent the contact force exerted by the road surface as a normal force $\vec{F}_n$ and a frictional force $\vec{f}_s$ directed to oppose the motion of the block. Select a coordinate system with the y axis in the normal direction and the x axis in the direction of motion.

• Apply Newton's second law to the car. Write out the equation for Newton's second law ($\Sigma \vec{F} = m\vec{a}$) in vector form. The acceleration vector is directed both parallel to the road and opposite to the direction of motion of the car.

• On the free-body diagram sketch in the x and y components of the weight. Relate the angle between the weight vector and the y axis with the angle of the incline. Use trigonometry to express the x and y components of the weight in terms of the magnitude of the weight (mg) and the angle of the incline.

• Solve the y component equation for F_n, substitute this result for F_n in the relation $f_{s, max} = \mu_s F_n$, and substitute this result for $f_{s, max}$ in the x component equation. Solve this for the maximum acceleration. Use the one-dimensional kinematic relations to determine the minimum stopping distance.

8. A block is projected at an initial speed of 3.5 m/s straight up at a ramp which makes a 26° angle with the horizontal. The coefficient of kinetic friction between the block and the incline is 0.3. How far up the incline will it slide before stopping?

How to Solve It
• Draw a sketch of the block and the ramp. Also draw a free-body diagram of the block. You should represent the contact force exerted by the ramp as a normal force $\vec{F}_n$ and a frictional force $\vec{f}_k$ directed to oppose the motion of the block. Select a coordinate system with the y axis in the normal direction and the x axis in the direction of motion.

• Apply Newton's second law to the block. Write out the equation for Newton's second law ($\Sigma \vec{F} = m\vec{a}$) in vector form. The acceleration vector is directed parallel to the incline and opposite to the direction of motion of the block.

• On the free-body diagram sketch in the x and y components of the weight. Relate the angle between the weight vector and the y axis with the angle the incline makes with the horizontal. Use trigonometry to express the x and y components of the weight in terms of the magnitude (mg) and the angle of the incline.

• Solve the y component equation for F_n, substitute this result for F_n in the relation $f_k = \mu_k F_n$, and substitute this result for f_k in the x component equation. Solve for the acceleration. Use the one-dimensional kinematic relations to determine the distance the block slides.

9. In Figure 5-7, a block of wood of unknown mass m_1 rests on a 5-kg block, which in turn rests on a tabletop. The blocks are connected by a light string that passes over a frictionless peg. The coefficient of kinetic friction at both surfaces is $\mu_k = 0.33$. The upper block is pulled to the left by a 60-N force, and the lower block is pulled to the right by the string that passes around

Figure 5-7

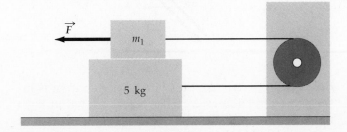

the peg. The blocks are moving at constant speed. Determine the mass m_1 of the upper block.

How to Solve It
• Draw a free-body diagram of the upper block. You should represent the contact force exerted by the upper surface of the lower block as a normal force $\vec{F}_{n1}$ and a frictional force $\vec{f}_{k1}$. Relative to the lower block, the upper block moves toward the left. We know $\vec{f}_{k1}$ is directed toward the right since kinetic frictional forces are always directed to oppose the relative motion of the surfaces in contact.

• Apply Newton's second law to the upper block. Write out the equation for Newton's second law ($\Sigma \vec{F} = m\vec{a}$) in vector form and write out the component equations. The peg is frictionless so the tension T is the same throughout the string. The blocks move at constant speed so their accelerations are both zero.

• Solve the y component equation for F_{n1}, then calculate F_{k1} using the relation $f_{k1} = \mu_k F_{n1}$, and next substitute this result for f_{k1} in the x component equation.

• Draw a free-body diagram of the 5-kg block. You should represent the contact force exerted by the upper block as a normal force $\vec{F}_{n2}$ and a frictional force $\vec{f}_{k2}$ directed opposite to $\vec{f}_{k1}$. Also, represent the contact force exerted by the tabletop as a normal force $\vec{F}'_{n2}$ and a frictional force $\vec{f}'_{k2}$ directed to oppose the block's motion relative to the tabletop. Again let the y axis be in the normal direction and the x axis be in the direction of the block's motion.

• Write out the equation for Newton's second law ($\Sigma \vec{F} = m\vec{a}$) in vector form for the 5-kg block and write out the component equations.

• The forces $\vec{F}_{n1}$ and $\vec{F}_{n2}$ form an action–reaction pair as do $\vec{f}_{k1}$ and $\vec{f}_{k2}$. That means $F_{n1} = F_{n2}$ and $f_{n1} = f_{n2}$ (their magnitudes are equal). Solve the y-component equation for F'_{n2}, then solve for f'_{k2} using the relation $f'_{k2} = \mu_k F'_{n2}$, and next substitute this result for f'_{k2} in the x component equation.

• Eliminate T from the two x-component equations and solve for m_1.

10. Assume that mass of the upper block in the previous problem is 4.53 kg. If the force pulling the upper block to the left is equal to 75 N, determine the acceleration of each block.

How to Solve It
• Follow the same procedure used to solve the previous problem except do not assume that the accelerations equal zero.

• Because the string doesn't stretch appreciably, the motion of the upper block to the left is identical to the motion of the lower block to the right. It follows that magnitudes of the accelerations of the two blocks are equal.

Solutions

1. Figure 5-8 is a free-body diagram of the block. There are three forces acting on the block, the gravitational force $\vec{w}$, the applied force $\vec{F}$, and the contact force exerted by the table on the block. This contact force is expressed as the sum of two forces, the normal force $\vec{F}_n$ and the frictional force $\vec{f}_k$. The acceleration vector is in the direction of the applied force.

Figure 5-8

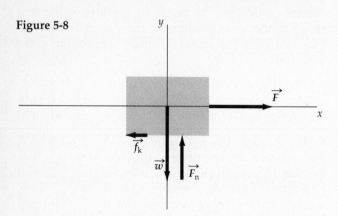

By applying Newton's second law to the block, we obtain the vector equation

$$\vec{w} + \vec{F}_n + \vec{f}_k + \vec{F} = m\vec{a}$$

This vector equation is a concise representation of the two component equations

$$w_x + F_{nx} + f_{kx} + F_x = m\,a_x$$
$$w_y + F_{ny} + f_{ky} + F_y = m\,a_y$$

We first consider the equation relating the y components. Because $\vec{w}$ is in the negative y direction, we know that w_y is equal to $-mg$. Because the motion of the block is horizontal, we know that a_y equals zero. Therefore,

$$w_y + F_{ny} + f_{ky} + F_y = m\,a_y$$
$$-mg + F_n + 0 + 0 = m\,(0)$$
$$F_n = mg$$

We next solve for the frictional force by substituting this result for F_n in the relation $f_k = \mu_k F_n$.

$$f_k = \mu_k F_n = \mu_k mg$$

Next we substitute this result for f_k in the equation relating the x components. Because both $\vec{w}$ and $\vec{F}_n$ are perpendicular to the x axis we know that w_x and F_{nx} are equal to zero. Therefore,

$$w_x + F_{nx} + f_{kx} + F_x = ma_x$$
$$0 + 0 - f_k + F = ma$$
$$-\mu_k mg + F = ma$$

Solving for μ_k we have

$$\mu_k = \frac{F - ma}{mg}$$
$$= \frac{(16\text{ N}) - (2\text{ kg})(3\text{ m/s}^2)}{(2\text{ kg})(9.8\text{ m/s}^2)} = \underline{0.51}$$

2. The acceleration is $\underline{3.91\text{ m/s}^2}$.

3. The velocities $\vec{v}_1$ and $\vec{v}_2$ of the ball just before and just after it reaches its highest point, respectively, are shown in Figure 5-9.

Figure 5-9

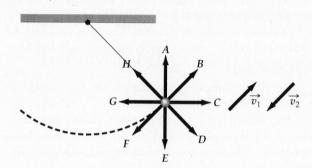

Next we draw (Figure 5-10) the change in velocity vector $\Delta\vec{v}$ using the head-to-tail method of vector addition and the relation $\vec{v}_1 + \Delta\vec{v} = \vec{v}_2$ (the initial velocity plus the change in velocity equals the final velocity). To do this we first draw $\vec{v}_1$ and $\vec{v}_2$ from the same point. Then we draw $\Delta\vec{v}$ from the tip of $\vec{v}_1$ to the tip of $\vec{v}_2$.

Figure 5-10

Because $\vec{a}_{av} = \Delta\vec{v}/\Delta t$, the average acceleration is in the same direction as $\Delta\vec{v}$. The arrow shown in Figure 5-3 that is in the direction of the acceleration vector is labeled $\underline{F}$.

Remark: To check our result we have drawn a free-body diagram of the ball, Figure 5-11, that shows that it is quite plausible that the net force vector is in the same direction as the acceleration vector just previously found. To draw the net force vector we use the head to tail method of vector addition.

Figure 5-11

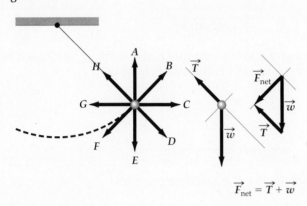

4. Arrow G

5. We sketch a free-body diagram of the ball, shown in Figure 5-12, when it is at the lowest point in its arc. There are two forces acting on the ball, the weight $\vec{w}$ and the tension force $\vec{T}$ exerted by the string. The weight, as always, acts downward. We know that strings can only pull along their length, so the tension force points straight up. On this diagram we also draw a coordinate system with the y axis in the centripetal direction and the x axis in the direction of the motion. Because we do not know the direction of the acceleration vector $\vec{a}$, it is drawn in an arbitrary direction. The acceleration vector is drawn as a shaded arrow in order to differentiate it from the force vectors.

Figure 5-12

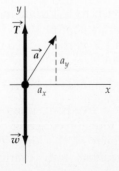

By applying Newton's second law to the ball, we obtain the vector equation

$$\vec{w} + \vec{T} = m\vec{a}$$

This vector equation is a concise representation of the two component equations

$$w_x + T_x + ma_x$$
$$w_y + T_y + ma_y$$

We will first consider the equation relating x components. Because both $\vec{w}$ and $\vec{T}$ are perpendicular to the x axis, we know that w_x and T_x are both equal to zero. Therefore,

$$w_x + F_x = ma_x$$
$$0 + 0 = ma_x$$
$$a_x = 0$$

(The acceleration vector's x component is also its tangential component. That it is zero tells us that the ball's rate of change of speed is momentarily zero. If we do not neglect air drag, both F_x and a_x would be negative, indicating that the ball's speed is decreasing.)

We now consider the equation relating the y components. Because $\vec{T}$ is in the positive y direction and $\vec{w}$ is in the negative y direction we know that T_y is equal to T and that w_y is equal to $-mg$. Also, the ball is moving in a circle with a radius r equal to the length of the string. When the ball passes through the lowest point of the arc the centripetal direction coincides with the $+y$ direction so $a_y = v^2/r$. Therefore,

$$w_y + T_y = ma_y$$
$$-mg + T = m\frac{v^2}{r}$$

Solving for T gives

$$T = m\left(g + \frac{v^2}{r}\right)$$
$$= (0.1 \text{ kg})\left(9.8 \text{ m/s}^2 + \frac{(0.8 \text{ m/s})^2}{0.4 \text{ m}}\right)$$
$$= \underline{1.14 \text{ N}}$$

Remark: The tension exceeds the weight of the ball by $mv^2/r = 0.16$ N. You may not have expected the tension to exceed the weight of the ball. The ball is accelerating upward so the net force acting on it must also be upward. For this to occur the tension must exceed the weight.

6. The tension in the string is <u>1.0 N</u> and the yo-yo's speed is changing at a rate of <u>−9.8 m/s²</u>.

7. A sketch of the car on the incline and a free-body diagram for the car are shown in Figure 5-13. There are two forces acting on it, the gravitational force $\vec{w}$ and the contact force exerted by the road on the tires. This contact force will be expressed as the sum of two forces, the normal force $\vec{F}_n$ and the static frictional force $\vec{f}_s$. Because the car is slowing down, we know the acceleration vector $\vec{a}$ is opposite to the velocity vector.

Figure 5-13

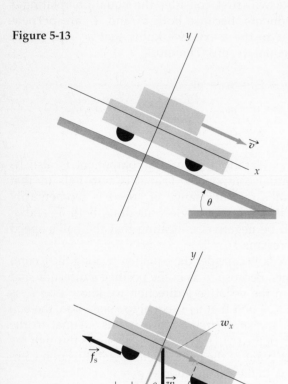

Applying Newton's second law to the car gives

$$\vec{w} + \vec{F}_n + \vec{f}_s = m\vec{a}$$

We have selected a coordinate axis with the y axis in the normal direction and the x axis in the direction of motion. Putting this vector equation into component form gives the equations

$$w_x + F_{nx} + f_{sx} = ma_x$$

$$w_y + F_{ny} + f_{sy} = ma_y$$

We first consider the equation relating the y components. Because the acceleration vector is in

the negative x direction, we know that a_y equals zero. Therefore,

$$w_y + F_{ny} + f_{sy} = ma_y$$

$$-mg\cos\theta + F_n + 0 = m(0)$$

$$F_n = mg\cos\theta$$

The magnitude of the acceleration will be a maximum when $f_s = f_{s,\,max}$. Substituting our previous result for F_n in the relation $f_{s,\,max} = \mu_s F_n$ gives

$$f_{s,\,max} = \mu_s mg\cos\theta$$

Next we substitute this result for $f_{s,\,max}$ in the equation relating x components to obtain

$$w_x + F_{nx} + f_{sx} = ma_x$$

$$mg\sin\theta + 0 - f_{s,\,max} = ma_x$$

$$mg\sin\theta - \mu_s mg\cos\theta = ma_x$$

Solving for the acceleration gives

$$a_x = -g(\mu\cos\theta - \sin\theta)$$

The acceleration is constant so the minimum stopping distance can be determined using the kinematic equation

$$v^2 = v_0^2 + 2a_x\,\Delta x$$

Solving for Δx gives

$$\Delta x = \frac{v^2 - v_0^2}{2a_x}$$

which shows that when the acceleration is a maximum, the stopping distance Δx is a minimum. For that case,

$$\Delta x = \frac{0 - v_0^2}{-2g\,(\mu_s\cos\theta - \sin\theta)}$$

$$= \frac{(33\text{ m/s})^2}{2(9.8\text{ m/s}^2)(0.58\cos 20° - \sin 20°)} = \underline{274\text{ m}}$$

Remark: If you wish to test your mettle, try solving this problem using a coordinate system with a horizontal x axis and a vertical y axis. The algebra is more complex, but it certainly can be done.

8. The stopping distance is <u>0.883 m</u>.

9. Figure 5-14 is a free-body diagram of the upper block. There are four forces acting on this block, the weight $\vec{w}_1$, the force $\vec{F}$, the force $\vec{T}_1$ exerted by the string, and the contact force exerted by the lower block. This contact force will be expressed as the sum of two forces, the normal force $\vec{F}_{n1}$ and the kinetic frictional force $\vec{f}_{k1}$. The speed is constant so the acceleration is zero. The pin is frictionless so we know the tension T is the same throughout the string.

Figure 5-14

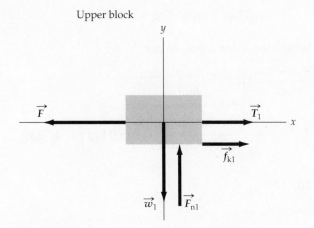

Upper block

Applying Newton's second law to the upper block results in the equation

$$\vec{F} + \vec{w}_1 + \vec{T}_1 + \vec{f}_{k1} + \vec{F}_{n1} = m_1\vec{a}_1$$

Putting this vector equation into component form results in the equations

$$F_x + w_{1x} + T_{1x} + f_{k1x} + F_{n1x} = m_1a_{1x}$$
$$F_y + w_{1y} + T_{1y} + f_{k1y} + F_{n1y} = m_1a_{1y}$$

We first consider the equation relating the y components. This gives

$$F_y + w_{1y} + T_{1y} + f_{k1y} + F_{n1y} = m_1a_{1y}$$
$$0 - m_1g + 0 + 0 + F_{n1} = m_1(0)$$
$$F_{n1} = m_1g \qquad (1)$$

Next we substitute this result for F_{n1} in the relation $f_{k1} = \mu_k F_{n1}$ and solve for the friction force. This gives

$$f_{k1} = \mu_k m_1g \qquad (2)$$

We next substitute this result for f_{k1} in the equation relating x components. This gives

$$F_x + w_{1x} + T_{1x} + f_{k1x} + F_{n1x} = m_1a_{1x}$$
$$F + 0 - T - f_{k1} + 0 = m_1(0)$$
$$F - T - \mu_k m_1g = 0 \qquad (3)$$

Figure 5-15 is a free-body diagram of the lower block. There are four forces acting on this block, the gravitational force $\vec{w}_2$, the force $\vec{T}_2$ exerted by the string, the contact force exerted by the table, and the contact force exerted by the upper block. These contact forces will each be expressed as the sum of two forces, a normal force and a kinetic frictional force. The acceleration is zero.

Figure 5-15

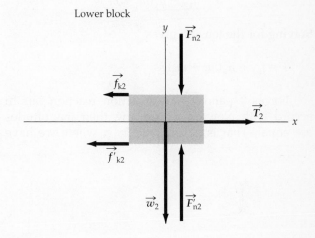

Lower block

Applying Newton's second law to this block results in the equation

$$\vec{w}_2 + \vec{T}_2 + \vec{f}_{k2} + \vec{F}_{n2} + \vec{f}'_{k2} + \vec{F}'_{n2} = m_2\vec{a}_2$$

Putting this vector equation into component form results in the equations

$$w_{2x} + T_{2x} + f_{k2x} + F_{n2x} + f'_{k2x} + F'_{n2x} = m_2a_{2x}$$
$$w_{2y} + T_{2y} + f_{k2y} + F_{n2y} + f'_{k2y} + F'_{n2y} = m_2a_{2y}$$

We first consider the equation relating the y components. Therefore,

$$-m_2g + 0 + 0 - F_{n2} + 0 + F'_{n2} = m_2(0)$$
$$F'_{n2} = F_{n2} + m_2g \qquad (4)$$

$\vec{F}_{n1}$ and $\vec{F}_{n2}$ are an action–reaction pair. In accord with Newton's third law, their magnitudes are equal. That is, $F_{n2} = F_{n1} = m_1g$, where we have substituted from Equation (1). Substituting m_1g for F_{n2} in Equation (4) we obtain

$$F'_{n2} = m_1g + m_2g = (m_1 + m_2)g \qquad (5)$$

(Wouldn't you have guessed that F'_{n2} equals the weight of the two blocks?)

Next we substitute our Equation (5) result for F'_{n2} in the relation $f'_{k2} = \mu_k F'_{n2}$ and solve for the f'_{k2}

$$f'_{k2} = \mu_k(m_1 + m_2)g \qquad (6)$$

Now we substitute our Equation (6) result for f'_{k2} in the equation relating x components. This gives

$$0 + T_2 - f_{k2} + 0 - f'_{k2} + 0 = m_2(0)$$
$$T - f_{k2} - \mu_k(m_1 + m_2)g = 0$$

Solving for the tension gives

$$T = f_{k2} + \mu_k(m_1 + m_2)g \qquad (7)$$

Forces $\overrightarrow{f_{k1}}$ and $\overrightarrow{f_{k2}}$ are an action–reaction pair. In accord with Newton's third law, their magnitudes are equal. That is, $f_{k2} = f_{k1} = \mu_k m_1 g$, where we have

substituted from Equation (2). Substituting $\mu_k m_1 g$ for f_{k2} in Equation (7) we obtain

$$T = \mu_k m_1 g + \mu_k(m_1 + m_2)g = \mu_k(2m_1 + m_2)g$$

Substituting this expression for T in Equation (3) we obtain

$$F - \mu_k(2m_1 + m_2)g - \mu_k m_1 g = 0$$

which we solve for m_1 to get

$$m_1 = \frac{1}{3}\left(\frac{F}{\mu_k g} - m_2\right)$$
$$= \frac{1}{3}\left(\frac{60 \text{ N}}{(0.33)(9.8 \text{ m/s}^2)} - (5.0 \text{ kg})\right) = \underline{4.52 \text{ kg}}$$

10. $\underline{1.55 \text{ m/s}^2}$

Chapter 6

Work and Energy

I. Key Ideas

Under certain conditions, energy is transferred between objects through the forces they exert on each other. This kind of energy transfer is called work. Work—like force, mass, and numerous other terms that have both a scientific and an everyday usage—has a precise scientific meaning which, like kinetic energy, arises from the work–kinetic energy theorem. Kinetic energy is energy associated with motion. Another form of energy is potential energy, energy associated with the configuration of a system. Potential energy changes are related to work that depends on the initial and final positions, but not on the path traveled.

6-1   Work and Kinetic Energy: Motion in One Dimension with Constant Forces   The **work** done by a constant force on an object is the product of the force and the displacement of the point of application of the force. If the force and the displacement are in different directions (as shown in Figure 6-1), then only the component of the force in the direction of the displacement is multiplied by the displacement.

Work in one dimension by a constant force

$$W = F \cos \theta \, \Delta x = F_x \, \Delta x$$

Figure 6-1

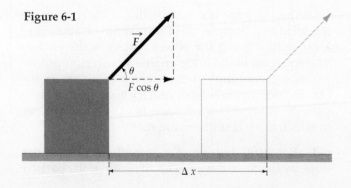

Work (which can be either positive or negative) is a scalar quantity and thus has no direction. When the force acts at right angles to the displacement, the work done by the force is zero.

When an object undergoes a displacement, each force acting on it may, in principle, do work. The **total work** W_{total} done on an object is the algebraic sum of the works done by each of the forces acting on it. An extended object can be modeled as a particle only if all its parts undergo *identical* displacements during any and all time intervals. That is, an object is a particle if it is perfectly rigid and moves without rotating. The following discussion applies only to objects that can be modeled as particles. It follows that for a particle, the total work done on it equals the product of its displacement and the net force acting on it. The **work–kinetic energy theorem**, which follows from Newton's second law ($\Sigma \vec{F} = m\vec{a}$), states that the total work W_{total} done on a particle equals its change in kinetic energy K.

Work–kinetic energy theorem

$$W_{total} = \Delta K = \tfrac{1}{2}mv_f^2 - \tfrac{1}{2}mv_i^2$$

When a variable force does work, we can determine this work by dividing the net displacement into a large number of very short displacements. We can estimate the work done in each short displacement by multiplying that displacement and the value of the force at the beginning of that displacement. The overall work done by a force acting over a finite displacement equals the sum of the works done by the force acting over a sequence of infinitesimal displacements that make up the finite displacement. That is,

Work done by a variable force

$$W = \int_{x_1}^{x_2} F_x \, dx = \text{area under the } F_x\text{-versus-}x \text{ curve}$$

6-2   Work and Kinetic Energy in Three Dimensions and the Dot Product   Next we discuss how to calculate the work done by a force on a particle that moves along a curved path. The work done by the force $\vec{F}$ acting over the small displacement $d\vec{s}$ is equal to ds times the component of $\vec{F}$ in the direction of $d\vec{s}$. That is,

Work

$$dW = F_s\,ds = F\cos\phi\,ds$$

where ϕ is the angle between $\vec{F}$ and $d\vec{s}$ as shown in Figure 6-2.

Figure 6-2

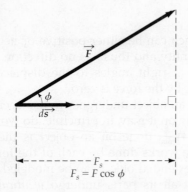

Multiplying the vectors $\vec{F}$ and $d\vec{s}$ to get $F_s ds = F\cos\phi\,ds$ gives their **scalar** or **dot product.** The scalar product of two vectors $\vec{A}$ and $\vec{B}$ is written $\vec{A}\cdot\vec{B}$ and is defined as

The dot product

$$\vec{A}\cdot\vec{B} = AB\cos\phi$$

where ϕ is the angle between them. This product is related to the x, y, and z components of the two vectors by

The dot product

$$\vec{A}\cdot\vec{B} = A_xB_x + A_yB_y + A_zB_z$$

In the dot product notation, the work W done by a force $\vec{F}$ on a particle moving from point 1 to point 2 is

Work—a general definition

$$W = \int_{\vec{s}_1}^{\vec{s}_2}\vec{F}\cdot d\vec{s}$$

The dot product of a unit vector and a vector $\vec{A}$ gives the component of $\vec{A}$ in the direction of the unit vector. For example,

Projecting $\vec{A}$ onto the x axis

$$\hat{i}\cdot\vec{A} = A\cos\phi = A_x$$

$A\cos\phi$ is the projection of $\vec{A}$ onto the x axis, where ϕ is the angle between $\vec{A}$ and $\hat{i}$. Consequently, taking the dot product of a unit vector and a vector can be thought of as projecting the vector onto a line in the direction of the unit vector. This may not seem particularly significant now, but thinking of the dot product of a unit vector and a vector in this way will prove useful in later chapters.

6-3   Power   What we call **power** is the rate at which a force does work, the work per unit time. The power P associated with a force $\vec{F}$ acting on an object is

Power

$$P = \frac{dW}{dt} = \frac{\vec{F}\cdot d\vec{s}}{dt} = \vec{F}\cdot\frac{d\vec{s}}{dt} = \vec{F}\cdot\vec{v}$$

where $\vec{v}$ is the velocity of the point of application of the force. The SI unit of power, one joule per second, is called the watt (W). More generally, the term power refers not just to the rate of doing work but to any rate of transfer of energy.

The power associated with the net force on a particle equals the rate of change of the particle's kinetic energy. This relation is derived below for a particle constrained to move in one dimension. Starting with the power relation $P = \vec{F}\cdot\vec{v}$ we have

$$P = \vec{F}_{net}\cdot\vec{v} = F_{net\,x}v_x = ma_xv_x$$

$$= mv_x\frac{dv_x}{dt} = \frac{d}{dt}\left(\tfrac{1}{2}mv_x^2\right) = \frac{dK}{dt}$$

where we have substituted ma_x for $F_{net\,x}$ and dv_x/dt for a_x. Rearranging this relation gives $dK = P\,dt$, which for constant power becomes

Kinetic energy and power

$$dK = P\,dt \quad \text{(constant power)}$$

While a complex system like a car cannot be modeled as a particle, this formula can be applied to a car with P being the power output of the engine. The equation holds if the engine's power output remains constant, no external forces do work on the car, and no internal or external frictional forces dissipate the energy. This equation is valid to the degree to which these conditions are met.

6-4   Potential Energy   For most forces, the work done by the force depends on the path of the particle it acts on. For example, if you push your chair from point A to point B, the amount of work you do depends on the path over which you push the chair. Other things being equal, the longer the path, the more work you do. Yet, even when the path lengths are equal, the work done is not necessarily equal. For example, if one path is over a slick newly waxed floor while the other path is over rough concrete, more work is required to push the chair from A to B over the rough concrete than over the waxed floor. That is, pushing harder (exerting a greater force) over the same distance, means you do more work.

There are situations, however, when the work done by a force depends only on the initial and final positions and not on the path taken. Gravitational forces are an example of this. The work done by the force of gravity on a particle of mass m is $-mg$ $(y_f - y_i)$, where y_i and y_f are the initial and final heights. For example, when you lift a book of mass m off your desk and place it on a shelf, the work done by the gravitational force on the book is $-mg$ $(y_s - y_d)$, where y_s and y_d are the heights of the shelf and of the desktop. Whether you pick up the book and place it directly on the shelf or carry it around with you all day before placing it on the shelf, the work done on the book by gravity remains $-mg$ $(y_s - y_d)$. **Conservative forces,** forces such as gravity, are forces where the work done depends only on the initial and final positions and not on the path taken. If a force is conservative, then the work done by the force on a particle that returns to its initial position must be zero. To be conservative a force must be a function of position only.

The change in **potential energy** U associated with a conservative force is the negative of the work done by the force. That is

Potential energy

$$\Delta U = U_f - U_i = -W = -\int_{s_i}^{s_f} \vec{F} \cdot d\vec{s}$$

Consequently, near the surface of the earth, the change in gravitational potential energy is related to the change in position by the relation $U - U_0 = +mg(y - y_0)$, where y_0 and y are the initial and final heights of the particle. U, a function of y, is called a potential energy function. If we define U_0 as mgy_0, this may be expressed as

Gravitational potential energy

$$U = U_0 + mgy$$

Your physics book and the earth attract each other via gravitational forces. When objects attract each other, the farther apart they are, the greater potential energy they have. Where is this potential energy? Potential energy is an abstract concept, so that's a challenging question. You should think of the potential energy as an inherent part of the earth–book system. Physicists sometimes talk as if this potential energy is associated only with the book and its position; but actually the potential energy is always shared between the book and the earth. The potential energy of the book–earth system depends on the position of the book relative to the earth (not just on the book's position relative to an arbitrary reference frame).

The potential energy of the book–earth system increases when you raise the book. To raise the book you must do positive work on it. However, while you are doing this positive work the gravitational force on the book is doing negative work. It is the negative of the work done by the gravitational force that equals the increase in the system's gravitational potential energy.

Another example of a conservative force is the force exerted by a spring that obeys Hooke's law ($F_x = -kx$). The potential energy function associated with such a spring is

Potential energy of a spring

$$U = \tfrac{1}{2}kx^2$$

In one dimension, the change in potential energy associated with a conservative force $\vec{F} = F_x \hat{i}$ is $dU = -F_x \, dx$. Thus the force can be expressed as the negative of the derivative of the potential energy function

Force and potential energy

$$F_x = -\frac{dU}{dx}$$

It follows that on a graph of the potential energy function U versus the position x, the conservative force F_x is the negative of the slope of the curve. On a U-versus-x graph, a positive slope indicates a negatively directed force, and a negative slope indicates a positively directed force. The important thing to note is that *the force is in the direction such as to accelerate the particle toward lower potential energy.* If the conservative force F_x is the only force acting on a particle, then the particle is in **equilibrium** when the slope of the potential energy versus position curve is zero, that is, when F_x is zero.

A particle is in **stable equilibrium** if small displacements away from equilibrium result in a restoring force that pushes the particle back toward its equilibrium position. Consequently, when a particle is in stable equilibrium it is at a potential energy minimum. For example, a ball resting at the lowest point in a valley is in stable equilibrium. A particle is in **unstable equilibrium** if small displacements from equilibrium result in a force that pushes the particle even further away from equilibrium. Thus when a particle is in unstable equilibrium it is at a potential energy maximum, like a ball poised at the crest of a hill. A particle is in **neutral equilibrium** if the force remains zero following any small displacement from an equilibrium position.

A conservative force always tends to accelerate a particle toward a position of lower potential energy.

II. Numbers and Key Equations

Numbers

$1\,J = 1\,N{\cdot}m = 1\,kg{\cdot}m^2/s^2 = 0.738\,ft{\cdot}lb$

$1\,W = 1\,J/s$

$1\,hp = 550\,ft{\cdot}lb/s = 746\,W$

$1\,kW{\cdot}h = 3.6\,MJ$

Key Equations

Work in one dimension by a constant force

$$W = F \cos \phi\, \Delta x = F_x\, \Delta x$$

Work in one dimension by a variable force

$$W = \int_{x_1}^{x_2} F_x\, dx$$

The dot product

$$\vec{A}\cdot\vec{B} = AB \cos \phi = A_x B_x + A_y B_y + A_z B_z$$

Projecting $\vec{A}$ onto the x axis

$$\hat{i}\cdot\vec{A} = A \cos \phi = A_x$$

Work in three dimensions by a variable force

$$W = \int_{s_1}^{s_2} F_s\, ds = \int_{s_1}^{s_2} F \cos \phi\, ds = \int_{\vec{s}_1}^{\vec{s}_2} \vec{F}\cdot d\vec{s}$$

Work-kinetic energy theorem

$$W_{\text{total}} = \Delta K = \tfrac{1}{2}mv_f^2 - \tfrac{1}{2}mv_i^2$$

Increment of work

$$dW = F_s\, ds = F \cos \phi\, ds = \vec{F}\cdot d\vec{s}$$

Power

$$P = \frac{dW}{dt} = \vec{F}\cdot\vec{v}$$

Kinetic energy and power

$$\Delta K = P\, \Delta t \quad \text{(constant power)}$$

Conservative force defined

If $\displaystyle\int_{\vec{s}_1}^{\vec{s}_2} \vec{F}\cdot d\vec{s}$ is path independent then $\vec{F}$ is conservative.

Potential energy

$$\Delta U = U_2 - U_1 = -W = -\int_{\vec{s}_1}^{\vec{s}_2} \vec{F}\cdot d\vec{s}$$

(where $\vec{F}$ is conservative)

Force and potential energy

$$F_x = -\frac{dU}{dx} \quad \text{(where } \vec{F} = F_x\hat{i} \text{ is conservative)}$$

Gravitational potential energy

$$U = U_0 + mgy$$

Potential energy of a spring

$$U = \tfrac{1}{2}kx^2$$

III. Potential Pitfalls

This chapter contains many words whose technical meanings differ from their meaning in everyday usage. These include *work, kinetic energy, potential energy,* and *power.* Be sure to understand the technical definitions.

Don't think the work has to be nonzero just because the force and the displacement are both nonzero. The work done by a force is equal to the dot product of the force and the displacement of the point of application of the force. This work is zero when the displacement of the point of application of the force is either zero or is directed perpendicular to the force.

Don't think of work as having a direction just because the force and the displacement do. Work is a scalar quantity, and thus it has no direction. The sign of a work term represents the sense of the energy change. When the work is positive, energy is transferred to the object that the force acts on.

Don't think that the work–kinetic energy theorem can always be applied to a block sliding along a flat surface. If there is kinetic friction between the block and the surface then the block cannot be modeled as a particle. It follows that the theorem does not apply. Only if the surface is frictionless can the block be modeled as a particle.

Only changes in potential energy have physical meaning. The choice of the reference point for potential energy is always arbitrary, so choose a reference point that is convenient for a particular situation. Although kinetic energy can never be negative, potential energy can be either positive or negative.

Don't associate potential energy with a single particle. Potential energy depends on the system's configuration and is associated with the entire system. If only one particle in a system moves, it is common parlance to say that the particle's potential energy has changed, but in reality it is the potential energy of a larger system that has changed.

Be aware of the sign in the definition of potential energy. When the displacement is in the direction of a conservative force, the change in potential energy is negative. For example, when an object moves downward, in the direction of the gravitational force, the gravitational potential energy decreases.

IV. True and False and Responses

True and False

_____ 1. Like kinetic energy, work is necessarily a positive quantity.

_____ 2. The work–kinetic energy theorem is best applied to problems where the force on a particle is known explicitly as a function of time.

_____ 3. When several forces act on a *particle*, the total work done by all of them is always equal to

$$\int_{\vec{s}_1}^{\vec{s}_2} \vec{F}_{net} \cdot d\vec{s}$$

where $\vec{F}_{net}$ is the net force acting on the particle.

_____ 4. When several forces act on an *object*, the total work done by all of them is always equal to

$$\int_{\vec{s}_1}^{\vec{s}_2} \vec{F}_{net} \cdot d\vec{s}$$

where $\vec{F}_{net}$ is the net force acting on the object.

_____ 5. The scalar or dot product of any two perpendicular vectors is always zero.

_____ 6. When a particle moves in the direction of a conservative force the potential energy of the particle increases.

_____ 7. Kinetic energy and work have the same dimensions.

_____ 8. The reference point for gravitational potential energy must be mean sea level.

_____ 9. An athlete throws a ball. To throw the ball again so that it moves with twice the speed requires that the athlete do twice as much work on the ball.

_____ 10. A particle is in stable equilibrium at a point where its potential energy is at a maximum.

_____ 11. Particles that attract via conservative forces have more potential energy when they are close together than when they are far apart.

_____ 12. Particles that repel via conservative forces have more potential energy when they are close together than when they are far apart.

Responses

1. False. Work is negative when the force and displacement are oppositely directed.

2. False. It is best applied to problems where the force on a particle is known explicitly as a function of position. Work is computed by evaluating an integral where the integration variable is an increment of position, not time.

3. True

4. False. The statement is true only if the points of application of each force undergo equal displacements. Although this condition must hold for a particle, it does not necessarily hold for an object. For example, if you compress a spring by pushing on opposite ends with your hands, the two ends do not move through identical displacements. (Even if they are identical in magnitude they will not be identical in direction.)

5. True

6. False. If a particle moves in the direction of a conservative force its potential energy decreases.

7. True. The dimensions are $\dfrac{[M][L]^2}{[T]^2}$.

8. False. The reference point is arbitrary as long as it is stationary relative to the earth.

9. False. For the ball to have twice its original speed, it must have four times its original kinetic energy.

Because the total work done on the ball equals the change in its kinetic energy, the athlete must do four times as much work.

10. False. When a particle is at a potential energy maximum, small displacements of the particle from the equilibrium position will result in a force that pushes the particle farther from the equilibrium position. Thus, the particle is in unstable equilibrium.

11. False. If one of the particles moves in the direction of the conservative force the potential energy will, as always, decrease. They will also become closer together. Thus they have less potential energy when they are closer together.

12. True

V. Questions and Answers

Questions

1. You pick a book up off your desk and place it on a book shelf on the wall above the desk. What is the total work done by all forces acting on the book? Explain.

2. A pendulum consists of a 1-inch-diameter steel ball of mass m suspended from the ceiling by a string. When the ball swings from its lowest position (where it has speed v_0) up to its highest position (where it reverses direction), what is the total work done on the ball by all the forces acting on it? How much work is done by each force acting on it? Explain.

3. A block of mass m, released from rest, slides down a frictionless incline and reaches the bottom with speed v_f. What is the total work done on the block? How much work is done by each force acting on it? Explain.

4. When you get up from your chair and start walking toward the door, your kinetic energy increases. What force does work on you to cause this increase in your kinetic energy? Where does this kinetic energy come from?

5. A car of mass m travels up a long, straight hill of height h at constant speed. Assume air drag is negligible. How much work is done by each force acting on the car? What is the total work done on it by external forces? Explain.

6. Your roommate solves an assigned physics problem and gets a negative value for the change in potential energy of the pendulum bob in Question 2. Is this possible? Explain.

Answers

1. The total work done on the book is zero. The work–kinetic energy theorem relates the total work done on a particle with the change in the particle's kinetic energy. The total work done on the book equals the change in kinetic energy of the book. The initial speed of the book is zero and the final speed of the book is zero, thus the change in kinetic energy is zero minus zero which equals zero. The gravitational force does negative work on the book, but your hand does positive work. The work–kinetic energy theorem tells us that the sum of the two works equals zero.

2. The total work done on the ball equals its change in kinetic energy. Its final kinetic energy is zero and its initial kinetic energy is $\frac{1}{2}mv_0^2$, where m is its mass and v_0 is its velocity at its lowest point. Thus the total work done on the ball is $-\frac{1}{2}mv_0^2$. Two forces act on the ball, the gravitational force and the tension force exerted by the string. The work done by the tension force is zero because the ball's velocity, and therefore the infinitesimal increments of its displacement, is always perpendicular to this force. The power delivered by the tension force $\vec{T}$ is given by $P = \vec{T}\cdot\vec{v} = Tv\cos 90° = 0$. If the direction of a force and the velocity of its point of application are perpendicular, the rate at which that force is doing work is zero. It follows that the total work done on the ball equals the work done by only the gravitational force.

3. The total work done on the block equals its change in kinetic energy. Its initial kinetic energy is zero and its final kinetic energy is $\frac{1}{2}mv_f^2$, where m is its mass and v_f is its velocity as it reaches the bottom of the incline. Thus the total work is $\frac{1}{2}mv_f^2$. Because the normal force is perpendicular to the displacement, the work done by it is zero. The work done by the gravitational force W_g is equal to the total work done on the block; that is, $W_g = \frac{1}{2}mv_f^2$.

4. Two forces act on you when you start walking: the force of gravity and the contact force of the floor on your feet. Assuming that you are walking on a horizontal floor, the force of gravity does no work because it acts at right angles to your displacement. The contact force of the floor only acts on a foot when that foot is in contact with the floor. Whenever this force acts on it, the foot is at rest so the displacement of the point of application of the contact force is zero. Therefore the contact force does no work. Because neither force acting on you does any work, the total work done on you is zero. Your increased kinetic energy comes from the chemical potential energy that is stored in your muscles. These chemical changes constitute changes in your internal structure. An object whose structure changes *cannot* be modeled as a particle. Because the work–kinetic energy theorem only holds for particles, this theorem does not apply to your motion. This type of problem is analyzed using another theorem, the work–energy theorem, which is presented in Chapter 7.

5. Because the structure of the car does not remain fixed as the car moves, the car cannot be modeled as a particle. This means the work–kinetic energy theorem does not apply. Neglecting air resistance, there are two forces acting on the car, the force of gravity and the contact force of the road on the tires. The force of gravity does work $-mgh$, where m is the mass of the car and h is the altitude gained by the car. The work done by the contact force of the road on the car is zero because the displacement of the point of application of this contact force is zero. (Unless the car is skidding, the part of the tire in contact with the road is not moving.) Thus the total work done on the car equals $-mgh$. If air drag were not negligible it would do additional negative work on the car.

6. It is possible but only if your roommate makes a mistake. The pendulum bob moves upward and the conservative force acting on it, its weight, is downward. The potential energy must increase if the bob moves either directly or obliquely upward. The value of the potential energy depends on the reference point chosen. If the chosen reference point is on the ceiling, then the bob's potential energy will remain negative so long as the ball remains below the ceiling. However, the *change* in potential energy is unaffected by the choice of reference point; in this case, it will still be positive since the potential energy is increasing.

VI. Problems and Solutions

Problems

1. A 2-kg particle is subjected to a single force F_x that varies with position as shown in Figure 6-3.

Figure 6-3

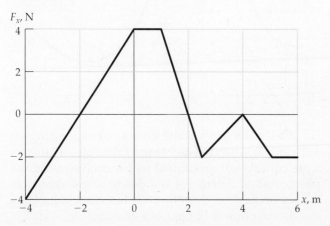

From the graph, determine the work done by the force when the particle moves from $x = 0$ to (*a*) $x = 4$ m; and (*b*) $x = -3$ m. (*c*) If the particle is projected from the origin with a speed of 1 m/s, determine its speed when it is at $x = 4$ m.

How to Solve It

• The work done by a force is equal to the area under the F_x-versus-x curve. Areas below the x axis are negative. To find the area, first divide the region between the curve and the x axis into rectangles and triangles. Then for each region multiply the average force and the displacement.

• To find the work done by the force when the particle moves from $x = 0$ to $x = 4$ m, take the sum of the areas for the individual displacements. Use the same technique to determine the work done by the force when the particle moves from $x = 0$ to $x = -3$ m.

• To find the speed of the particle at $x = 4$ m, use the work–kinetic energy theorem to relate the work done on the particle to the change in its kinetic energy.

2. A 2-kg particle is subjected to a single force F_x that varies with position as shown in Figure 6-4. From the graph, determine the work done by the force when the particle moves from $x = 0$ to (*a*) $x = 6$ m; and (*b*) $x = -4$ m. (*c*) If the particle is projected from $x = -4$ m with a velocity of 5 m/s in the $+x$ direction, determine its speed when it is at $x = 6$ m.

Figure 6-4

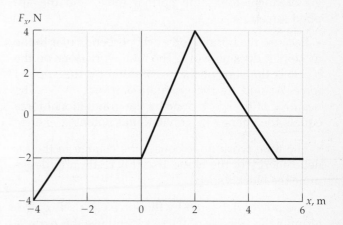

How to Solve It

• The work done by a force is equal to the area under the F_x-versus-x curve. Areas below the x axis are negative. To find the area, first divide the region between the curve and the x axis into rectangles and triangles. Then for each region multiply the average force and the displacement.

• To find the work done by the force when the particle moves from $x = 0$ to $x = 6$ m, take the sum of the areas for the individual displacements. Use the same technique to determine the work done by the force when the particle moves from $x = 0$ to $x = -4$ m.

• To find the speed of the particle at $x = 6$ m, use the work–kinetic energy theorem to relate the work done on the particle to the change in its kinetic energy.

3. Starting from rest, a 2-kg block is pulled up a frictionless 37° incline by a force $\vec{F}$ of 15 N directed up the incline, as shown in Figure 6-5. (*a*) Determine the speed of the block after it has traveled a distance d. (*b*) Determine the power being delivered by the force when d equals 0.5 m.

Figure 6-5

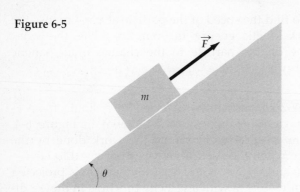

How to Solve It

• There are three forces acting on the block: the gravitational force, the normal force, and the applied force $\vec{F}$.

• Because the normal force acts perpendicular to the motion, it does no work. The total work done on the block is the sum of the work done by the applied force W_F and by the gravitational force W_{grav}. The negative of the work done by the gravitational force equals the change in gravitational potential energy.

• The total work done equals the change in the kinetic energy of the block. Use this relation to determine the block's speed.

• The power delivered by the force equals $\vec{F} \cdot \vec{v}$. Determine the velocity of the block and use this expression to determine the power.

4. A 5-kg block rests on a frictionless 53° incline, as shown in Figure 6-6. The block is pushed by a constant horizontal force $\vec{F}$ of 96 N. (*a*) Determine the speed of the block after it has traveled a distance d of 0.5 m. (*b*) Determine the power being delivered by the force when d equals 0.5 m.

Figure 6-6

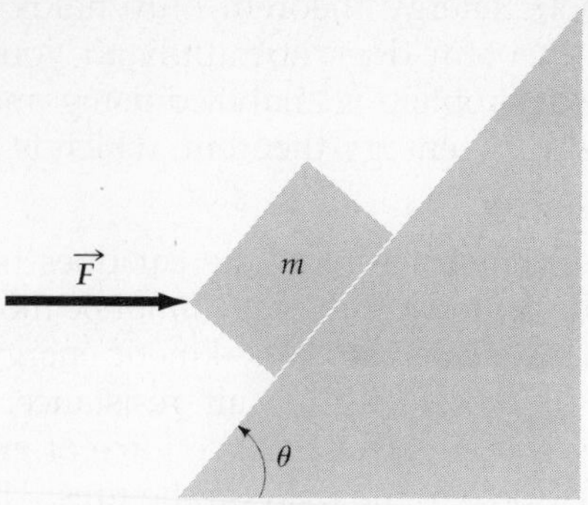

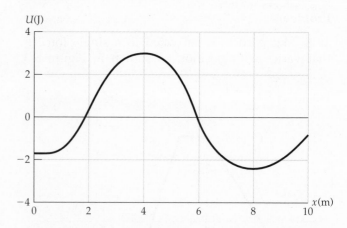

How to Solve It

• There are three forces acting on the block. The gravitational force, the normal force, and the applied force $\vec{F}$.

• Because the normal force acts perpendicular to the block, it does no work. The total work done on the block is the sum of the works done by the applied force W_F and by the gravitational force W_{grav}. The negative of the work done by the gravitational force equals the change in gravitational potential energy.

• The total work done equals the change in the kinetic energy of the block. Use this relation to determine the block's speed.

• The power being delivered by the force equals $\vec{F} \cdot \vec{v}$. Determine the velocity of the block and use this expression to determine the power.

5. The potential-energy function for a conservative force $\vec{F} = F_x \hat{i}$ acting on a 2.00-kg particle is shown in Figure 6-7.

Figure 6-7

For what values of x is the force (*a*) zero; (*b*) directed leftward; (*c*) directed rightward? (*d*) What values of x are equilibrium positions? (*e*) For each equilibrium position state whether it is a position of stable, neutral, or unstable equilibrium. (*f*) For what value(s) of x is the magnitude of the force greatest?

How to Solve It
- Use $F_x = -dU/dx$ to relate the force to the slope of the $U(x)$-versus-x plot. Equilibrium positions are locations where the force is zero. Look up the conditions for stable, neutral, and unstable equilibrium.

6. The potential-energy function for a conservative force $\vec{F} = F_x \hat{i}$ acting on a 2.00-kg particle is shown in Figure 6-8.

Figure 6-8

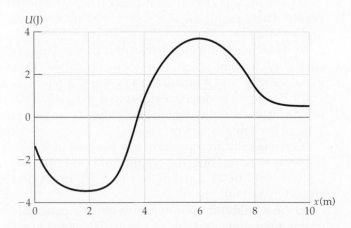

For what values of x is the force (*a*) zero; (*b*) directed leftward; (*c*) directed rightward. (*d*) What values of x are equilibrium positions? (*e*) For each equilibrium position state whether it is a position of stable, neutral, or unstable equilibrium. (*f*) For what value(s) of x is the magnitude of the force greatest?

How to Solve It
- Use $F_x = -dU/dx$ to relate the force to the slope of the $U(x)$-versus-x plot. Equilibrium positions are locations where the force is zero. Look up the conditions for stable, neutral, and unstable equilibrium.

Solutions

1. The work done by the force equals the area under the F_x-versus-x curve. To calculate the area, we divided up the region between the curve and the x axis into triangles and rectangles, as shown in Figure 6-9.

(*a*) The work done by the force as the particle moves from $x = 0$ to $x = 4$ m is found by adding areas 3, 4, and 5. Each area may be computed by taking the product of the average force and the displacement:

$$W = A_3 + A_4 + A_5$$
$$= (4\text{ N})(1\text{ m}) + \tfrac{1}{2}(4\text{ N})(1\text{ m}) + \tfrac{1}{2}(-2\text{ N})(2\text{ m})$$
$$= 4\text{ J} + 2\text{ J} - 2\text{ J}$$
$$= \underline{4\text{ J}}$$

Figure 6-9

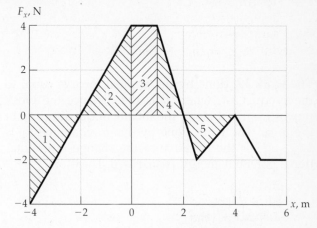

(*b*) The work done by the force as the particle moves from $x = 0$ to $x = -3$ m is found by adding areas 1 and 2. Each area may be computed by taking the product of the average force and the displacement:

$$W = A_1 + A_2$$
$$= \tfrac{1}{2}(4\text{ N})(-2\text{ m}) + \tfrac{1}{2}(-2\text{ N})(-1\text{ m})$$
$$= -4\text{ J} + 1\text{ J}$$
$$= \underline{-3\text{ J}}$$

(*c*) The work–kinetic energy theorem states that the total work done on the particle as it moves from $x = 0$ to $x = 4$ m equals the change in its kinetic energy. Using K_i for the kinetic energy at the origin where its speed $v_i = 1$ m/s and K_f for its kinetic energy at $x = 4$ m where its speed is v_f, we have

$$W = \Delta K = K_f - K_i$$
$$= \tfrac{1}{2}mv_f^2 - \tfrac{1}{2}mv_i^2$$

Solving this equation for the speed v_f yields

$$v_f = \sqrt{\frac{2W}{m} + v_i^2}$$
$$= \sqrt{\frac{2(4\text{ J})}{(2\text{ kg})} + (1\text{ m/s})^2}$$
$$= \underline{2.24\text{ m/s}}$$

2. (*a*) 3 J; (*b*) 9 J; (*c*) 4.36 m/s

3. The total work done on the block equals the sum of the works done by the applied force and by the force of gravity. The work–kinetic energy theorem states that the total work equals the change in the kinetic energy:

$$W_{\text{total}} = \Delta K$$
$$W_F + W_{\text{grav}} = \tfrac{1}{2}mv_f^2 - \tfrac{1}{2}mv_i^2$$

The work W_F done by the applied force is equal to the product Fd, where d is the distance the block moves from rest along the incline in the direction of the force. The work done by the gravitational force W_{grav} equals the negative of the change in gravitational potential energy. Also, $v_i = 0$. Therefore,

$$W_F = -W_{\text{grav}} + \tfrac{1}{2}mv_f^2$$
$$Fd = \Delta U + \tfrac{1}{2}mv_f^2$$
$$= mg\,\Delta h + \tfrac{1}{2}mv_f^2$$

where $mg\,\Delta h$ is the change in gravitational potential energy when the block undergoes a change in height Δh.

The relation between d and Δh ($\Delta h = d \sin\theta$) is illustrated in Figure 6-10.

Figure 6-10

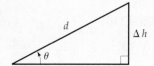

By substituting $d \sin\theta$ for Δh in the previous equation we obtain

$$Fd = mgd \sin\theta + \tfrac{1}{2}mv_f^2$$

Solving this equation for the final velocity v_f we get

$$v_f = \sqrt{\frac{2Fd}{m} - 2gd \sin\theta}$$

$$= \sqrt{\frac{2(15\ \text{N})(0.5\ \text{m})}{(2\ \text{kg})} - 2(9.8\ \text{m/s}^2)(0.5\ \text{m})\sin(37°)}$$

$$= 1.28\ \text{m/s}$$

The power delivered by the force is

$$P = \vec{F} \cdot \vec{v} = Fv \cos\phi$$
$$= (15\ \text{N})(1.28\ \text{m/s})\cos(0)$$
$$= 19.1\ \text{W}$$

4. (a) 1.92 m/s; (b) 110 W

5. The force and the potential energy are related by the equation $F_x = -dU/dx$. dU/dx is the slope of the $U(x)$-versus-x curve.

(a) The force is zero at each point where the slope is zero. These points are at $0 < x < \approx 0.7$ m, $x = 4$ m, and $x = 8$ m.

(b) The force is directed leftward if $F_x < 0$. This occurs at each point where the slope is positive. That is, for $0.7\ \text{m} < x < 4\ \text{m}$ and for $8\ \text{m} < x < 10\ \text{m}$.

(c) The force is directed rightward if $F_x > 0$. This occurs at each point where the slope is negative. That is, for $4\ \text{m} < x < 8\ \text{m}$.

(d) An equilibrium position is a position at which the force is zero. Thus, the equilibrium positions are $0 < x < \approx 0.7$ m, at $x = 4$ m and at $x = 8$ m, as discussed in part (a).

(e) An equilibrium position is stable at a potential-energy minimum, unstable at a potential-energy maximum, and neutral where the potential energy is constant. It follows that $0 < x < \approx 0.7$ m are points of neutral stability, at $x = 4$ m the equilibrium is unstable, and at $x = 8$ m it is stable.

(f) The force equals the negative of the slope, so the magnitude of the force is greatest where the slope is the steepest. This occurs at both $x \approx 2.0$ m and $x \approx 5.8$ m.

6. (a) $x = 2$ m, $x = 6$ m, and $\approx 9\ \text{m} < x < 10\ \text{m}$
(b) $2\ \text{m} < x < 6\ \text{m}$
(c) $0 < x < 2$ m and $6\ \text{m} < x < \approx 9\ \text{m}$
(d) $x = 2$ m, $x = 6$ m, and $\approx 9\ \text{m} < x < 10\ \text{m}$
(e) stable at $x = 2$ m, unstable at $x = 6$ m, and neutral for $\approx 9\ \text{m} < x < 10\ \text{m}$
(f) $x \approx 3.5$ m

Chapter 7

Conservation of Energy

I. Key Ideas

Energy is a central concept in all of science. As far as anyone can tell, energy is never created or destroyed; rather, it is transformed from one form to another and transferred from one location to another. Forms of energy include kinetic energy (energy associated with motion); potential energy (energy associated with the relative positions of interacting objects); thermal energy (energy associated with random molecular motions and configurations); and others. Under certain conditions, energy is transferred between objects through the forces they exert on each other. This kind of energy transfer is called work. Work, like force, mass, and numerous other terms that have both a scientific and an everyday usage, has a precise scientific definition. The energy and mass of a system are directly proportional, and the proportionality constant is the speed of light squared. For bound systems, like a stable atomic nucleus, the energy can only take on specific discrete values. We then say that the system's energy spectrum is quantized.

7-1 The Conservation of Mechanical Energy Kinetic energy and potential energy together constitute what is called **mechanical energy.** The total kinetic energy K of a system of particles is the sum of the kinetic energies of the individual particles. Potential energy is energy associated with the configuration of a system. For any particular configuration the actual value of the potential energy is arbitrary since only changes in potential energy are physically meaningful. For a specific configuration, called the reference configuration, the potential energy is set equal to zero. The forces exerted on each particle in the system by the other particles in the system are called internal forces. If a system's configuration changes, the accompanying change in its potential energy is the negative of the total work done by the internal conservative forces. For a given configuration of a system, the potential energy U is its change

in potential energy as it moves from its reference configuration to its present one.

For example, consider a system consisting of only a crate and the earth, and the initial configuration has the crate resting on the floor. We specify this as the reference configuration where the system's potential energy is zero. You now lift the crate some distance above the floor. (You are not part of this system.) The gravitational forces of the earth on the crate and the crate on the earth are both conservative internal forces. As the crate rises, the gravitational force on it does negative work (the force and the displacement are oppositely directed) and the gravitational force of the crate on the earth does no work (the earth's displacement is zero). Thus, during the lift the change in the earth–crate system's potential energy equals the negative of the work done on the crate by the earth's gravitational pull. This potential-energy change equals the system's new potential energy.

The **total mechanical energy** E of a system is its total kinetic energy plus its total potential energy. That is,

Total mechanical energy

$$E_{\text{mech}} = K + U = K + U_{\text{s}} + U_{\text{g}} + \cdots$$

where U_{s} is the potential energy associated with elastic deformations, such as that associated with a stretched spring, and U_{g} is the gravitational potential energy, such as that associated with the water behind a dam. Other forms of potential energy will be added in later chapters.

If, while a system changes, no external forces do work on it and if the internal forces doing work are all conservative (the work done by a conservative force doesn't depend on the path of the system), its total mechanical energy remains constant. That is,

Conservation of mechanical energy

$$E_{mech} = \text{constant} \quad or \quad K_f + U_f = K_i + U_i$$

This result is called the **conservation of mechanical energy.**

7-2 The Conservation of Energy   There are two ways that the total energy of a system can change. One way is if one or more external forces does work on it and the other is if heat is transferred to or from the system. A discussion of energy transfer via heat is explored in Chapter 19. This chapter will consider only energy transferred by work. In this context the relation between work and energy can be expressed

Work–energy theorem

$$W_{ext} = \Delta E_{mech} + \Delta E_{therm} + \Delta E_{chem} + \cdots$$

where W_{ext} is the total work done on the system by external forces; E_{mech} is the system's mechanical energy; E_{chem} is its chemical energy, such as the energy stored in a battery; and E_{therm} is its thermal energy, the energy associated with the random motion of molecules. The equation is not complete as there are other forms of energy, like nuclear energy that is stored in the nucleus of atoms.

Thermal energy is increased if two surfaces slide on each other. It can be shown that if both surfaces are within a system

Frictional dissipation of mechanical energy

$$\Delta E_{therm} = f_k d$$

where ΔE_{therm} is the system's thermal-energy change, f_k is the frictional force on one of the surfaces, and d is the distance that surface slides relative to the other surface.

Consider a system consisting of planet Earth, a massless spring, a ramp, and a block of mass m as shown in Figure 7-1. Initially the block is at rest and the spring is neither stretched nor compressed.

Figure 7-1

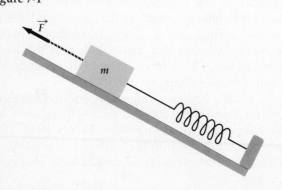

A student is pulling on the string (which is part of the block) with a force F. The block gains height h as it slides a distance L up the ramp. Applying the work–energy theorem to the system we have

$$W_{ext} = \Delta E_{mech} + \Delta E_{therm} + \Delta E_{chem}$$
$$= \Delta K + \Delta U_s + \Delta U_g + \Delta E_{therm} + \Delta E_{chem}$$

It is useful for bookkeeping purposes to make a table of the work done by each external force and a second table for the various energy terms. These tables are

External forces	W_{ext}
F	FL

	K	U_g	U_s	E_{therm}	E_{chem}
initial	0	0	0	n.a.	n.a.
final	$\frac{1}{2}mv^2$	mgh	$\frac{1}{2}kL^2$	n.a.	n.a.
change (Δ)	$\frac{1}{2}mv^2$	mgh	$\frac{1}{2}kL^2$	$f_k L$	0

n.a. = Not applicable.

We don't know the initial or final thermal energies, but we do know that the final thermal energy exceeds the initial by $f_k L$. The change in chemical energy is zero. Substituting these values into the work–energy theorem gives

$$FL = \tfrac{1}{2}mv^2 + \tfrac{1}{2}kL^2 + mgh + f_k L + 0$$

Had the system included the student, the change in chemical energy would not be zero because chemical energy is transformed into mechanical and thermal energy by the student's muscles. The decrease in the student's chemical energy has to be at least as large as the work FL done by her on the system. However, the student was not included in the system so this change does not appear in this application of the work–energy theorem.

7-3 Mass and Energy   Fundamental subatomic particles come in pairs. For each particle there is a corresponding antiparticle. If a particle-antiparticle pair combine, they are transformed from material particles into photons (electromagnetic energy), and the energy of the photons can be accounted for only if we accept the fact that each material particle has an intrinsic energy called its **rest energy** E_0. A particle's rest energy is given by the famous equation

Rest energy

$$E_0 = mc^2$$

where m is its mass and c is the speed of light in a vacuum.

Fundamental particles combine to form atomic nuclei, so we can treat an atomic nucleus as a system of these particles. Since the particles that are bound together in a nucleus attract each other, energy must be supplied to separate them. The minimum energy required to separate a nucleus into its constituent particles is called its **binding energy.** Atomic and nuclear energies are usually measured in units called **electron volts (eV),** where $1.00 \text{ eV} = 1.60 \times 10^{-19}$ J.

Energy has mass. Since the energy of a bound nucleus is less than the energy of the same particles widely separated, the mass of the system in the nucleus, with its particles bound together, is less than the sum of the masses of the widely separated constituent particles.

7-4 Quantization of Energy   For a bound system, like an atomic nucleus, an atom, or a molecule, the total energy of the system is found to be quantized. That is, the smallest increases or decreases of a bound system's total energy occur in finite amounts, so the system's energy levels are discrete (quantized). It is common practice to label each energy level of a system with one or more integers called quantum numbers. Because these energy changes are small in comparison to the changes that typically occur in our everyday life, the finite size of the smallest increments in energy were not noticed until about a century ago.

Electromagnetic energy (radio waves, microwaves, light waves, X rays, and gamma rays) is always absorbed or emitted by an amount directly proportional to the frequency of the radiation. That is,

Energy quantum for electromagnetic radiation

$$E = hf$$

where the proportionality constant h, called Planck's constant, equals 6.626×10^{-34} J·s.

II. Numbers and Key Equations

Numbers

$$1 \text{ eV} = 1.60 \times 10^{-19} \text{ J}$$

$$h = 6.626 \times 10^{-34} \text{ J·s}$$

Key Equations

Total mechanical energy

$$E_{\text{mech}} = K + U = K + U_{\text{s}} + U_{\text{g}} + \cdots$$

Conservation of mechanical energy

$$E_{\text{mech}} = \text{constant} \qquad or \qquad K_{\text{f}} + U_{\text{f}} = K_{\text{i}} + U_{\text{i}}$$

Work–energy theorem

$$W_{\text{ext}} = \Delta E_{\text{mech}} + \Delta E_{\text{therm}} + \Delta E_{\text{chem}} + \cdots$$

Frictional dissipation of mechanical energy

$$\Delta E_{\text{therm}} = f_{\text{k}} d$$

Rest energy

$$E_0 = mc^2$$

Energy quantum for electromagnetic radiation

$$E = hf$$

III. Potential Pitfalls

Don't confuse a system's total mechanical energy with its total energy. The total energy includes the total mechanical energy and any additional forms of energy.

Only changes in potential energy have physical meaning. The choice of the reference point for potential energy is arbitrary, so choose a reference configuration that is convenient for a particular situation. Although kinetic energy can never be negative, potential energy can be either positive or negative.

Keep in mind that the amount of mechanical energy that is transformed into thermal energy when two surfaces slide across each other is $f_{\text{k}} d$, where f_{k} is the kinetic frictional force and d is the distance one surface moves relative to the other surface. Don't mistake $f_{\text{k}} d$ for the work done by the kinetic frictional force. It isn't. The work done by f_{k} cannot be directly calculated because the displacements of the points where this force is applied are not directly observable.

Don't think that the mass of a system of particles is always equal to the sum of the individual masses of the constituent particles. A bound system has less energy than a system where the particles are widely separated. Since the mass of a system is proportional to its energy, a bound system has less mass than a system of the same particles that are widely separated. The difference is called the binding energy. These distinctions are always valid, but they are most significant for subatomic systems.

IV.　True and False and Responses

True and False

_____1. The work done by a kinetic frictional force f_k equals $f_k d$, where d is the distance one surface slides relative to the other surface.

_____2. The mechanical energy dissipated by a kinetic frictional force f_k equals $f_k d$, where d is the distance one surface slides relative to the other surface.

_____3. A block sliding on a horizontal floor is brought to rest by a kinetic frictional force. All of the dissipated kinetic energy appears as the thermal energy of the block.

_____4. The mass of a nucleus is exactly equal to the sum of the masses of its constituent protons and neutrons.

_____5. The initial and final energies of a system are the sums of all types of energy (kinetic, potential, chemical, thermal, . . .) at two specific times, the initial and final times.

_____6. External work refers to a process that occurs over a time interval as a result of external forces that act on the system doing work between the initial and final times.

_____7. A ball rolling on a horizontal surface will eventually come to rest when it completely runs out of energy.

_____8. A wooden ball that is not moving has no energy.

_____9. Energy is a force.

Responses

1. False. The displacement of the point of application of the frictional force is not equal to the bulk displacement of the object that the force acts on.

2. True

3. False. Some of the dissipated kinetic energy appears as the thermal energy of the block, almost all the rest as the thermal energy of the floor.

4. False. For a bound system like an atomic nucleus, energy has to be added to separate it into widely separated protons and neutrons. Energy and mass are proportional so the mass of the nucleus is less than the sum of the masses of the constituent protons and neutrons.

5. True

6. True

7. False. It will eventually come to rest but it doesn't run out of energy. When the ball finally stops rolling, all of its initial kinetic energy has been transformed into thermal energy, some of which resides in the ball, some in the surface, and the rest in the air.

8. False. It stops rolling when it has no kinetic energy, but it still has other forms of energy including chemical and thermal.

9. False. Energy is not a force.

V.　Questions and Answers

Questions

1. A block of mass m, released from rest, slides down a frictionless incline and reaches the bottom with speed v_f. What is the change in the mechanical energy of the *block-slide-Earth* system? Explain.

2. When you step on the accelerator (gas pedal) the car's kinetic energy increases. What external force, if any, does work on the car to cause this increase in its kinetic energy? Where does this kinetic energy come from?

3. A boy pulls a wagon up a long, straight hill at constant speed. What is the change in total energy of the *boy-wagon-hill-Earth* system? Explain.

4. A boy pulls a wagon up a long, straight hill at constant speed. Does the total energy of the *wagon-hill-Earth* system change? Explain.

5. A blob of putty falls on the floor (plop). Neglecting air resistance, does the total energy of the *putty-floor-Earth* system change? Explain.

Answers

1. Zero. There are no external forces acting on the block-slide-Earth system, so the total energy of the system remains constant. The slide is frictionless so no mechanical energy is dissipated by friction. The only force doing work on the block is the conservative force of gravity. Therefore, for this system mechanical energy is conserved. The decrease in gravitational potential energy equals the increase in kinetic energy.

2. No external force does work on the car. The static frictional force on the tires by the pavement does no work because that part of each tire in contact with the pavement is at rest. (We are assuming the tires do not slip on the pavement.) The increase in the car's kinetic energy equals the decrease in the chemical energy stored in the gasoline (and the oxygen in the atmosphere). The kinetic energy comes from the gasoline–oxygen mixture.

3. The total energy of this system does not change. There are no external forces acting on this system so its total energy remains unchanged.

4. The boy is external to the wagon-hill-Earth system. As the boy pulls the wagon up the hill, the

force he exerts on the wagon does positive work. This is the only external work done on the system so the total external work done on the system is positive. This means the system's total energy increases, in accord with the work–energy theorem. Where does this energy come from? From the chemical energy stored in the boy's muscles.

5. No. The total energy of the putty-floor-Earth system does not change because there are no external forces on the system. Thus, no work is done on it by external forces. During the fall, the system's gravitational potential energy is transformed into the kinetic energy of the ball and during the plop this kinetic energy is dissipated via friction within the putty and between the putty and the floor.

VI. Problems and Solutions

Problems

1. A pendulum consists of a compact 0.50-kg particle suspended from a 2.00-m-long string that is attached to a mount on the ceiling. If the pendulum is released from rest with the string taut and horizontal, how fast will the particle be moving when the string makes an angle of 30 degrees with the vertical while the particle is on its downward arc? Air drag is negligible.

How to Solve It
• Draw a figure. Choose a system for which the mechanical energy is constant. This system should not have any external forces doing work on it and should not have any dissipative internal forces.

• Equate the initial and final mechanical energies and solve for the speed of the particle at the final height.

2. A 3-kg block sliding along a horizontal surface at 2.00 m/s begins to slide up a 20-cm high hill. Assuming friction is negligible on all surfaces, what is the speed of the block after it reaches the top of the hill?

How to Solve It
• Draw a figure. Choose a system for which the mechanical energy is constant. This system should not have any external forces doing work on it and should not have any dissipative internal forces.

• Equate the initial and final mechanical energies and solve for the speed of the particle at the final height.

3. A pendulum consists of a compact particle of mass m suspended from a string of length L attached to a mount on the ceiling, shown in Figure 7-2. The

Figure 7-2

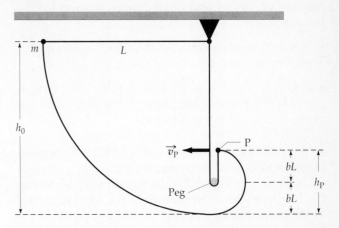

particle is released from rest with the string horizontal. As the particle passes through the lowest point in its path, the string strikes a skinny peg a distance bL above this lowest point. As the particle passes through point P (a distance bL directly above the peg), determine an expression for the tension in the string in terms of m, g, L, and b. Find the value of b in the limit that the tension approaches zero. Assume air drag is negligible. For certain values of b, the equation yields a negative value for the tension. Is that physically possible? Explain.

How to Solve It
• Following its release, the particle swings through a circular arc of radius L and then a circular arc of radius bL. Neglecting air resistance, as it swings along these arcs two forces act on the particle, the force exerted by the string and the gravitational force. As the particle passes through point P, both of these forces, and thus the acceleration, are directed downward. Obtain an expression for the acceleration in terms of the speed v and the radius bL using the kinematic formula for centripetal acceleration and apply Newton's second law ($\Sigma \vec{F} = m\vec{a}$) to the block to obtain an expression for the tension in the string.

• Select as a system the string, the particle, the peg, the ceiling and string mount, and the earth. Everything affecting the motion is included in this system. There are no external forces on it, so the system's total energy remains constant. Because there is no friction, no mechanical energy is dissipated, and since the length of the string remains constant, the system has no potential energy. Thus mechanical energy is conserved—the sum of the potential and kinetic energies remains constant. Write an expression equating the system's initial mechanical energy with its mechanical energy when the particle is at point P.

• Eliminate the speed of the particle at point P from the two equations and obtain an expression for the tension in terms of m, g, and b.

• When b equals a certain value, the tension equals zero. Determine this value of b.

• A negative value of tension implies the string is pushing the particle. We know that a string can not push, so this can not be physical.

4. A small block of mass m slides down a frictionless incline and around a circular loop-the-loop of radius R, as shown in Figure 7-3. In terms of R, determine the height h above the bottom of the loop such that as the block passes through the highest point on the loop-the-loop the force of the track on the block equals half the weight of the block.

Figure 7-3

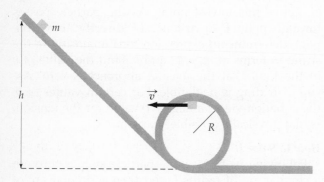

How to Solve It
• Neglecting friction, two forces act on the block as it travels along the track, the normal force exerted by the track and the gravitational force. As the block passes through the top of the loop-the-loop, both of these forces, and thus the acceleration, are directed downward. Obtain an expression for the acceleration in terms of the speed v and the radius R using a kinematic formula for centripetal acceleration and apply Newton's second law $(\Sigma \vec{F} = m\vec{a})$ to the block to obtain an expression for the normal force.

• No external forces act on the block-track-Earth system so its total energy must remain constant. The track is frictionless so no mechanical energy is dissipated by friction. Thus, we know that the system's mechanical energy remains constant. Write an expression equating the initial mechanical energy with the mechanical energy at the top of the loop.

• Eliminate the speed at the top of the loop from the two equations and obtain an expression for the normal force in terms of the initial height. Find the value for the initial height h for which the normal

force at the top of the loop equals half of the block's weight.

5. Blocks of mass m_1 and m_2 ($m_1 > m_2$) are hung over a pulley using a string of negligible mass, as shown in Figure 7-4. Both friction and the mass of the pulley can be neglected. When the system is released from rest, block 2 is in contact with the floor. Following release, block 1 falls a distance h to the floor as block 2 is pulled upward through the same distance. To what maximum height does block 2 rise above its starting position if $m_1 = 6$ kg, $m_2 = 5$ kg, and $h = 1.5$ m?

Figure 7-4

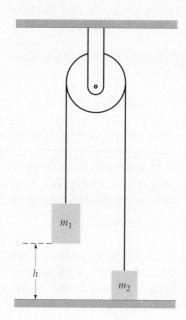

How to Solve It
• Choose as a system the two blocks, the string, the pulley, the pulley mount and ceiling, and the earth. The system includes everything affecting the motion. There are no external forces on this system so the work done on it by external forces is zero, which means its total energy must remain constant. The pulley is massless and its axis bearing is frictionless, so no mechanical energy is dissipated. The string's length remains constant so it stores no potential energy. Thus, the system's mechanical energy must remain constant, at least until block 1 strikes the floor. Equate the system's initial mechanical energy with its mechanical energy immediately prior to the impact of block 1 with the floor to obtain an expression for the speed of the blocks immediately prior to the impact.

• After impact, the string goes slack and block 2 continues to rise. Now choose as a system block 2 and the earth. There are no external forces on this system and no friction to dissipate mechanical energy, so after impact its mechanical energy remains constant. Equate its mechanical energy at the time of block 1's impact with its mechanical energy when block 2 reaches its highest point to obtain an expression for the maximum height attained by block 2.

6. Blocks of mass m_1 and m_2 are hung over a pulley by a light string, as shown in Figure 7-5. The mass of the pulley can be neglected, as can friction both in the pulley and between the block and the incline. When the system is released from rest in the position shown, block 1 descends a distance h to the ground as the string pulls block 2 along the incline through the same distance. What is the maximum total displacement along the incline that block 2 undergoes when $m_1 = m_2 = 5$ kg, $h = 1.0$ m, and $\theta = 30°$?

Figure 7-5

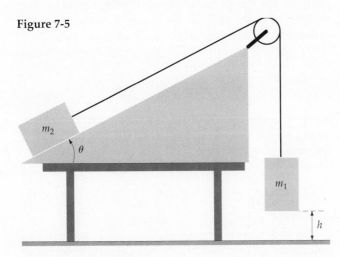

How to Solve It
• Choose as a system the two blocks, the string, the pulley, the pulley mount, the table, the ramp, and the earth. The system includes everything affecting the motion. There are no external forces on this system so the work done on it by external forces is zero, which means its total energy must remain constant. The pulley is massless, its bearings are frictionless as is the inclined surface, so no mechanical energy is dissipated. The string's length remains constant so it stores no potential energy. Thus, the system's mechanical energy remains constant, at least until block 1 strikes the floor. Equate the initial mechanical energy with the mechanical energy immediately prior to the impact of block 1 with the floor to obtain an expression for the speed of the blocks immediately prior to the impact.

• After impact, the string goes slack and block 2 continues to slide up the incline. Now choose as a system block 2, the incline and table, and the earth.

There are no external forces on this system and no friction to dissipate mechanical energy, so after impact its mechanical energy remains constant. Equate the mechanical energy at the time of block 1's impact with the mechanical energy when block 2 reaches its highest point. Use this equation to obtain an expression for the maximum total displacement of block 2.

7. You are inside a crate that is released from rest at the top of a ramp inclined 30 degrees with the horizontal. After sliding 4.00 m down the ramp the crate runs into a spring bumper which it compresses as it slows. You and the crate have a mass of 80 kg, the spring is massless and has a constant of 500 N/m, and the coefficient of kinetic friction between the block and the ramp is 0.30. What is the maximum distance that the spring is compressed?

How to Solve It
• Draw a sketch of the situation. Then choose a system and apply the work–energy theorem to it.

• List each external force acting on the system if any and, if possible, write an expression for the work done by each external force. Do this in tabular form.

• Make a table of the expressions for the initial and final energy terms. These terms are kinetic energy, gravitational potential energy, elastic potential energy, and thermal energy. Also, list the change (final minus initial) in these terms.

• Substitute terms from the table into the work–energy equation. Check for the number of unknowns. If there is more than one unknown you will have to find an additional equation so it can be eliminated. One way to obtain an additional equation is to apply Newton's second law.

8. Starting from rest, Buck the sled dog drags a 45-kg sled up a 5.00-m long ramp that is inclined 35° with the horizontal. The kinetic coefficient of friction between the ramp and the sled is 0.35. Using the work–energy theorem, determine how much work Buck must do on the sled just to drag it up the ramp. Assume Buck's paws do not slip on the ramp.

How to Solve It
• Draw a sketch of the situation. Then choose a system and apply the work–energy theorem to it. Choose a system that Buck does external work on.

• List each external force acting on the system if any and, if possible, write an expression for the work done by each external force. Do this in tabular form.

• Make a table of the expressions for the initial and final energy terms. These terms are kinetic energy, gravitational potential energy, and thermal energy. Also, list the change (final minus initial) in these terms.

• Substitute from the table into the work–energy equation. Check for the number of unknowns. If there is more than one unknown you will have to find an additional equation so it can be eliminated. One way to obtain an additional equation is to apply Newton's second law.

Solutions and Answers

1. First we draw Figure 7-6, showing the initial and final positions.

Figure 7-6

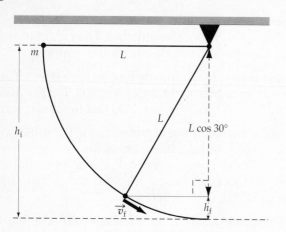

Next we choose a system that consists of the particle, the string, the ceiling mount and ceiling, and the earth. The system includes everything that affects the motion and there are no external forces on it. Also, air drag is negligible so there are no forces to dissipate the mechanical energy. Thus, mechanical energy is conserved. Choosing the lowest point in the arc as the reference point for potential energy we equate the initial and final mechanical energy and solve for the speed of the particle.

$$E_i = E_f$$

$$U_i + K_i = U_f + K_f$$

$$mgh_i + \tfrac{1}{2}mv_i^2 = mgh_f + \tfrac{1}{2}mv_f^2$$

$$mgL + 0 = mg\,(L - L\cos 30°) + \tfrac{1}{2}mv_f^2$$

Solving for the speed gives

$$v_f^2 = 2gL\cos 30°$$

$$= 2(9.81\text{ m/s}^2)(2.00\text{ m})\cos 30° = 34.0\text{ m}^2/\text{s}^2$$

so

$$v_f = \underline{5.83\text{ m/s}}$$

2. $\underline{0.28\text{ m/s}}$

3. Following its release, the particle will swing through a circular arc of radius L. After the string strikes the peg the particle will swing in a near circular arc of radius bL. (In the skinny peg approximation, we treat this motion as if the particle swings through a circular arc of radius bL.) Neglecting air resistance, two forces act on the particle as it swings through its arcs, the force exerted by the string and the gravitational force. The motion of the particle will obey Newton's second law $\vec{T} + \vec{w} = m\vec{a}$, where $\vec{T}$ is the force exerted by the string, $\vec{w}$ is the gravitational force, and $\vec{a}$ is the acceleration of the particle. As the particle passes through point P the vectors $\vec{T}$, $\vec{w}$, and $\vec{a}$ are each directed straight downward. Therefore, at P

$$T + mg = m\frac{v_P^2}{r}$$

where r is the radius of the circular path. By substituting bL for r, and subtracting mg from both sides, we obtain the equation

$$T = m\left(\frac{v_P^2}{bL} - g\right) \tag{1}$$

In order to determine the tension T, we need to determine an expression for the speed v_P in terms of m, g, L, and b. We will equate the particle's initial potential energy with its mechanical energy at P. Neglecting air drag, as the particle swings from its release position to point P only two forces act on it, the force exerted by the string and the gravitational force. The force exerted by the string always acts at right angles to the velocity of the particle, so this force does not do any work. The only force doing work on the particle is the gravitational force so we know the mechanical energy E must be conserved.

By equating the initial mechanical energy with the mechanical energy of the particle when it is at P we obtain the equations

$$E_0 = E_P$$

$$mgh_0 + \tfrac{1}{2}mv_0^2 = mgh_P + \tfrac{1}{2}mv_P^2$$

$$gL + 0 = g\,(2bL) + \tfrac{1}{2}v_P^2$$

The lowest point in the path of the particle is a convenient reference point for determining potential energies, so all heights are measured with respect to that point. Rearranging the last equation to obtain an expression for v_P^2 yields

$$v_P^2 = 2gL(1 - 2b)$$

Substituting this expression for v_P^2 into Equation (1) results in

$$T = m\left(\frac{2gL(1 - 2b)}{bL} - g\right) = mg\left(\frac{2 - 5b}{b}\right)$$

$$= mg\left(\frac{2}{b} - 5\right) \qquad (2)$$

Rearranging Equation (2) for b gives

$$b = \frac{2}{\dfrac{T}{mg} + 5}$$

so

$$\operatorname*{Lim}_{T \to 0} b = \operatorname*{Lim}_{T \to 0}\left(\frac{2}{\dfrac{T}{mg} + 5}\right) = \frac{2}{0 + 5} = \underline{0.4}$$

If the peg is placed so $b > 0.4$, Equation (2) yields negative values for the tension. In reality, this does not occur. Our analysis to determine an expression for T assumes the particle moves in the circular arcs shown in Figure 7-2. This assumption holds as long as the string remains taut, that is, as long as the tension remains greater than zero. An exception to this occurs if $b = 0.4$. In this case the tension remains greater than zero except for the single instant that the particle passes through the topmost point of the arc.

If the peg is placed so $b > 0.4$, the tension still never becomes negative because the string will become slack before the particle reaches point P.

4. $\underline{2.75\,R}$

5. No external forces do work on the blocks-string-pulley-ceiling-Earth system and there is no friction so none of the system's mechanical energy is dissipated. Thus the mechanical energy of the system remains constant. We equate the system's initial mechanical energy to its mechanical energy just prior to impact. Thus,

$$E_i = E_f$$

$$U_{1i} + U_{2i} + K_{1i} + K_{2i} = U_{1f} + U_{2f} + K_{1f} + K_{2f}$$

$$m_1gh_{1i} + m_2gh_{2i} + \tfrac{1}{2}m_1v_{1i}^2 + \tfrac{1}{2}m_2v_{2i}^2 =$$
$$m_1gh_{1f} + m_2gh_{2f} + \tfrac{1}{2}m_1v_{1f}^2 + \tfrac{1}{2}m_2v_{2f}^2$$

$$m_1gh + 0 + 0 + 0 = 0 + m_2gh + \tfrac{1}{2}m_1v_f^2 + \tfrac{1}{2}m_2v_f^2$$

where $h_{1i} = h_{2f} = h$. By rearranging terms in this equation we obtain an expression for the speed v_f of either block just before impact:

$$v_f = \sqrt{\frac{2(m_1 - m_2)gh}{m_1 + m_2}}$$

Block 2 continues to rise after impact. During this time no external forces act on the block 2-Earth system so its mechanical energy is conserved. By equating its mechanical energy E_a at impact with its mechanical energy E_b when block 2 reaches the top of its trajectory we obtain the equation

$$E_a = E_b$$

$$m_2gh_a + \tfrac{1}{2}m_2v_a^2 = m_2gh_b + \tfrac{1}{2}m_2v_b^2$$

Dividing through by m_2 and substituting using $h_a = h_{2f} = h$, $v_a = v_f$, $v_b = 0$, and $h_b = h_{max}$ we obtain

$$gh + \frac{1}{2}\left(\frac{2(m_1 - m_2)gh}{m_1 + m_2}\right) = gh_{max} + 0$$

where we have substituted for v_f the expression obtained for the velocity just before impact. Solving for h_{max} gives

$$h_{max} = \frac{2h}{1 + \dfrac{m_2}{m_1}} = \frac{2(1.5\text{ m})}{1 + \dfrac{5\text{ kg}}{6\text{ kg}}}$$

$$= \underline{1.64\text{ m}}$$

Remarks: Before moving on we will evaluate the maximum height for two limiting values of the ratio m_2/m_1. The physical equivalent of taking the first of these limits is to consider the system when block 1 is a blacksmith's anvil and block 2 is a small pea. Limiting evaluations of obtained expressions are extremely useful because they provide a check as to the correctness of the result. First we will determine h_{max} in the limit that the ratio m_2/m_1 approaches zero. That is,

$$\operatorname*{Lim}_{m_2/m_1 \to 0}(h_{max}) = \operatorname*{Lim}_{m_2/m_1 \to 0}\left(\frac{2h}{1 + m_2/m_1}\right)$$

$$= \frac{2h}{1 + 0}$$

$$= 2h$$

We expect that being tied to a pea via a light string will have little affect on the rate of fall of the anvil. That is, we expect that the anvil will fall with an acceleration almost equal to the free-fall acceleration g. This means the pea will accelerate upward with the same acceleration $\sim g$. After the anvil hits the floor, the magnitude of the pea's acceleration will again be g, but now the acceleration will be directed downward. Because both the acceleration and deceleration of the pea have the same magnitude, it follows that the distances the pea rises while accelerating and while decelerating will be equal. Thus we expect the pea to have a maximum height of $2h$ which is in agreement with our calculation by determining h_{max} in the limit that m_2/m_1 approaches zero.

We now consider the situation in the limit where the mass ratio m_2/m_1 approaches one. Physically this is the limit in which block 2 is an anvil and block 1 an identical anvil with a small mosquito sitting on it. Here we expect that following release, block 1 will descend with a very small acceleration and thus will strike the floor with a miniscule speed. Block 2 will have that same small speed, and thus it will rise only a miniscule distance above the height h.

By evaluating our expression for h_{max} in the limit that the ratio m_2/m_1 approaches zero, we have

$$\lim_{m_2/m_1 \to 1} (h_{max}) = \lim_{m_2/m_1 \to 1} \left(\frac{2h}{1 + m_2/m_1} \right)$$

$$= \frac{2h}{1 + (1)}$$

$$= h$$

which is in agreement with our expectations.

6. 1.5 m

7. First we draw a sketch, Figure 7-7, of the situation. In this sketch, L is the distance the crate slides before hitting the spring, x_{max} is the spring's maximum compression, $\theta = 30°$, k is the spring constant, m is the mass of you and the crate, $L' = L + x_{max}$, and h is the crate's initial height relative to its final height.

Figure 7-7

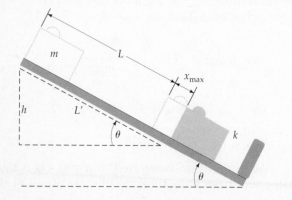

Next we select a system. We choose it to consist of you and the crate, the spring, the ramp, and planet Earth. That way everything affecting the motion is within the system. The work–energy theorem is

$$W_{ext} = \Delta K + \Delta U_g + \Delta U_s + \Delta E_{therm} \qquad (1)$$

In order to organize the data we will make two tables. One for the work done by each external force and a second for the initial and final values of the various energy terms as well as the change (final minus initial) in each of these values.

External forces	W_{ext}
none	0

	K	U_g	U_s	E_{therm}
initial	0	mgh	0	n.a.
final	0	0	$\frac{1}{2}kx_{max}^2$	n.a.
change (Δ)	0	$-mgh$	$\frac{1}{2}kx_{max}^2$	$f_k L'$

n.a. = Not applicable.

Substituting from the tables into the work–energy formula (Equation 1) gives

$$0 = 0 - mgh + \tfrac{1}{2}kx_{max}^2 + f_k L'$$

Since $h = L' \sin \theta$ and $L' = L + x_{max}$ this can be written

$$0 = -mg(L + x_{max})\sin \theta + \tfrac{1}{2}kx_{max}^2 + f_k(L + x_{max}) \qquad (2)$$

In order to solve for x_{max} we first have to find the kinetic frictional force f_k. To do this we draw a free-body diagram, Figure 7-8, and use Newton's second law to find the normal force F_n, and then use $f_k = \mu_k F_n$ to find the kinetic frictional force.

Figure 7-8

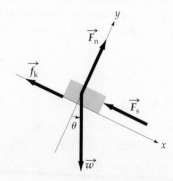

To find F_n we apply Newton's second law in the y direction. This gives

$$F_{n\,y} + w_y + f_{k\,y} + F_{s\,y} = ma_y$$
$$F_n - mg \cos \theta + 0 + 0 = 0$$

so

$$F_n = mg \cos \theta$$

and

$$f_k = \mu_k F_n = \mu_k mg \cos \theta$$

Substituting this into Equation (2) gives

$$0 = -mg(L + x_{max})\sin \theta$$
$$+ \tfrac{1}{2}kx_{max}^2 + \mu_k mg (\cos \theta)(L + x_{max})$$

This is a quadratic equation in x_{max}. Rearranging terms to put it in standard quadratic form gives

$$0 = \tfrac{1}{2}kx_{max}^2 + mgx_{max}(\mu_k \cos \theta - \sin \theta)$$
$$+ mgL(\mu_k \cos \theta - \sin \theta)$$
$$= ax_{max}^2 + bx_{max} + c$$

where

$$a = \frac{k}{2} = \frac{500 \text{ N/m}}{2} = 250 \text{ N/m}$$
$$b = mg(\mu_k \cos \theta - \sin \theta)$$
$$= (80 \text{ kg})(9.81 \text{ N/kg})(0.3 \cos 30° - \sin 30°)$$
$$= -189 \text{ N}$$
$$c = mgL(\mu_k \cos \theta - \sin \theta) = bL$$
$$= (-189 \text{ N})(4.00 \text{ m})$$
$$= -756 \text{ N·m}$$

so

$$x_{max} = \frac{-b \pm \sqrt{b^2 - 4ac}}{2a}$$
$$= \frac{189 \pm \sqrt{(189 \text{ N})^2 - 4(250 \text{ N/m})(-754 \text{ N·m})}}{2(250 \text{ N/m})}$$
$$= 2.16 \text{ m}, -1.40 \text{ m}$$

x_{max} is defined as the distance the block moves while compressing the spring, and distance is never negative. Consequently, $x_{max} = \underline{2.16 \text{ m}}$.

8. $\underline{1.90 \times 10^3 \text{ J}}$

Chapter 8

Systems of Particles
and Conservation of Momentum

I. Key Ideas

In our study of Newton's laws, we have treated extended objects—objects like blocks, balls, and automobiles—as particles. The justification of these treatments is presented in this chapter. We will see that the total force exerted on a system of particles equals the product of the total mass of the system and the acceleration of a point called the center of mass. The motion of a system of particles will be divided in two terms: the motion of the center of mass and the motion of various parts of the system relative to the center of mass.

We will introduce the concept of momentum, which is the product of the mass of a particle and its velocity. Momentum is considered a fundamental concept in physics because the total momentum of an isolated system always remains constant.

8-1 The Center of Mass
The position of the **center of mass** $\vec{R}_{cm}$ of a system is related to the positions of the particles that make up the system by the equation

Center of mass

$$M\vec{R}_{cm} = m_1\vec{r}_1 + m_2\vec{r}_2 + \cdots = \sum_i m_i\vec{r}_i$$

where m_i is the mass of the ith particle, $M (= \Sigma m_i)$ is the total mass of the system, and $\vec{r}_i$ and $\vec{R}_{cm}$ are the position vectors of the ith particle and the center of mass, respectively. (For a continuous object, the summation in the above expression is replaced by the integral

$$M\vec{R}_{cm} = \int \vec{r}\,dm$$

where dm is the mass of an arbitrary infinitesimal piece of the object and $\vec{r}$ is the position vector of that piece.) The vector $\vec{r}_i$ can be expressed in terms of its components as $\vec{r}_i = x_i\hat{i} + y_i\hat{j} + z_i\hat{k}$, and the vector $\vec{R}_{cm}$ can be expressed in terms of its components as $\vec{R}_{cm} = X_{cm}\hat{i} + Y_{cm}\hat{j} + Z_{cm}\hat{k}$.

The center-of-mass equation can be expressed in component form as the three equations

Center of mass (component form)

$$M X_{cm} = \sum_i m_i x_i$$

$$M Y_{cm} = \sum_i m_i y_i$$

$$M Z_{cm} = \sum_i m_i z_i$$

It can be shown that the center of mass of a system consisting of a number of objects, like a collection of bricks, is determined by the expression

Center of mass for a composite system

$$M\vec{R}_{cm} = m_a\vec{r}_{cm,\,a} + m_b\vec{r}_{cm,\,b} + \cdots$$

where a, b, etc., refer to the individual objects (bricks).

The gravitational potential energy of a system of particles is expressed as $U = \Sigma m_i g y_i$. Factoring out the g from this sum leaves us with the expression

Gravitational potential energy

$$U = \Sigma(m_i g y_i) = g\Sigma(m_i y_i) = M g Y_{cm}$$

(The gravitational potential energy of a complex system depends only on the height of the center of mass of the system.)

8-2 Finding the Center of Mass by Integration
Sometimes the mass distribution of a system can be treated as a continuous smear of mass, as opposed to a discrete collection of point particles. In these cases the center of mass is calculated by doing an integral. Integrals are sums. When you use an integral

to find the center of mass what you are doing is dividing up the continuous mass distribution into infinitely many *infinitesimal* pieces and then adding up the contribution of each piece, just as you would do for a discrete mass distribution. If the mass is distributed along a line, then each infinitesimal length dl has mass $dm = \lambda dl$, where λ is the mass per unit length. If the mass is distributed on a surface then each infinitesimal piece of the surface has area dA and mass $dm = \sigma dA$, where σ is the mass per unit area. Lastly, if the mass is distributed throughout a volume then each infinitesimal piece has volume dV and mass $dm = \rho dV$, where ρ is the mass per unit volume, also known as density.

8-3   *Motion of the Center of Mass of a System*   As an object moves, the bulk motion of the object is often of particular importance. For example, when a car moves along the highway, the complex motion of the pistons, valves, and other parts may be of interest, but it is the bulk motion of the car that is of foremost interest. After all, a car is primarily a transportation vehicle. It is the motion of the center of mass that best represents the bulk motion. The **velocity of the center of mass** $\vec{V}_{cm}$ equals the rate of change of the position of the center of mass. To find the velocity of the center of mass we differentiate both sides of the center-of-mass equation with respect to time. This gives

Velocity of center of mass

$$M \vec{V}_{cm} = m_1\vec{v}_1 + m_2\vec{v}_2 + \cdots = \sum_i m_i\vec{v}_i$$

By differentiating again we get

Acceleration of center of mass

$$M \vec{A}_{cm} = m_1\vec{a}_1 + m_2\vec{a}_2 + \cdots = \sum_i m_i\vec{a}_i$$

where $\vec{A}_{cm}$ is the **acceleration of the center of mass.**

Newton's second law ($\vec{F}_{net} = m\vec{a}$) holds for individual particles, so we can substitute the net force on each particle in the previous equation. Thus we have

$$M \vec{A}_{cm} = \vec{F}_{1,net} + \vec{F}_{2,net} + \cdots = \sum_i \vec{F}_{i,net}$$

All forces come in action–reaction pairs. The forces of an action–reaction force pair, in accord with Newton's third law, are equal in magnitude and oppositely directed. It follows that the vector sum of any two forces forming an action–reaction pair is zero. The right-hand term in the above equation is the sum over all forces acting on the system, including all internal forces (forces exerted by one part of the system on another part). However, the internal

forces sum to zero because this sum includes only action–reaction force pairs. Because the internal forces sum to zero, the sum on the right side of the equation equals the sum over only the external forces acting on the system. External forces (forces exerted on the system by things not included in the system) also come in action–reaction pairs, but only one of each of these action–reaction pairs is included in the sum. It follows that the net force exerted on a system is the sum of all external forces acting on the system. Substituting the sum over only the external forces for the right-hand term in the above equation, we have

Newton's second law for a system

$$\vec{F}_{net,ext} = \sum_i \vec{F}_{i,ext} = M \vec{A}_{cm}$$

where the net external force on the system $\vec{F}_{net,ext}$ is the sum of all the external forces acting on the system. The implication of this equation is that the center of mass of a system moves like a particle of mass M ($= \Sigma m_i$) under the influence of the net external force acting on the system.

8-4   *The Conservation of Momentum*   The **momentum** of a particle is defined as the product of its mass and its velocity:

Momentum of a particle

$$\vec{p} = m\vec{v}$$

It is useful to write Newton's second law in terms of momentum. This can be done by realizing that the derivative of the momentum with respect to time equals the product of the mass and the acceleration, that is,

$$\frac{d\vec{p}}{dt} = \frac{d(m\vec{v})}{dt} = m\frac{d\vec{v}}{dt} = m\vec{a}$$

Thus we can express Newton's second law as

Newton's second law for a particle

$$\vec{F}_{net} = \frac{d\vec{p}}{dt}$$

The concept of momentum is important because if the net external force acting on a system is zero, the total momentum of the system $\vec{P}$ ($= \Sigma \vec{p}_i$) remains constant. The velocity of the center of mass is related to the momentum by the equation

Momentum of a system

$$\vec{P} = \sum_i \vec{p}_i = \sum_i m_i \vec{v}_i = M \vec{V}_{cm}$$

Differentiating both sides of this equation with respect to time results in an equation relating the acceleration of the center of mass of the system $\vec{A}_{cm}$ and its total momentum $\vec{P}$. Substituting from this equation into Newton's second law for particles results in

Newton's second law for systems

$$\vec{F}_{net,ext} = \frac{d\vec{P}}{dt}$$

So long as the net external force on a system is zero, the total momentum of the system remains constant (as does the velocity of the center of mass). This result

Conservation of momentum

If $\vec{F}_{net,ext} = 0$, then $\vec{P} = \sum_i \vec{p}_i = M\vec{V}_{cm} =$ constant

is known as the **law of conservation of momentum.**

8-5 Kinetic Energy of a System of Particles The kinetic energy of a system can be expressed as the sum of two terms: the kinetic energy associated with the center-of-mass motion and the kinetic energy associated with the motions relative to the center of mass

Kinetic energy for systems

$$K = \tfrac{1}{2} MV_{cm}^2 + K_{rel}$$

where $K = \sum \tfrac{1}{2} m_i v_i^2$ and $K_{rel} = \sum \tfrac{1}{2} m_i u_i^2$ in which v_i is the speed of the ith particle and u_i is its speed relative to the center of mass.

Consider two identical bicycle wheels, one in your hand and another on the unicycle you are riding. The velocities of the centers of mass of these wheels are the same since they must both move along with the unicycle. However the unicycle wheel is spinning (because it is rolling) and the wheel in your hand is not. Consequently, the spinning wheel has additional kinetic energy due to its spinning motion. This additional kinetic energy is K_{rel}, the kinetic energy due to motion relative to the center of mass.

8-6 Collisions In a collision two objects approach and interact strongly for a short time. During the brief time of the collision, any external forces are

much smaller than the forces of interaction between the objects. Thus, during the collision the only significant forces are the forces of interaction between the colliding objects. These forces are equal and opposite, so the total momentum of the system remains unchanged. Also, the collision time is usually so small that any displacement of the objects during the collision can be neglected.

Impulse and Average Force Forces transfer momentum to the objects they act on. The measure of the momentum transferred by a force is called **impulse** $\vec{I}$. The impulse imparted by the force $\vec{F}$, during the time interval from t_i to t_f, is defined by the equation

Impulse

$$\vec{I} = \int_{t_i}^{t_f} \vec{F} \, dt$$

Integrating Newton's second law $\vec{F}_{net} = d\vec{p}/dt$ and applying the definition of impulse gives

Impulse–momentum theorem

$$\vec{I}_{net} = \int_{t_i}^{t_f} \vec{F}_{net} \, dt = \vec{p}_f - \vec{p}_i = \Delta\vec{p}$$

where $\vec{I}_{net}$ is the impulse associated with the net force $\vec{F}_{net}$ acting on the particle.

The **average force** is defined as

Time average of a force

$$\vec{F}_{av} = \frac{1}{\Delta t} \int_{t_i}^{t_f} \vec{F} \, dt$$

where t is the elapsed time $t_f - t_i$. Using the definition of impulse it follows that

Impulse and average force

$$\vec{I} = \vec{F}_{av} \, \Delta t$$

Collisions in One Dimension When two objects collide the momentum of the two-object system remains constant. In one dimension, directions are indicated by signs, so the explicit use of vectors can be avoided. Thus we can write the conservation of momentum equation for two objects as

Conservation of momentum

$$m_1 v_{1f} + m_2 v_{2f} = m_1 v_{1i} + m_2 v_{2i}$$

For **perfectly inelastic collisions,** where the objects stick together, both objects have the same final velocity—the velocity of their center of mass. Combining this condition with the previous equation gives

Perfectly inelastic collision ($v_{1f} = v_{2f} = V_{cm}$)

$$(m_1 + m_2)V_{cm} = m_1 v_{1i} + m_2 v_{2i}$$

For **elastic collisions,** the total initial and final kinetic energies are the same. Thus,

Elastic collision

$$\tfrac{1}{2}m_1 v_{1f}^2 + \tfrac{1}{2}m_2 v_{2f}^2 = \tfrac{1}{2}m_1 v_{1i}^2 + \tfrac{1}{2}m_2 v_{2i}^2$$

Using this equation combined with the conservation of momentum equation, it can be shown that, for a collision, the magnitude of the relative velocity of approach equals the magnitude of relative velocity of recession. That is,

Relative velocities in an elastic collision

$$v_{2f} - v_{1f} = -(v_{2i} - v_{1i})$$

In the center-of-mass reference frame, the total momentum of the system is zero, that is, the momenta of the two objects are equal in magnitude and oppositely directed. Thus, for an elastic collision, the velocity of each object relative to the center of mass is merely reversed:

$$u_{1f} = -u_{1i} \quad \text{and} \quad u_{2f} = -u_{2i}$$

Most collisions fall somewhere between being elastic and perfectly inelastic. The degree of elasticity varies with the **coefficient of restitution** e, which is defined as the ratio of the relative speed of recession to the relative speed of approach:

Coefficient of restitution

$$v_{2f} - v_{1f} = -e(v_{2i} - v_{1i})$$

For an elastic collision, $e = 1$ and for a perfectly inelastic collision, $e = 0$.

Collisions in Three Dimensions For two and three dimensions, the vector nature of momentum is important. Perfectly inelastic collisions can be analyzed in a straightforward manner using the equation

Perfectly inelastic collision ($\vec{v}_{1f} = \vec{v}_{2f} = \vec{V}_{cm}$)

$$(m_1 + m_2)\vec{V}_{cm} = m_1 \vec{v}_{1i} + m_2 \vec{v}_{2i}$$

For all collisions, the total momentum is conserved in each dimension. However, given the masses and initial velocities, conservation of momentum alone is not sufficient to determine the final velocities. The final velocities will not be the same for elastic collisions as for inelastic collisions.

8-7 The Center-of-Mass Reference Frame Any reference frame in which the center of mass of a system remains stationary is called a **center-of-mass reference frame** for that system. Relative to this frame both the velocity and acceleration of the center of mass are always zero, and measurements are usually made from a coordinate system attached to this frame with its origin at the center of mass. In this frame the total momentum of the system, which equals the product of the total mass of the system and the velocity of the center of mass, equals zero. Consequently, a center-of-mass reference frame is also a **zero-momentum reference frame.**

In the analysis of problems it is frequently useful to transform back and forth between a given reference frame and the center-of-mass reference frame. To do this we make use of the relative velocity formula $\vec{v}_{AC} = \vec{v}_{AB} + \vec{v}_{BC}$ developed in Chapter 3. In words this states: the velocity of particle A relative to reference frame C equals the velocity of A relative to frame B plus the velocity of frame B relative to frame C. If A is the ith particle of a system, B is a center-of-mass reference frame, and C is an arbitrary reference frame, then this formula can be written $\vec{v}_i = \vec{u}_i + \vec{V}_{cm}$, where $\vec{v}_i$ is the velocity of particle i relative to an arbitrary frame, $\vec{u}_i$ is the velocity of the same particle relative to a center-of-mass frame, and $\vec{V}_{cm}$ is the velocity of the center-of-mass frame relative to the arbitrary frame. It is often useful to rearrange the terms and write this formula as

Velocity relative to center of mass

$$\vec{u}_i = \vec{v}_i - \vec{V}_{cm}$$

8-8 Rocket Propulsion An interesting form of propulsion occurs when a system ejects a fluid in one direction and the rest of the system accelerates in the opposite direction. This kind of propulsion is known as **rocket propulsion.** When a balloon is inflated, and then released, the balloon propels itself through the air in this manner.

When a rocket engine is firing, exhaust gases are ejected out the back of the engine. The engine exerts a force on these gases causing them to eject out the back. It is the reaction force, the force of the gases on the engine, that propels the engine forward. If a rocket engine, initially at rest in empty space where no external forces act on it, begins to burn its fuel,

the center of mass of the fuel–engine system will remain at rest as the engine and the exhaust gases
move off in opposite directions.

Applying conservation of momentum and Newton's second law to a rocket results in an equation
known as the *rocket equation*:

Rocket equation

$$m \frac{dv}{dt} = u_{ex} \left| \frac{dm}{dt} \right| + F_{ext}$$

where m is the instantaneous mass of the rocket and
the remaining fuel, u_{ex} is the speed of the exhaust
relative to the rocket, and F_{ext} is the net external
force acting on the rocket. The mass m decreases as
the burned fuel is exhausted, thus dm/dt, the time
rate of change of the mass of the rocket and remain-
ing fuel, is negative. The quantity $u_{ex}|dm/dt|$ is called
the **thrust**. It is the contact force exerted by the com-
bustion products on the rocket.

II. Numbers and Key Equations

Numbers

There are no new numbers for this chapter.

Key Equations

Center of mass vector (for discrete particles)

$$M\vec{R}_{cm} = m_1\vec{r}_1 + m_2\vec{r}_2 + \cdots = \sum_i m_i\vec{r}_i$$

or

$$M X_{cm} = \sum_i m_i x_i$$

$$M Y_{cm} = \sum_i m_i y_i$$

$$M Z_{cm} = \sum_i m_i z_i$$

Center of mass vector (for continuous objects)

$$M \vec{R}_{cm} = \int \vec{r}\, dm$$

or

$$M X_{cm} = \int dm$$

$$M Y_{cm} = \int dm$$

$$M Z_{cm} = \int dm$$

Gravitational potential energy

$$U = g_i\, m_i y_i = M\, g Y_{cm}$$

Velocity of center of mass

$$M \vec{V}_{cm} = m_1\vec{v}_1 + m_2\vec{v}_2 + \cdots = \sum_i m_i\vec{v}_i$$

Acceleration of center of mass

$$M \vec{A}_{cm} = m_1\vec{a}_1 + m_2\vec{a}_2 + \cdots = \sum_i m_i\vec{a}_i$$

Newton's second law for systems

$$\vec{F}_{net,ext} = \sum_i \vec{F}_{i,ext} = M \vec{A}_{cm}$$

Momentum of a particle

$$\vec{p} = m\vec{v}$$

Newton's second law for particles

$$\vec{F}_{net} = \frac{d\vec{p}}{dt}$$

Momentum of a system

$$\vec{P} = \sum_i \vec{p}_i = \sum_i m_i \vec{v}_i = M \vec{V}_{cm}$$

Newton's second law for systems

$$\vec{F}_{net,ext} = \frac{d\vec{P}}{dt}$$

Velocity relative to center of mass

$$\vec{u}_i = \vec{v}_i - \vec{V}_{cm}$$

Kinetic energy for systems

$$K = \sum_i \tfrac{1}{2} m_i v_i^2 = \tfrac{1}{2} M V_{cm}^2 + K_{rel}$$

Conservation of momentum

$$m_1\vec{v}_{1f} + m_2\vec{v}_{2f} = m_1\vec{v}_{1i} + m_2\vec{v}_{2i}$$

Perfectly inelastic collision ($\vec{v}_{1f} = \vec{v}_{2f} = \vec{V}_{cm}$)

$$(m_1 + m_2)\vec{V}_{cm} = m_1\vec{v}_{1i} + m_2\vec{v}_{2i}$$

Elastic collision

$$\tfrac{1}{2}m_1 v_{1f}^2 + \tfrac{1}{2}m_2 v_{2f}^2 = \tfrac{1}{2}m_1 v_{1i}^2 + \tfrac{1}{2}m_2 v_{2i}^2$$

or

$$v_{2f} - v_{1f} = -(v_{2i} - v_{1i})$$

Coefficient of restitution (head-on collision)

$$v_{2f} - v_{1f} = -e(v_{2i} - v_{1i})$$

Impulse–momentum theorem

$$\vec{I} = \int_{t_i}^{t_f} \vec{F}\, dt = \vec{p}_f = \vec{p}_i = \Delta\vec{p}$$

Time average of a force

$$\vec{I} = \vec{F}_{av}\, \Delta t = \int_{t_i}^{t_f} \vec{F}\, dt$$

Rocket equation

$$m\frac{dv}{dt} = u_{ex}\left|\frac{dm}{dt}\right| + F_{ext}$$

III. Potential Pitfalls

The center of mass of a system does not necessarily coincide with the position of any of the particles or objects that make up a system.

Don't think that if momentum isn't conserved in one direction it isn't conserved in any direction. The law of conservation of momentum is a vector statement. If the component of the net external force on a system in a given direction remains zero, the component of the system's momentum in that direction remains constant.

It is often the case that during collisions, impulses associated with external forces acting on the system are negligible compared to impulses associated with the internal forces between the colliding objects. Thus even for systems that are not isolated, it is usually an excellent approximation to equate the system's pre- and post-collision momenta if the collision occurs in a sufficiently short time. During a collision, the momentum of the system remains constant to the degree that impulses associated with external forces are negligible.

When a perfectly inelastic one-dimensional collision occurs, the kinetic energy of the center-of-mass motion remains constant while the kinetic energy relative to the center of mass becomes zero. Thus the final kinetic energy of the system cannot be zero unless the velocity of the center of mass is zero.

IV. True and False and Responses

True and False

_____1. The total momentum of a system is the product of its total mass and the velocity of its center of mass.

_____2. If the sum of the internal forces in a system remains zero, the momentum of the system necessarily remains constant.

_____3. If the *x* component of the total external force on a system remains zero, the *x* component of the momentum of the system necessarily remains constant.

_____4. During a collision, the kinetic energy relative to the center of mass must remain constant.

_____5. Following a head-on, perfectly inelastic collision, the kinetic energy of the colliding particles must equal zero.

_____6. When a head-on collision is perfectly elastic, the coefficient of restitution is zero.

_____7. The impulse associated with a force is a measure of the mechanical energy transferred.

_____8. In an attempt to knock a coconut off the top of a coconut tree, a soldier tosses a live hand grenade at it. If the grenade explodes before reaching the coconut, it is possible that none of the fragments strike the coconut even though the grenade's center of mass passes right through it.

_____9. In a one-dimensional elastic collision between two objects, the velocity of each particle relative to the center of mass is reversed.

Responses

1. True

2. False. Internal forces necessarily come in action–reaction pairs, so they always sum to zero. The rate of change of the momentum of the system equals the impulse imparted by the net external force on the system.

3. True

4. False. For example, in a perfectly inelastic collision the kinetic energy relative to the center of mass is greater than zero before the collision and zero following the collision.

5. False. The velocity of the center of mass, along with the associated kinetic energy, is the same after the collision as before. If it is nonzero before the collision it is nonzero following the collision.

6. False. In a one-dimensional perfectly elastic collision the coefficient of restitution is one.

7. False. The impulse associated with a force is the measure of the momentum transferred.

8. True. If a thrown grenade explodes in flight, after the explosion no fragments will remain near its center of mass. Consequently, it is possible for an object to go unscathed even though the grenade's center of mass passes through it.

9. True

V. Questions and Answers

Questions

1. When a large truck collides head on with a small car, the force exerted by the car on the truck is equal in magnitude and oppositely directed to the force exerted by the truck on the car. Would the momentum of the car–truck system be conserved if this were not so?

2. When a space ship in empty gravity-free space fires a rocket, does the momentum of the ship–fuel system change? Does the momentum of the ship change? Does the momentum of the fuel change?

3. The longer it takes for a specific momentum change to occur, the smaller the average force required to cause it. Does a longer time interval also mean the maximum force that acts during the momentum change must also be smaller?

4. Can a system have zero kinetic energy and non-zero momentum? Can it have zero momentum and nonzero kinetic energy?

Answers

1. If these forces were not equal in magnitude and oppositely directed, the changes in momentum of the car and the truck would not be equal in magnitude and oppositely directed, and thus the momentum of the car–truck system would not remain constant.

2. The momentum of the ship–fuel system remains constant. Any changes in the momentum of the ship are balanced by oppositely directed changes in the momentum of the exhausted fuel.

3. No. The momentum change always equals the product of the average force and the time. This says nothing about how large the maximum force can be.

4. When a system has zero kinetic energy, nothing is moving. (There are no negative terms in the kinetic energy sum.) Thus the momentum must be zero. When a system has zero momentum, the center of mass is at rest. However, the system can still have kinetic energy due to the motion of parts of the system relative to the center of mass. A spinning ball necessarily has kinetic energy, but may have zero momentum.

VI. Problems and Solutions

Problems

1. A 2-kg particle is located at the origin and a 1-kg particle is located on the x axis at $x = 3$ m, as shown in Figure 8-1. Determine the location of the center of mass of this two-particle system.

Figure 8-1

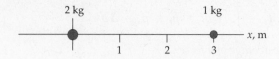

How to Solve It
- Use the formula for the x component of the center of mass to determine X_{cm}.

2. As shown in Figure 8-2, a 3-kg particle is located at the origin, a 1-kg particle is located on the x axis at $x = 2$ m, and a 2-kg particle is located in the xy plane at the point (2 m, 3 m). Determine the location of the center of mass of this three-particle system.

Figure 8-2

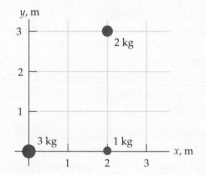

How to Solve It
- Use the formula for the x component of the center of mass to determine X_{cm}.

- Use the formula for the y component of the center of mass to determine Y_{cm}.

3. Find the location of the center of mass of a uniform thin rod with one end at the origin and the other end on the x axis at $x = L$.

How to Solve It
- Use the integral form of the center-of-mass formula $MX_{cm} = \int x\,dm$, where dm is the mass between x and $x + dx$.

4. Find the center of mass of a thin rod of length L whose density increases linearly with distance from one end according to the formula $\lambda = dm/dx = kx$; where $k = 2$ kg/m^2, and x is the distance from one end of the rod.

How to Solve It
- Use the integral form of the center-of-mass formula $MX_{cm} = \int x\,dm$, where dm is the mass between x and $x + dx$.

• Determine M by using the formula $M = \int dm = \int_0^L \dfrac{dm}{dx}\, dx.$

5. As shown in Figure 8-3, a uniform 0.3-kg meter stick, free to rotate about a frictionless horizontal axis perpendicular to the stick through its 25-cm mark, is released from rest from a horizontal position. Determine the kinetic energy of the stick as it rotates through the vertical position, with the 100-cm mark directly below the rotation axis.

Figure 8-3

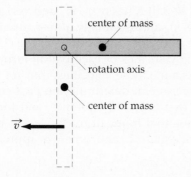

center of mass

rotation axis

center of mass

$\vec{v}$

How to Solve It
• Use the conservation of mechanical energy to determine the kinetic energy of the stick.

6. As shown in Figure 8-4, a uniform 0.3 kg meter stick, free to rotate about a frictionless horizontal axis, perpendicular to the stick through its 20-cm mark, has a compact 0.2-kg weight fastened to the stick at the 80-cm mark. Determine (*a*) the location of the center of mass of the stick–weight system, and (*b*) the change in the gravitational potential energy of the system when the stick rotates from a horizontal position to a vertical position with the weight directly above the axis.

Figure 8-4

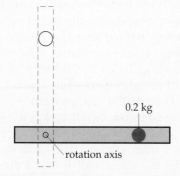

0.2 kg

rotation axis

How to Solve It
• Use the formula for the center of mass of a composite system to locate the center of mass of the stick–weight system. Choose the origin to be at the zero-cm mark of the stick, with the positive x axis along the stick, and solve for the x component of the center of mass.

• Relate the change in the gravitational potential energy to the change in the position of the center of mass.

7. A uniform 3.0-kg meter stick is rotating freely about a fixed horizontal axis through its 25-cm mark. At the instant the stick's center of mass passes directly below the rotation axis (as shown in Figure 8-5), it has a speed of 1.0 m/s. Determine the force exerted on the stick by the axis at this instant.

Figure 8-5

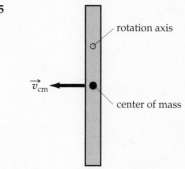

rotation axis

$\vec{v}_{cm}$

center of mass

How to Solve It
• Draw a free-body diagram of the stick. Be sure to include each force acting on it.

• Determine the acceleration, magnitude, and direction of the center of mass.

• Apply Newton's second law for systems ($\vec{F}_{net,ext} = M\vec{A}_{cm}$) and solve for the desired force.

8. As the meter stick referred to in problem 6 rotates through the vertical position, with the 100-cm mark directly below the rotation axis, the stick–weight system's center of mass has a speed of 1.3 m/s. Determine the force exerted on the stick by the rotation axis at this instant. Compare this force with the weight of the stick.

How to Solve It
• Follow part (*a*) of Problem 6 to determine the location of the center of mass of this system. It is at the 62-cm mark of the stick.

• Draw a free-body diagram of the stick–weight system.

• Determine the acceleration, magnitude and direction, of the center of mass as the stick swings through the vertical position.

• Apply Newton's second law for systems ($\vec{F}_{net,ext} = M\vec{A}_{cm}$) and solve for the desired force. Compare this force with the weight.

9. A 3-kg rifle, initially at rest, fires a bullet of mass 10 g at a muzzle speed of 650 m/s. Assuming the rifle is free to move, at what speed would it recoil?

How to Solve It
• The momentum of the rifle–bullet system is the same before and after firing the rifle. This momentum is zero.

10. A 900-kg car traveling north at 60 km/h collides with a 1200-kg light truck traveling west at 50 km/h. The vehicles stick together following the collision, as shown in Figure 8-6. Determine the velocity (magnitude and direction) of the wreck immediately following the collision. (Neglect friction between the vehicles and the ground.)

Figure 8-6

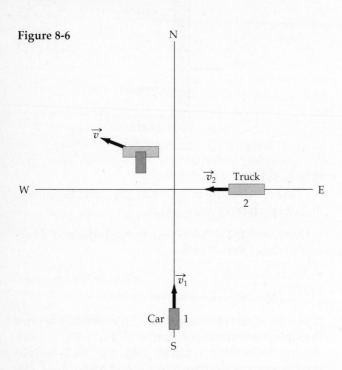

How to Solve It
• The momentum of the car–truck system is conserved during the collision. Thus, the velocity of the system's center of mass is the same before and after the collision. The velocity of the wreck equals the velocity of the center of mass.

• This is a two-dimensional problem. Solve for the x and y components of the center-of-mass velocity separately. Let east be the positive x direction and let north be the positive y direction.

11. Consider a head-on elastic collision between a cue ball and an object ball. Assuming the cue ball has an initial velocity $\vec{v}_0$, the object ball is initially at rest, and the balls have equal masses m, determine

the velocities of each ball immediately following impact. To determine these velocities, transfer to the center-of-mass system, determine the after-impact velocities in that frame of reference, and then transfer back to the initial frame of reference.

How to Solve It
• Determine the velocity of the center of mass.

• Recognize that in a one-dimensional elastic collision, the velocities relative to the center of mass get reversed.

12. Consider a head-on elastic collision between ball A and ball B. Ball A has mass m_A and velocity $\vec{v}_0$, ball B has mass m_B and is initially at rest, and the ratio of the masses is $m_B/m_A = \alpha$. Following the collision, determine the velocities of each ball in terms of α and $\vec{v}_0$. To determine these velocities, transfer to the center-of-mass system, determine the post-collision velocities in that frame of reference, and then transfer back to the initial frame of reference. Check your results for α equals 1, and when α approaches zero and infinity.

How to Solve It
• Determine the velocity of the center of mass.

• Recognize that in a one-dimensional elastic collision, the velocities relative to the center of mass get reversed.
• When α equals 1, the two masses are equal. This is the situation in Problem 11 above, and we expect the same results. When α approaches infinity, the collision is analogous to that of a Ping-Pong ball colliding with a stationary bowling ball. The velocity of the center of mass of this system is essentially zero. During the collision, the velocity of ball A (the Ping-Pong ball) is reversed and the velocity of ball B (the bowling ball) remains essentially zero.

  When α approaches zero, the collision is analogous to that of a bowling ball colliding with a stationary Ping-Pong ball. The velocity of the center of mass of this system equals the initial velocity of ball A (the bowling ball). From the perspective of the center-of-mass reference frame (the one moving with the bowling ball), the velocity of ball A (the bowling ball) remains zero while the velocity of ball B (the Ping-Pong ball) changes from $-\vec{v}_0$ to $\vec{v}_0$; thus following the collision, ball B moves away from ball A with a velocity of $\vec{v}_0$. In the original frame of reference, ball A (the bowling ball) continues to move with velocity $\vec{v}_0$, while ball B (the Ping-Pong ball) moves with a velocity of $2\vec{v}_0$.

13. A 2-kg sphere with a velocity of 8 m/s $\hat{i}$ runs into a stationary 8-kg sphere. Following the collision the 2-kg sphere moves with a velocity of -6 m/s $\hat{j}$. Determine the magnitude and direction of velocity of the 8-kg sphere following the collision. Is this collision elastic?

How to Solve It
• The momentum of the system is conserved. Make a sketch illustrating the motion of the spheres prior to and following the collision.

• Write out the conservation of momentum equation in component form for both the x and y components.

• Express the velocity vector of the 8-kg sphere in terms of its magnitude v and the angle it makes with the x axis.

• Compare the kinetic energy of the system prior to the collision with the kinetic energy following the collision.

14. A 2-kg sphere with a velocity of 8 m/s $\hat{i}$ runs into a stationary 8-kg sphere. Following the collision the 2-kg sphere moves with a velocity of 3 m/s $\hat{i}$ − 4 m/s $\hat{j}$. Determine the velocity of the 8-kg sphere following the collision.

How to Solve It
• The momentum of the system is conserved. Make a sketch illustrating the motion of the spheres prior to and following the collision.

• Write out the conservation of momentum equation in component form for both the x and y components.

• Express the velocity vector of the 8-kg sphere in terms of its magnitude v and the angle it makes with the x axis.

15. Rainwater is falling straight down with a velocity of 15 m/s at the rate of 10 cm of rain per hour. An electronic scale is placed in the rain. The scale's pan consists of a 25-cm-diameter circular disk. If the weight of the water on the platform is negligible, determine the reading of the scale.

How to Solve It
• This is an impulse–momentum problem. The impulse exerted on the rainwater by the pan equals the change in momentum of the rainwater.

• Determine the change in the momentum of the rainwater for each kilogram of water stopped by the pan.

• Determine the time required for a single kilogram of water to be stopped by the pan. Recall that the density of water is 1000 kg/m^3.

• Use the impulse–momentum theorem and solve for the average force exerted by the pan on the rainwater during the time it takes for a single kilogram of water to be stopped by the pan.

16. An empty, open railroad car of mass M with frictionless wheels is rolling along a horizontal track with speed v_0 when it begins to rain. There is no wind so the rain falls straight down. The mass of the rainwater that accumulates in the car is m. (*a*) Determine the speed of the car. (*b*) After the rain stops, a workman opens a drain hole in the bottom of the car and lets the water out. Determine the speed of the car when it is once again empty.

How to Solve It
• Make a sketch illustrating the motion of the car and rainwater prior to and following the collision.

• This is a conservation of momentum problem. There are two external forces acting on the car–rainwater system: the downward force of gravity and the upward contact force of the track on the car. Because this track is frictionless, this contact force is directed upward and the horizontal component of the external force exerted on the car–rainwater system is zero, therefore the horizontal component of the momentum of the system is conserved.

• The collision between the car and the rainwater is perfectly inelastic. Using the component form of the conservation of momentum, write down an equation relating the horizontal components of the momentum before and after the rainwater has accumulated in the car and solve for the speed of the car and accumulated rainwater.

• Again use the conservation of the horizontal component of momentum to determine the speed of the car after the water has drained out. (*Hint:* What is the horizontal velocity of the water just after it leaves the rail car?)

Solutions

1. The formula for the x component of the center of mass is $MX_{cm} = \Sigma m_i x_i$. Thus

$$X_{cm} = \frac{1}{M} \sum_i m_i x_i = \frac{(2 \text{ kg})0 + (1 \text{ kg})(3 \text{ m})}{1 \text{ kg} + 2 \text{ kg}} = \underline{1 \text{ m}}$$

As one might expect, the center of mass of the system is between the two particles, twice as far from the 1 kg mass as from the 2 kg mass.

2. $(X_{cm}, Y_{cm}) = \underline{(1 \text{ m}, 1 \text{ m})}$. (Is this where you expected the center of mass to be?)

3. Because the rod is uniform, the mass dm of an arbitrary segment of the rod of length dx (see Figure 8-7) is given by the formula $dm = M/L\, dx$.

Figure 8-7

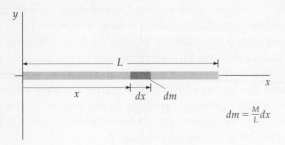

$$dm = \frac{M}{L}dx$$

Thus

$$X_{cm} = \frac{1}{M}\int x\, dm = \frac{1}{M}\int_0^L x\frac{dm}{dx}dx$$

$$= \frac{1}{M}\int_0^L x\frac{M}{L}dx = \frac{1}{L}\int_0^L x\, dx = \frac{1}{L}\frac{x^2}{2}\bigg|_0^L = \frac{L}{2}$$

This is where you should have expected the mass of a uniform rod to be, at the midpoint of the rod.

4. $X_{cm} = \dfrac{2}{3}L$

5. The total mechanical energy is constant, thus

$$K_0 + U_0 = K + U \quad \text{or} \quad 0 + 0 = K + MgY_{cm}$$

where we have chosen the rotation axis as the reference point where the gravitational potential energy is zero. It follows that

$$K = -MgY_{cm} = -(0.3\text{ kg})(9.8\text{ m/s}^2)(-0.25\text{ m})$$
$$= 0.74\text{ J}$$

6. (a) The center of mass is at the 62-cm mark.
(b) 2.06 J

7. There are two forces acting on the stick, the weight of the stick $\vec{w}$ and the force of the axis $\vec{F}$. The free-body diagram is shown in Figure 8-8.

Figure 8-8

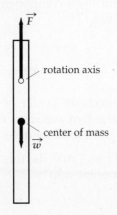

rotation axis

center of mass

$\vec{w}$

The center of mass moves in a 25-cm-radius circle centered at the rotation axis. The acceleration of the center of mass equals V_{cm}^2/r toward the rotation axis (upwards).

Applying Newton's second law for systems yields the equation

$$\vec{F}_{net,ext} = \vec{w} + \vec{F} = M\vec{A}_{cm}$$

For the y components this gives

$$w_y + F_y = MA_{y,cm}$$

$$-Mg + F_y = M\frac{v^2}{r}$$

Solving for the F_y results in

$$F_y = M\left(\frac{v^2}{r} + g\right) = 0.3\text{ kg}\left(\frac{(1\text{ m/s})^2}{0.25\text{ m}} + 9.8\text{ m/s}^2\right)$$
$$= 4.14\text{ N}$$

The weight of the stick is $(0.3\text{ kg})(9.8\text{ m/s}^2) = 2.94\text{ N}$, thus the force of the axis on the stick is about 41% greater than the weight of the stick. The force of the axis includes the net external force needed to produce the upward acceleration of the stick's center of mass.

8. 6.91 N

9. The momentum of the rifle–bullet system is conserved. Thus

$$\vec{P}_0 = \vec{P} \quad \text{or} \quad 0 = m_r\vec{v}_r + m_b\vec{v}_b$$

where $\vec{P}_0$ is the momentum of the system before the rifle is fired, and $\vec{P}$ is the momentum of the system after the rifle is fired. Solving the equation on the right for the velocity of the rifle gives

$$\vec{v}_r = -\frac{m_b}{m_r}\vec{v}_b$$

Because we are asked to find the speed and not the velocity, we equate the magnitudes of the vectors on both sides of this equation. Thus,

$$v_r = \frac{m_b}{m_r}v_b = \frac{0.01\text{ kg}}{3\text{ kg}}(650\text{ m/s}) = 2.17\text{ m/s}$$

10. 38.4 km/h at 42° north of west

11. As shown in Figure 8-9, the cue ball with initial velocity $\vec{v}_0$ collides head-on with a stationary object ball.

Figure 8-9

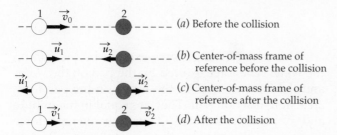

(a) Before the collision

(b) Center-of-mass frame of reference before the collision

(c) Center-of-mass frame of reference after the collision

(d) After the collision

The velocity of the system's center of mass is

$$\vec{V}_{cm} = \frac{1}{M} \sum_i m_i \vec{v}_i = \frac{m_1 \vec{v}_1 + m_2 \vec{v}_2}{m_1 + m_2}$$

$$= \frac{m\vec{v}_0 + 0}{m + m} = \frac{\vec{v}_0}{2}$$

In elastic collisions in two and three dimensions, velocities relative to the center of mass are not necessarily reversed. However, when all motion takes place along a single straight line, as it would in a head-on collision between spheres, these velocities do get reversed. The velocity of either ball in the center-of-mass reference frame is given by the equations $u_i = v_i - V_{cm}$, where $i = 1$ refers to the cue ball and $i = 2$ refers to the object ball. Prior to the collision, $v_1 = v_0$, $v_2 = 0$, $u_1 = +v_0/2$, and $u_2 = -v_0/2$. During the collision, the center-of-mass reference frame velocities become reversed, so $u_1' = -v_0/2$ and $u_2' = +v_0/2$, where the prime (') is used to denote velocities following the collision. Transferring back to the initial reference frame via the equations $v_i' = u_i' + V_{cm}$ yields the results $v_1' = 0$ and $v_2' = v_0$. Thus the cue ball is at rest and the object ball moves with the initial velocity of the cue ball.

12. $v_A' = \left(\dfrac{1 - \alpha}{1 + \alpha}\right) v_0, \quad v_B' = \dfrac{2v_0}{1 + \alpha}$

Substitute the values $\alpha = 0$, 1, and ∞ in these equations and compare your results with the predictions given with the problem.

13. The 2-kg sphere collides with an 8-kg sphere, as shown in Figure 8-10.

Figure 8-10

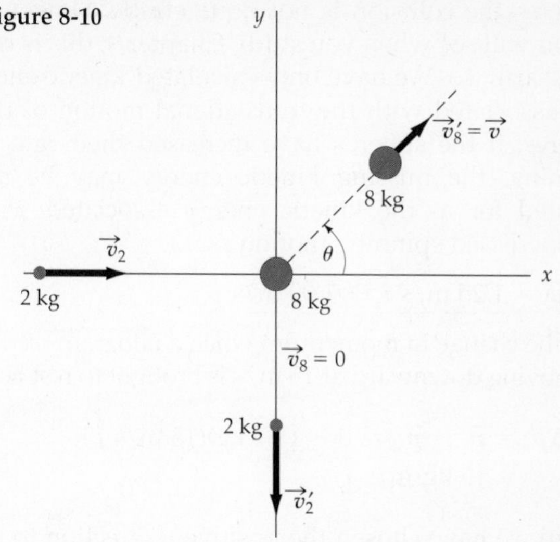

The law of conservation of momentum for this collision is

$$m_2 \vec{v}_2 + m_8 \vec{v}_8 = m_2 \vec{v}_2' + m_8 \vec{v}_8'$$

Putting this in component form for the x and y components gives

$$m_2 v_{2,x} + m_8 v_{8,x} = m_2 v_{2,x}' + m_8 v_{8,x}' \quad \text{and}$$
$$m_2 v_{2,y} + m_8 v_{8,y} = m_2 v_{8,y}' + m_8 v_{8,y}'$$

Substituting in values yields

$$(2 \text{ kg})(8 \text{ m/s}) + 0 = 0 + (8 \text{ kg})v \cos\theta \quad \text{and}$$
$$0 + 0 = (2 \text{ kg})(-6 \text{ m/s}) + (8 \text{ kg})v \sin\theta$$

or

$$v \cos\theta = 2 \text{ m/s} \quad \text{and} \quad v \sin\theta = 1.5 \text{ m/s}$$

Solving these simultaneously for v and θ yields the results

$$v = \sqrt{(1.5 \text{ m/s})^2 + (2 \text{ m/s})^2} = \underline{2.5 \text{ m/s}}$$

$$\tan\theta = 1.5/2 \Rightarrow \theta = \tan^{-1}\left(\frac{3}{4}\right) = \underline{37°}$$

To determine whether or not the collision is elastic, we compare the kinetic energy of the system prior to the collision with the kinetic energy following the collision.

Before the collision:

$$\tfrac{1}{2} m_2 v_2^2 + \tfrac{1}{2} m_8 v_8^2 = \tfrac{1}{2}(2 \text{ kg})(8 \text{ m/s})^2 + 0 = 64 \text{ J}$$

After the collision:

$$\tfrac{1}{2} m_2 v_2'^2 + \tfrac{1}{2} m_8 v_8'^2 = \tfrac{1}{2}(2 \text{ kg})(-6 \text{ m/s})^2$$
$$+ \tfrac{1}{2}(8 \text{ kg})(2.5 \text{ m/s})^2 = 61 \text{ J}$$

It seems the collision is not quite elastic. However, as you will see when you study Chapter 9, this is not necessarily so. We have only calculated kinetic energies associated with the translational motion of the spheres. If the spheres have increased their rate of spinning, the missing kinetic energy may be accounted for in the kinetic energy associated with this increased spinning motion.

14. $\vec{v} = 1.25 \text{ m/s } \hat{i} + 1.00 \text{ m/s } \hat{j}$

15. The change in momentum when a kilogram of water moving downward at 15 m/s is brought to rest is

$$\Delta \vec{p} = \vec{p} - \vec{p}_0 = 0 - (-1 \text{ kg})(15 \text{ m/s } \hat{j})$$
$$= 15 \text{ kg·m/s } \hat{j}$$

where we have chosen the positive y direction to be upward.

A kilogram of water has a volume of 10^{-3} m³. The formula for the volume V of a right circular cylinder with a height h and a base of radius r is $V = hr^2$. The height of the cylinder is $h = V/\pi r^2$.

The rain is falling at a rate of 10 cm per hour. Thus the time required for a column of water of height h to accumulate is

$$h = \frac{dh}{dt} \Delta t \quad \text{or} \quad \Delta t = \frac{h}{(dh/dt)} = \frac{V}{\pi r^2 (dh/dt)}$$

$$\vec{F}_{av} \Delta t = \Delta \vec{p} \quad \text{or}$$

$$\vec{F}_{av} = \frac{\Delta \vec{p}}{\Delta t} = \pi r^2 \frac{\Delta \vec{p}}{V} \frac{dh}{dt}$$

$$= \pi (0.125 \text{ m})^2 \frac{15 \text{ kg·m/s} \hat{j}}{10^{-3} \text{ m}^3} \left(0.1 \frac{\text{m}}{\text{h}}\right) \left(\frac{1 \text{ h}}{3600 \text{ s}}\right)$$

$$= 2.05 \times 10^{-2} \text{ N } \hat{j}$$

This is the average force exerted by the pan on the water. The force exerted by the pan on the water and the force exerted by the water on the pan form an action–reaction pair. They are equal in magnitude and oppositely directed. Thus the average force exerted on the pan by the rainwater is $-2.05 \times 10^{-2} \text{ N } \hat{j}$.

16. (a) $v = \dfrac{M}{M + m} v_0$

(b) $v = \dfrac{M}{M + m} v_0$ (same as for (a))

Chapter 9

Rotation

I. Key Ideas

In this chapter we'll continue to study the motion of many-particle systems, and emphasize the rotational aspects of such motion. For rigid-body motion, we'll look at the kinematic concepts of angular displacement, velocity, and acceleration and the dynamic concepts of torque and moment of inertia. This chapter treats the rotation of a rigid object either about a fixed rotation axis, or, in the case of a rolling object, about a moving rotation axis—one that keeps a fixed direction while moving.

9-1 Angular Velocity and Angular Acceleration
Suppose a disk is constrained to rotate about a fixed axis through its center and perpendicular to the disk as shown. A radial line drawn on the disk will sweep out angle θ (see Figure 9-1) as the disk rotates. This angle between the drawn line and a reference direction is called the **angular position** of the disk. The change in the angular position $\Delta\theta$ is called the **angular displacement,** and the rate of increase of angular position $d\theta/dt$ is the angular velocity.

Angular velocity $\vec{\omega}$ is a vector parallel to the rotation axis. The magnitude of $\vec{\omega}$ is the rotation rate and its direction is along the axis specified via a right-hand rule as shown in Figure 9-2. (If you curl

your fingers in the direction of rotation, $\vec{\omega}$ is in the same direction as your thumb—using your right hand of course!) **Angular speed** is the magnitude of the angular velocity vector. The **angular acceleration** vector is $\vec{\alpha} = d\vec{\omega}/dt$.

When a vector's direction remains parallel to an axis (not necessarily a rotation axis), the vector can be specified by a signed scalar. For example, if you make the z axis parallel to the given axis, the vector can be described by only its z component. One example of such a vector is the velocity vector of a particle undergoing one-dimensional linear motion (motion along a straight line). Another is the angular velocity vector of an object constrained to rotate such that the direction of its rotation axis remains fixed. This is called *one-dimensional rotational motion*.

For one-dimensional rotational motion, θ, ω, and α denote the angular position, velocity, and acceleration, respectively. The table below shows them with their one-dimensional linear-motion counterparts.

Figure 9-1

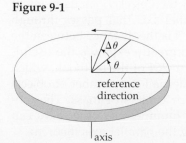

reference direction

axis

Figure 9-2

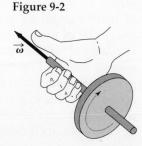

$\vec{\omega}$

	1-Dimensional (fixed-axis direction) Linear and Rotational Motion	
Linear motion	Rotational (angular) motion	
x	θ	Position
Δx	$\Delta\theta$	Displacement
$v_{av} = \dfrac{\Delta x}{\Delta t}$	$\omega_{av} = \dfrac{\Delta\theta}{\Delta t}$	Average velocity
$v = \dfrac{dx}{dt}$	$\omega = \dfrac{d\theta}{dt}$	Velocity
$a_{av} = \dfrac{\Delta v}{\Delta t}$	$\alpha_{av} = \dfrac{\Delta\omega}{\Delta t}$	Average acceleration
$a = \dfrac{dv}{dt} = \dfrac{d^2 x}{dt^2}$	$\alpha = \dfrac{d\omega}{dt} = \dfrac{d^2\theta}{dt^2}$	Acceleration

For constant angular acceleration the relations between the rotational kinematic parameters are

Kinematic formulas for constant α

$$\Delta\theta = \omega_0\Delta t + \tfrac{1}{2}\alpha(\Delta t)^2$$

$$\Delta\omega = \alpha\Delta t$$

$$\omega^2 = \omega_0^2 + 2\alpha\Delta\theta$$

$$\omega_{av} = \tfrac{1}{2}(\omega_0 + \omega)$$

For rotations about a fixed axis the relations between the linear kinematic parameters and the angular kinematic parameters are

Relations between linear and angular kinematic parameters

$$\Delta s = r\Delta\theta$$

$$v_t = r\omega$$

$$a_t = r\alpha$$

$$a_c = \frac{v^2}{r} = r\omega^2$$

where v_t and a_t are the tangential components of the linear velocity and acceleration, respectively, and a_c is the centripetal (radial) component of the acceleration.

The common units of angle are the degree, the radian, and the revolution. The radian is the SI unit of angle. By definition, one **radian (rad)** is the measure of the central angle of a circle whose intercepted arc length equals the radius. Examples of conversion factors are

Angular conversion factors

$$\frac{2\pi\,\text{rad}}{1\,\text{rev}}, \quad \frac{1\,\text{rev}}{360°}, \quad \text{and} \quad \frac{\pi\,\text{rad}}{180°}$$

9-2 Rotational Dynamics In motion, the quantity of fundamental dynamic importance is force. In rotational motion, the dynamic quantity associated with the force is the torque. *Torque τ* is the measure of a force's ability to produce a change in the rotational motion of an object. The torque τ produced by a force about an axis equals rF_t, where r, shown in Figure 9-3, is the distance from the rotation axis to the point of application of the force and F_t is the tangential component of the force. A little trigonometry will show that $\tau = rF_t = r(F\sin\theta) = (r\sin\theta)F = \ell F$, where the *lever arm ℓ* is the perpendicular distance from the axis to the *line-of-action* of the force.

Torque about an axis

$$\tau = rF_t = \ell F = rF\sin\theta$$

Figure 9-3

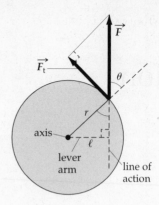

By mentally dividing up an object constrained to rotate about a fixed axis—a Ferris wheel for example—into infinitely many infinitesimal pieces, and by then applying Newton's laws for particle motion to each piece, it can be shown that

Newton's second law for rotations

$$\Sigma\tau = I\alpha$$

where $\Sigma\tau$ is the sum of the torques due to the external forces, α is the angular acceleration, and I is the moment of inertia. The moment of inertia is given by

Moment of inertia

$$I = \Sigma m_i r_i^2 \rightarrow \int r^2 dm$$

where m_i is the mass of the ith piece and r_i is its distance from the axis. For an object with a continuous mass distribution, the moment of inertia sum becomes an integral.

9-3 Calculating the Moment of Inertia The *moment of inertia I* is a measure of a body's inertial resistance to changes in angular velocity. It depends on the mass and the distribution of the mass about the axis. It can be shown that the moment of inertia I of an object about a given axis is equal to

Parallel-axis theorem

$$I = Mh^2 + I_{cm}$$

where M is the mass of the object, I_{cm} is its moment of inertia about a parallel axis that passes through the center of mass, and h is the distance between the two axes. This theorem greatly expands the usefulness of Table 9-1 on page 264 of the textbook, which gives the moments of inertia for a number of objects about axes through their centers of mass. The moments of inertia about innumerable other axes can be easily obtained using the parallel-axis theorem.

9-4 Applications of Newton's Second Law for Rotation For an application of Newton's second law for rotations about a fixed axis see the solution to Problem 11 at the end of this chapter. For an application of Newton's second law for rotations about a moving axis see the solution to Problem 15.

9-5 Rotational Kinetic Energy The kinetic energy K of any object is merely the sum of the kinetic energies of the individual particles of the object. For an object rotating about a fixed axis $v_i = r_i\omega$. Summing over the elements and using $v_i = r_i\omega$ gives

Kinetic energy for a rotation

$$K = \sum_i \tfrac{1}{2}m_i v_i^2 = \sum_i \tfrac{1}{2}m_i(r_i\omega)^2$$
$$= \tfrac{1}{2}\Big(\sum_i m_i r_i^2\Big)\omega^2 = \tfrac{1}{2}I\omega^2$$

Work is defined as $dW = F_t\,ds$, where dW is the work increment, ds is the magnitude of the displacement increment, and F_t is the component of the force in the direction of the displacement. The subscript t is used because the displacement points in the tangential direction. For an object rotating about a fixed axis $ds = r\,d\theta$, so

Work

$$dW = F_t\,ds = F_t\,r\,d\theta = \tau\,d\theta$$

where the torque τ equals $F_t r$. The rate at which the torque does work is the power input of the torque:

Power

$$P = \frac{dW}{dt} = \tau\frac{d\theta}{dt} = \tau\omega$$

9-6 Rolling Objects When a bicycle moves along a surface it can either follow a straight-line path or a curved path. In this section, discussion of rotation is restricted to the kind executed by the wheel of a bicycle traveling along a straight-line path. For this kind of rolling, the motion of each individual particle making up the wheel remains confined to a plane. It follows that the direction of the wheel's instantaneous rotation axis, which is perpendicular to the plane, remains fixed.

Rolling Without Slipping At each instant, each point on the wheel, shown in Figure 9-4, is moving with speed $v = r\omega$, where r is its distance from the rotation axis and ω is the wheel's angular speed. The rotation axis perpendicular to the plane of the page passes through the point(s) of contact between the wheel and the road surface.

Figure 9-4

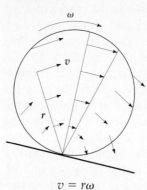

$v = r\omega$

Figure 9-5

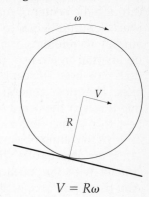

$V = R\omega$

A special case of this relation, shown in Figure 9-5, is $V = R\omega$, where R is the wheel's radius and V is the speed of its center. *Note:* Only if $V = R\omega$ is the wheel rolling *without* slipping.

Rolling-without-slipping condition

$$V = R\omega$$

By differentiating both sides of $v = r\omega$ with respect to time we get $a_t = r\alpha$, where a_t is the tangential acceleration and α is the angular acceleration, shown in Figure 9-6.
A special case of this relation, shown in Figure 9-7, is

Figure 9-6

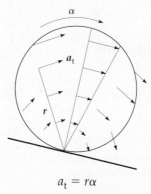

$a_t = r\alpha$

Figure 9-7

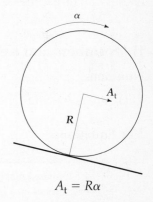

$A_t = R\alpha$

$A_t = R\alpha$, where A_t is the tangential acceleration of the wheel's center.

The most general method of solving for the motion of an object is to simultaneously apply both the linear and the rotational forms of Newton's second law. These are:

Newton's second laws for translation and rotation

$$\sum_i F_{i,\,\mathrm{ext}} = MA_{\mathrm{cm}}$$

$$\sum_i \tau_{i,\,\mathrm{cm}} = I_{\mathrm{cm}}\alpha$$

where each torque is calculated about an axis through the center of mass. When the center of mass is accelerating (a ball rolling down an incline, for example), its reference frame is not an inertial one. In a noninertial reference frame the equation $\Sigma\tau = I\alpha$ does not hold—except for certain specific rotation axes. The most general of these specific axes is the one through the center of mass. For this reason it is best to use the center of mass axis when applying $\Sigma\tau = I\alpha$ to rolling objects.

If an object rolls without slipping, the point of contact with the surface is instantaneously at rest. Any frictional force is that of static friction so no mechanical energy is dissipated. This suggests that many rolling-without-slipping problems can best be solved by using conservation of mechanical energy. The kinetic energy K of an object can be expressed as the sum of two terms, one for its translational motion K_{trans} and the other for its rotational motion K_{rot}. That is

Translational and rotational kinetic energy

$$K = K_{trans} + K_{rot} = \tfrac{1}{2}MV_{cm}^2 + \tfrac{1}{2}I_{cm}\omega^2$$

where I_{cm} is the moment of inertia about an axis through the center of mass. By substituting the rolling-without-slipping condition $V_{cm} = R\omega$ into this equation, ω can be eliminated and the kinetic energy can be expressed in terms of V_{cm} alone.

II. Numbers and Key Equations

Numbers

There are no new numbers for this chapter.

Key Equations

1-Dimensional (fixed-axis direction) Linear and Rotational Motion

Linear motion	Rotational (angular) motion	
x	θ	Position
Δx	$\Delta\theta$	Displacement
$v_{av} = \dfrac{\Delta x}{\Delta t}$	$\omega_{av} = \dfrac{\Delta\theta}{\Delta t}$	Average velocity
$v = \dfrac{dx}{dt}$	$\omega = \dfrac{d\theta}{dt}$	Velocity
$a_{av} = \dfrac{\Delta v}{\Delta t}$	$\alpha_{av} = \dfrac{\Delta\omega}{\Delta t}$	Average acceleration
$a = \dfrac{dv}{dt} = \dfrac{d^2 x}{dt^2}$	$\alpha = \dfrac{d\omega}{dt} = \dfrac{d^2\theta}{dt^2}$	Acceleration

---

Kinematic formulas for constant α

$$\Delta\theta = \omega_0\Delta t + \tfrac{1}{2}\alpha(\Delta t)^2$$

$$\Delta\omega = \alpha\Delta t$$

$$\omega^2 = \omega_0^2 + 2\alpha\Delta\theta$$

$$\omega_{av} = \tfrac{1}{2}(\omega_0 + \omega)$$

Relations between linear and angular kinematic parameters

$$\Delta s = r\Delta\theta$$

$$v_t = r\omega$$

$$a_t = r\alpha$$

$$a_c = \frac{v^2}{r} = r\omega^2$$

Angular conversion factors

$$\frac{2\pi\,\text{rad}}{1\,\text{rev}}, \quad \frac{1\,\text{rev}}{360°}, \quad \text{and} \quad \frac{\pi\,\text{rad}}{180°}$$

Torque about an axis

$$\tau = rF_t = \ell F = rF\sin\theta$$

Newton's second law for rotations

$$\Sigma\tau = I\alpha$$

Moment of inertia

$$I = \Sigma m_i r_i^2 \rightarrow \int r^2 dm$$

Parallel-axis theorem

$$I = Mh^2 + I_{cm}$$

Kinetic energy for a rotation

$$K = \sum_i \tfrac{1}{2}m_i v_i^2 = \tfrac{1}{2}I\omega^2$$

Work

$$dW = F_t\,ds = F_t\,r\,d\theta = \tau\,d\theta$$

Power

$$P = \frac{dW}{dt} = \tau\frac{d\theta}{dt} = \tau\omega$$

Newton's second laws for translation and rotation

$$\sum_i F_{i,\,\text{ext}} = MA_{\text{cm}}$$

$$\sum_i \tau_{i,\,\text{cm}} = I_{\text{cm}}\alpha$$

Rolling-without-slipping condition

$$V = R\omega$$

Translational and rotational kinetic energy

$$K = K_{\text{trans}} + K_{\text{rot}} = \tfrac{1}{2}MV_{\text{cm}}^2 + \tfrac{1}{2}I_{\text{cm}}\omega^2$$

III. Potential Pitfalls

What is a radian? If you cut a wedge of pumpkin pie and the length of one side of the wedge *equals* the length of the crust, the wedge angle equals one radian.

Don't think angular measure must be in radians in all the kinematic equations. Equations with *only* rotational kinematic parameters (e.g., θ, ω, and α) and time, such as $\Delta\theta = \omega\Delta t$, are valid with *any* consistent unit of angular measure. Thus, in the kinematic equations for rotational motion with constant α, any consistent unit for angular measure may be used. Equations with *both* linear and angular parameters, such as $\Delta s = r\Delta\theta$, are valid *only* when the unit of angular measure is the radian.

Don't be mislead into thinking that linear acceleration means the acceleration component in the direction of the velocity vector. In general, linear acceleration $\vec{a}$ has both a centripetal a_c and a tangential a_t component. Linear position, velocity, and acceleration are the same old friends that we called position, velocity, and acceleration prior to introducing the corresponding angular terms. When referring to angular position, angular velocity, or angular acceleration, it is best to explicitly use the "angular" descriptor.

Don't think that angular velocity is a scalar. Angular velocity is a vector quantity. The direction of the angular velocity vector is found by the right-hand rule. Your right thumb points in the direction of the angular velocity vector if you curl the fingers of your right hand in the direction of the rotational motion. Alternatively, the direction of the angular velocity vector is the direction a common (or right-handed) screw would advance if rotated in the same direction as the rotating body. The scalar quantity ω represents either the angular velocity vector's magnitude (the angular speed) or its component in the direction of some axis.

For one-dimensional rotational motion, the magnitude of the angular velocity is the instantaneous rate at which a body rotates. The direction of rotation, clockwise or counterclockwise, is indicated by a plus or minus sign. Counterclockwise is usually the positive direction. When a particle is moving along the arc of a circle, the angular velocity of the particle is the rate at which the line from the particle to the center of the circle sweeps out angle.

Don't think the rolling-without-slipping relation $V = R\omega$ is valid only for rolling *on flat surfaces*. It holds for rolling on curved surfaces also, like when cresting a hill.

For a rotating object, $r\omega$ is the linear velocity of a point at distance r from the rotation axis. Don't fall into the trap of thinking that for a rotating object $r\alpha$ is the linear acceleration of the point. This is only the tangential component of the linear acceleration vector. There is also a centripetal component given by $r\omega^2$.

Don't think of the *centripetal force* as a distinct force. It's not. Rather it's the component of the net force in the centripetal direction (the direction toward the center of curvature of the particle's path). A centripetal force is always present on a particle that is rotating about a fixed axis. It is needed to change the direction of the motion.

Don't think that if a particle is moving in a circle that the net force is directed toward the circle's center. That is so only if the particle moves with constant speed. For the particle to gain (or lose) speed, the net force must also have a component in the tangential direction.

Don't think the torque depends on the distance between the rotation axis and the point of application of the force. Instead it depends on the length of the moment arm (the perpendicular distance between the rotation axis and the line of action of the force). If the line of action intersects the rotation axis then the torque must be zero.

IV. True and False and Responses

True and False

_____ 1. The acceleration of the center of mass of an object equals the net external force acting on it divided by its total mass unless the body rotates about the center of mass.

_____ 2. Angular velocity and angular acceleration can be defined only in terms of angles measured in radians.

_____ 3. A circular disk of radius r rotates about a fixed axis perpendicular to the disk and through its center. If the disk's angular acceleration is α, the linear acceleration of a point on its rim is equal to $r\alpha$.

_____ 4. The second hand of a watch rotates with an angular velocity of roughly 0.1 rad/s.

_____ 5. All parts of a rotating rigid body have the same angular velocity.

_____ 6. All parts of a rotating rigid body have the same centripetal acceleration.

_____ 7. The moment of inertia of a rotating rigid body about the rotational axis is independent of the rate of rotation.

_____ 8. If the net external force acting on an object is zero, the net torque on it must also be zero.

_____ 9. The net force acting on a rigid body rotating about a fixed axis through its center of mass is always zero.

_____ 10. The lever arm of a force is the perpendicular distance from the axis of rotation to the point at which the force acts.

_____ 11. The kinetic energy of a rigid body rotating about a fixed axis depends only on its angular speed and the total mass.

_____ 12. You cannot specify a moment of inertia unless you also specify an axis.

_____ 13. The moment of inertia of a rigid body about an axis through its center of mass is smaller than that about any other parallel axis.

_____ 14. For the purpose of calculating its moment of inertia, all the mass of a rigid body can be considered to be concentrated at its center of mass.

_____ 15. If rigid body is rotating, it must be rotating about a fixed axis.

Responses

1. False. The center-of-mass acceleration equals the net external force divided by the total mass regardless of whether the body is rotating or not.

2. False. Angular velocity is often expressed in revolutions per minute, for example. However, many of the equations that include angular velocity or angular acceleration are valid only for angles measured in radians.

3. False. $r\alpha$ is only the tangential component of the linear acceleration of a point on the rim. The radial (or centripetal) component, which equals $\omega^2 r$, also needs to be considered.

4. True. It moves through $2\pi = 6.28$ radians every 60 seconds, or through 0.105 radians each second.

5. True

6. False. The centripetal acceleration of any part of the body is proportional to its distance from the axis of rotation.

7. True. The moment of inertia depends only upon the mass of the body and how it is distributed relative to the axis.

8. False. Consider two forces of equal magnitude and

opposite direction. The forces add to zero, but the net torque is zero only if they act along the same line.

9. True. If its center of mass isn't accelerating, the net force must be zero.

10. False. The lever arm is the perpendicular distance from the axis of rotation to the line along which the force acts.

11. False. It also depends on the distribution of the mass relative to the rotation axis. The kinetic energy depends on the angular speed and the moment of inertia.

12. True

13. True. This follows from the parallel-axis theorem.

14. False. The distribution of the body's mass in space plays an important role in determining the moment of inertia. If all the mass were at its center of mass, its moment of inertia about an axis through the center of mass would be zero.

15. False. In addition, the axis of rotation can change direction and undergo translations. Consider the motion of a bicycle wheel as the bike turns a corner.

V. Questions and Answers

Questions

1. Consider two points on a disk rotating with increasing speed about its axis, one at the rim and the other halfway from the rim to the center. Which point has the greater (a) angular acceleration, (b) tangential acceleration, (c) radial acceleration, and (d) centripetal acceleration?

2. Can there be more than one value for the moment of inertia for a given rigid body?

3. In Figure 9-8, a man is hanging onto one side of a Ferris wheel, which swings him down toward the ground. (The Ferris wheel's motor is disengaged so it is free to rotate about its axis.) Can you use constant-angular-acceleration formulas to calculate the time it takes him to reach the bottom?

Figure 9-8

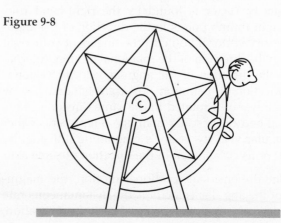

4. Does applying a positive (counterclockwise) net torque to a body necessarily increase its kinetic energy?

5. If the main rotor blade of the helicopter depicted in Figure 9-9 is rotating counterclockwise as seen from above, which way should the smaller tail rotor be pushing the air? Why?

Figure 9-9

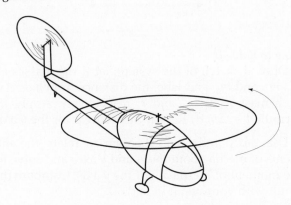

6. A solid ball, a solid disk, and a hoop, all with the same mass and the same radius, are set rolling without slipping up an incline, all with the same initial linear speed. Which goes farthest? If they are all set rolling with the same initial kinetic energy instead, which goes farthest?

Answers

1. (*a*) Both points have the same angular acceleration. This, of course, is the point of describing rotational motion in terms of angular quantities. (*b*) The tangential acceleration ($a_t = r\alpha$) of the point on the rim is larger because it is farther from the axis. (*c*) The radial acceleration ($a_{rad} = v^2/r = \omega^2 r$) of the point at the rim is larger for the same reason. (*d*) Since the radial acceleration and centripetal acceleration are the same ($a_c = a_{rad}$), the answer is the same as for (*c*).

Note: The centripetal direction is always defined as radially inward (toward the rotation axis). The radial direction is less definite. Sometimes radially inward is taken as the positive radial direction, but more often, radially outward is considered positive. Thus, $a_{rad} = \pm a_c$ depending whether radially inward or outward is taken as the positive direction.

2. The moment of inertia may be different about different axes.

3. No, the gravitational torque on the system decreases. As shown in Figure 9-10, the weight force

remains constant but its lever arm ℓ decreases, as he swings down toward the bottom. This decrease in torque results in a decrease in the angular acceleration. The constant-angular-acceleration formulas are valid only if the angular acceleration remains constant.

Figure 9-10

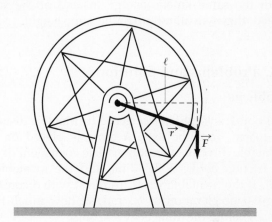

4. Not necessarily. It will increase the kinetic energy and therefore the angular speed of the body only if it does positive work, that is, if the motion is also counterclockwise. If the motion is clockwise then a counterclockwise torque will do negative work and reduce the body's kinetic energy.

5. The net torque about the helicopter's center of mass must equal zero, otherwise it will start to rotate. The rotational motion of the main rotor is counterclockwise. Thus the drag force of the air, which opposes this motion, produces a clockwise torque on the helicopter. To counter this clockwise torque, there must be a counterclockwise torque of equal magnitude acting on the helicopter. This is accomplished if the tail rotor pushes air to the helicopter's left (port). According to Newton's third law, if the tail rotor pushes air toward the left, then the air pushes the tail rotor to the helicopter's right (starboard), thus producing a counterclockwise torque on the helicopter. (The tail rotor's axis is oriented both horizontally and at right angles to the helicopter's forward direction.)

6. It is helpful to separately consider the translational ($K_{trans} = \frac{1}{2}MV_{cm}^2$) and rotational ($K_{rot} = \frac{1}{2}I\omega^2$) kinetic energies. All the objects have the same linear speed so they all have the same translational kinetic energy. Also, they all have the same angular speed so the larger the moment of inertia, the larger the rotational kinetic energy. The hoop ($I = MR^2$) has the largest moment of inertia, the disk ($I = \frac{1}{2}MR^2$) the next largest, and the sphere ($I = \frac{2}{5}MR^2$) the smallest, so the hoop has the largest rotational kinetic energy and the sphere has the least. Each will roll up the hill until all its kinetic energy is transformed into gravitational potential energy, so the greater the kinetic energy, the higher up the hill it will roll. This means the hoop rolls the highest, with the disk second, and the sphere last. Of course, if all three start with the same kinetic energy instead of the same speed, they will all roll up to the same height.

VI. Problems and Solutions

Problems

1. Point P is located on a record turntable a distance of 10 cm from the axis. A dime is placed on the turntable directly over point P. The coefficient of static friction between the dime and the turntable is 0.21. If the turntable starts from rest with a constant angular acceleration of 1.2 rad/s², how much time passes before the dime begins to slip?

How to Solve It
• The linear acceleration of P has both a tangential and a centripetal component. The tangential component is constant but the centripetal component increases as the angular speed increases.

• The dime will not slip as long as the static frictional force is sufficient to give it the same acceleration as P.

2. A phonograph turntable with a mass of 1.8 kg and a radius of 15 cm is being braked to a stop. After 5 s its initial angular speed has decreased by 15%. Assuming constant angular acceleration, (*a*) how long does it take to come to rest? (*b*) If the initial angular speed is 33⅓ rpm, through how many revolutions does it turn while coming to rest? (*c*) How much work is done on the turntable in order to bring it to rest?

How to Solve It
• This is a kinematics problem with constant angular acceleration.

• For part (*a*) assume that the total braking time can be determined without being given the initial angular velocity. Assign the symbol ω_0 to it and proceed.

3. A system consists of four point masses connected by rigid rods of negligible mass as shown in Figure 9-11. (*a*) Calculate the moment of inertia about the *A* and *y* axes by direct application of the formula $I = \Sigma m_i r_i^2$. (*b*) Use the parallel-axis theorem to relate the

moments of inertia about the *A* and *y* axes and solve for the moment of inertia about the *y* axis. Compare this result with your result in part (*a*). (*c*) Calculate the moments of inertia about axes *x* and *z*.

Figure 9-11

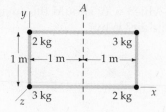

How to Solve It
• Draw a sketch of the system, label the masses m_1, m_2, m_3, and m_4, and label the distance from axis *A* to each mass r_1, r_2, r_3, and r_4 respectively. Apply the formula $I = \Sigma m_i r_i^2$. Repeat this process for the *y* axis.

• Use the parallel-axis theorem to relate the moments of inertia about the *A* and *y* axes and solve for the moment of inertia about the *y* axis. Compare this result with your result in part (*a*).

• Use the formula $I = \Sigma m_i r_i^2$ for finding the moments of inertia in part (*c*).

4. A system consists of three point masses connected by rigid rods of negligible mass as shown in Figure 9-12. (*a*) Calculate the moment of inertia about the *A* and *y* axes by direct application of the formula $I = \Sigma m_i r_i^2$. (*b*) Use the parallel-axis theorem to relate the moments of inertia about the *A* and *y* axes and solve for the moment of inertia about the *y* axis. Compare this result with your result in part (*a*). (*c*) Calculate the moments of inertia about axes *x* and *z*.

Figure 9-12

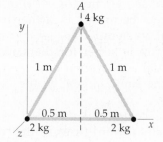

How to Solve It:
• Draw a sketch of the system, label the masses m_1, m_2, and m_3, and label the distance from axis *A* to each mass r_1, r_2, and r_3 respectively. Apply the formula $I = \Sigma m_i r_i^2$. Repeat this process for the *y* axis.

• Use the parallel-axis theorem to relate the moments of inertia about the *A* and *y* axes and solve for the moment of inertia about the *y* axis. Compare this result with your result in part (*a*).

• Use the formula $I = \Sigma m_i r_i^2$ for finding the moments of inertia in part (c).

5. A thin uniform rod of mass m and length L is free to pivot about an axis perpendicular to the rod a distance x from its center. (a) Find the moment of inertia about this axis as a function of x using the parallel-axis theorem and a formula, obtained from Table 9-1 in the text, for the moment of inertia of a thin uniform rod about an axis perpendicular to it and passing through its center. (b) Check your result by setting x equal to $L/2$, the value for an axis perpendicular to the rod and passing through one end. Then compare your result with the expression obtained from Table 9-1 for the same axis.

How to Solve It
• Use the parallel-axis theorem and the expression for I_{cm} from Table 9-1 in the textbook to find I.

• What is the value of x for an axis perpendicular to the rod and passing through one end? Check the validity of your part (a) result by substituting this value into it and comparing this expression with the appropriate expression from Table 9-1.

6. A uniform circular disk of mass M_d and radius R is free to pivot about an axis perpendicular to the plane of the disk a distance r from its center. (a) Find the moment of inertia about this axis as a function of r using the parallel-axis theorem and a formula, from Table 9-1 in the text, for the moment of inertia of a solid cylinder about its symmetry axis. (b) Find an expression for the moment of inertia if the axis is halfway between the disk's center and its perimeter.

7. As shown in Figure 9-13, a circular disk of mass M_1 and radius R is made from a uniform piece of thin sheet material. This disk is free to rotate about an axis perpendicular to it that passes through its center. A second disk with a radius of $R/2$ is cut from the same sheet and the two disks are glued together with the small disk's perimeter touching the perimeter of the large disk at a single point. Find an expression for the moment of inertia of the composite two-disk object about the given axis.

Figure 9-13

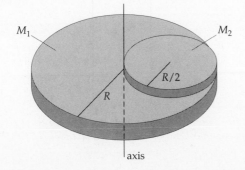

How To Solve It
• Separately calculate the moment of inertia of each disk about the specified axis. The total moment of inertia is then the sum of the two. (The moment of inertia of a composite of several objects is the sum of the moments of inertia of the individual objects.)

• The expression for the moment of inertia of a uniform disk about an axis perpendicular to the plane of the disk and passing through its center can be found in Table 9-1 of the textbook. To find the moment of inertia of the small disk you should first find its mass (in terms of M_1). Then use the parallel-axis theorem to find its moment of inertia about the given axis.

8. A 0.5-kg 0.25-m-radius disk is made from a uniform piece of thin sheet material. As shown in Figure 9-14, the disk is glued to a 0.200-kg meter stick, with the disk's center at the 75-cm mark. Find the moment of inertia of this composite object about an axis perpendicular to both the meter stick and the plane of the disk and passing through the stick's zero-cm mark.

Figure 9-14

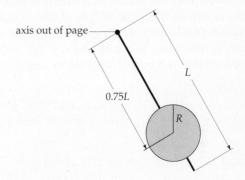

How to Solve It
• The moment of inertia of the composite object is the sum of the moments of inertia of the meter stick and of the disk.

• The expressions for the moments of inertia of a thin uniform rod about an axis perpendicular to the rod and passing through one end, and of a uniform disk about an axis perpendicular to the plane of the disk and passing through its center, can be found in Table 9-1 of the textbook. To find the moment of inertia of the disk about the given axis you also need to use the parallel-axis theorem.

9. Calculate the moment of inertia about each of the axes A, B, and C for the thin solid half-disk of mass M and radius R, shown in Figure 9-15 (on the next page). The C axis is perpendicular to the plane of the half-disk.

Figure 9-15

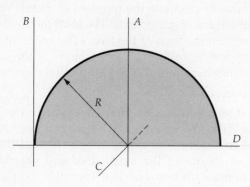

Figure 9-16

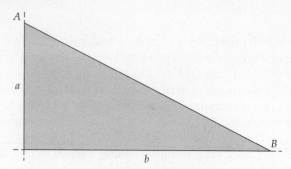

How to Solve It
• To find the moment of inertia about axis *A* first draw a strip of differential width at an arbitrary constant distance from the *A* axis. Set up a coordinate axis with its origin on the *A* axis and with the distance from the origin to the strip as *x*. Determine an expression for the differential moment of inertia *dI* of this strip about the *A* axis as a function of *x* times the strip width *dx*. Find the height of the rectangular strip as a function of *x*, and express the mass of the strip *dm* as the mass per unit area of the half-disk times the area of the strip. Integrate to determine the moment of inertia.

• Use the parallel-axis theorem and the result of part (*a*) to find the moment of inertia about the *B* axis.

• To find the moment of inertia about axis *C*, first draw a semicircular strip of differential width *dr* at an arbitrary constant distance *r* from the *C* axis. Determine an expression for the differential moment of inertia *dI* of this strip about the *C* axis as a function of *r* times the strip width *dr*. To do this you will have to find the length of the strip in terms of *r*, and will have to express the mass of the strip as the mass per unit area of the half-disk times the area of the strip. Integrate to determine the moment of inertia.

• Check your result. For the uniform half-disk the mass distribution from the *C* axis is the same as would be for a uniform full disk. (The formula for the moment of inertia for a uniform full disk, found in Table 9-1 of the textbook, is $I = \frac{1}{2}MR^2$.)

10. Calculate the moment of inertia about each of the axes *A* and *B* for the thin solid triangle of mass *M*, height *a*, and width *b* = 2*a*, shown in Figure 9-16.

How to Solve It
• To find the moment of inertia about axis *A*, first draw a strip of differential width at an arbitrary constant distance from the *A* axis. Set up a coordinate axis with its origin on the *A* axis and with the distance from the origin to the strip as *x*. Determine an expression for the differential moment of inertia *dI* of this strip about the *A* axis as a function of *x* times the strip width *dx*. Find the height *y* of the rectangular strip as a function of *x*, and express the mass of the strip *dm* as the mass per unit area of the triangle times the area of the strip. Integrate to determine the moment of inertia.

• To find the moment of inertia about axis *B* follow a procedure analogous to the one described for part (*a*).

11. In Figure 9-17 a 0.4-kg pulley has a 6-cm radius and blocks 1 and 2 have masses of 0.8 kg and 1.6 kg, respectively. The pulley wheel is a uniform disk, friction in the pulley axle is negligible, and the string neither slips nor stretches. (*a*) Find the acceleration of block 2 and the angular acceleration of the pulley. (*b*) Find the tensions in the string segments on either side of the pulley. Check the validity of your results by verifying that Newton's second law for rotations holds for the pulley wheel.

Figure 9-17

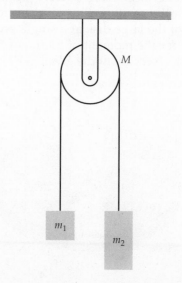

How to Solve It
- Draw a separate free-body diagram for each block and for the pulley wheel. Do not assume that the string's tension is the same on either side of the pulley. In order to exert a net torque on the pulley wheel these tensions must differ. Apply Newton's second law to each block and apply Newton's second law for rotations to the pulley wheel. This procedure will result in three independent equations. Count the number of unknowns in these equations to determine if you have enough equations to solve the problem.

- Write down the relation between the acceleration of each block and the angular acceleration of the pulley wheel. Also, write down an expression for the moment of inertia of the pulley wheel which is a uniform disk.

- Do the algebra. That is, solve the equations for the values requested.

- Check your work by substituting the values you got into the equation obtained by applying Newton's second law for rotations to the pulley wheel and seeing if opposite sides of the equations are numerically equal, as they should be.

12. In Figure 9-18, the pulley wheel has a moment of inertia of 1.0×10^{-3} kg·m² and radii of 4 and 8 cm. Blocks 1 and 2 have masses of 1.6 and 1.0 kg, respectively. Assume that friction in the pulley axle is negligible and the strings neither slip nor stretch. (*a*) Find the acceleration of each block and the angular acceleration of the pulley. (*b*) Find the tension in each string.

Figure 9-18

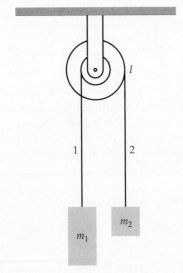

How to Solve It
- Draw a separate free-body diagram for each block and for the pulley wheel. Apply Newton's second

law to each block and apply Newton's second law for rotations to the pulley wheel. This procedure will result in three independent equations. Count the number of unknowns in these equations to see if you can now solve the equations for the values requested in the problem statement.

- Write down the relation between the acceleration of each block and the angular acceleration of the pulley wheel. The accelerations of the blocks are not equal so there are two relations to write down.

- Do the algebra. That is, solve the equations for the values requested.

13. A 2.00-kg 6.00-cm-radius wheel, shown in Figure 9-19, has a square hole through its center. The wheel's center of mass is at its geometric center. A string of negligible mass with one end attached to the ceiling is wrapped around its perimeter. As it falls, the string does not slip so the wheel rotates faster and faster as it gains speed. It is observed that the wheel is moving at 4.74 m/s after falling a distance of 2.00 m. Find its moment of inertia about an axis perpendicular to the page and passing through its center.

Figure 9-19

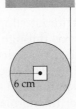

How to Solve It
- The string does not slip so no energy is dissipated via friction and the rolling-without-slipping condition holds. Choose a suitable system, apply the conservation of mechanical energy, and solve for the moment of inertia of the wheel.

14. A uniform sphere is rolling along a horizontal road with speed v_0 when it comes to a hill. It rolls up the hill, just making it to the top where it comes to rest. How high is the hill? Assume that it rolls without slipping and that air resistance is negligible.

How to Solve It
- The ball does not slip so no energy is dissipated via friction and the rolling-without-slipping condition holds. Choose a suitable system, apply the conservation of mechanical energy, and solve for the increase in height of the ball.

15. A bowling ball (mass 7 kg, diameter 20 cm) is set spinning about a horizontal axis at 120 rev/min. As the ball is set down on the floor, it is still spinning at 120 rev/min but its center of mass is at rest. If the coefficient of kinetic friction between ball and floor is 0.8 describe quantitatively the subsequent motion of the ball.

How to Solve It
- This is an example of rigid-body motion about an axis of fixed direction.

- Because a finite force cannot produce an infinite linear acceleration, the initial motion is that of rolling *with* slipping.

- During the initial rolling-*with*-slipping period calculate both the linear acceleration of the center of mass and the angular acceleration about the rotation axis through the center of mass. Using the kinematic formulas, calculate the time when slipping ceases.

- Once the rolling-without-slipping condition is reached, the force of friction remains zero.

16. A solid spherical ball rolls down an incline. If the coefficient of static friction between the ball and the incline is 0.33, what is the steepest angle of the incline for which rolling without slipping will occur?

How to Solve It
- The ball will roll without slipping as long as the net torque about an axis through the center of mass is sufficient to produce the needed angular acceleration to prevent skidding. The steeper the incline, the larger the linear acceleration and the smaller the normal force. The smaller the normal force, the smaller maximum static frictional force and, consequently, the smaller the torque exerted by the maximum static frictional force.

- You will need to apply both the linear and the rotational forms of Newton's second law to the ball.

Solutions

1. The dime moves in a circular path as shown in Figure 9-20a. It will not slip as long as its acceleration matches that of point P, the point on the turntable directly under the dime. As the speed of the dime increases, its linear acceleration increases. The net force on the dime must increase in order to provide sufficient force for this ever-increasing acceleration.

It is the static frictional force that provides the net force. (The normal and weight forces on the dime cancel.) The dime slips when the frictional force, shown in Figure 9-20b, reaches its maximum value and no longer provides sufficient force for the dime's acceleration to match that of point P. As shown in Figure 9-20c the acceleration has both tan-

gential and centripetal components. The magnitude of the acceleration is

$$a = \sqrt{a_t^2 + a_c^2} = \sqrt{(r\,\alpha)^2 + (r\,\omega^2)^2} = r\,\alpha\sqrt{1 + \alpha^2 t^4} \quad (1)$$

where we have substituted $r\alpha$ for a_t, $\omega^2 r$ for a_c, and αt for ω.

Figure 9-20

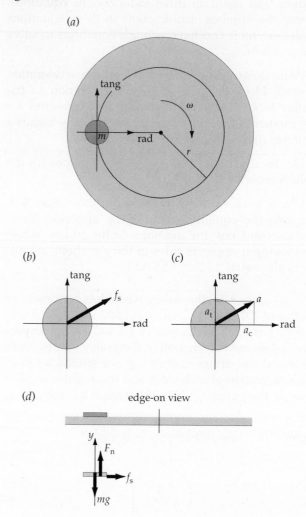

To find the maximum static frictional force we first find the normal force using Newton's second law. Figure 9-20d is an edge-on free-body diagram of the coin. Applying Newton's second law for the y components gives

$$F_n - w = ma_y$$

where $w = mg$ and $a_y = 0$. It follows that

$$F_n = mg$$

so

$$f_s \le \mu_s F_n = \mu_s mg \quad (2)$$

where f_s is shown in Figure 9-20b.

The weight force and the normal force are equal in magnitude and oppositely directed, and so add to zero. It follows then that $F_{net} = f_s$. Since $F_{net} = ma$, we can substitute ma for f_s in Equation 2. This gives

$$ma \leq \mu_s mg$$

Dividing by m and substituting for a from Equation 1 gives

$$ra\sqrt{1 + \alpha^2 t^4} \leq \mu_s g$$

The time when slipping begins is the time at which the left side of the equation above equals the right side. Equating the two sides and solving for the time gives

$$t = \frac{1}{\sqrt{\alpha}}\left[\left(\frac{\mu_s g}{r\,\alpha}\right)^2 - 1\right]^{1/4}$$

$$= \frac{1}{\sqrt{1.2\ \text{rad/s}^2}}\left[\left(\frac{0.21\ (9.81\ \text{m/s}^2)}{(0.1\ \text{m})(1.2\ \text{rad/s}^2)}\right)^2 - 1\right]^{1/4}$$

$$= \underline{3.78\ \text{s}}$$

2. (a) $\underline{33\ \text{s}}$
   (b) $\underline{9.3\ \text{rev}}$
   (c) $\underline{0.031\ \text{J}}$

3. (a) The masses in Figure 9-21a are $m_1 = 2$ kg, $m_2 = 3$ kg, $m_3 = 2$ kg, and $m_4 = 3$ kg, and the distances are $r_1 = r_2 = r_3 = r_4 = 1$ m.

Figure 9-21

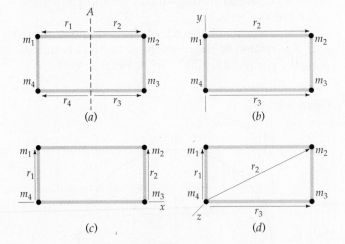

(a)

(b)

(c)

(d)

The moment of inertia about axis A is

$$I_A = \sum_i m_i r_i^2 = m_1 r_1^2 + m_2 r_2^2 + m_3 r_3^2 + m_4 r_4^2$$

$$= (2\ \text{kg})(1\ \text{m})^2 + (3\ \text{kg})(1\ \text{m})^2$$
$$\qquad\qquad + (2\ \text{kg})(1\ \text{m})^2 + (3\ \text{kg})(1\ \text{m})^2$$

$$= \underline{10\ \text{kg}\cdot\text{m}^2}$$

The distances in Figure 9-21b are $r_2 = r_3 = 2$ m and $r_1 = r_4 = 0$. Thus, the moment of inertia about the y axis is

$$I_y = \sum_i m_i r_i^2 = m_1 r_1^2 + m_2 r_2^2 + m_3 r_3^2 + m_4 r_4^2$$

$$= (2\ \text{kg})(0)^2 + (3\ \text{kg})(2\ \text{m})^2$$
$$\qquad\qquad + (2\ \text{kg})(2\ \text{m})^2 + (3\ \text{kg})(0)^2$$

$$= \underline{20\ \text{kg}\cdot\text{m}^2}$$

(b) The A axis passes through the center of mass, and is parallel to the y axis. Therefore, the moments of inertia about these axes are related by the parallel-axis theorem $I_y = Mh^2 + I_{cm}$, where $M = m_1 + m_2 + m_3 + m_4 = 10$ kg, $I_{cm} = I_A = 10$ kg·m², and h, the distance between the two axes, equals 1 m. Thus,

$$I_y = Mh^2 + I_{cm} = Mh^2 + I_A$$

$$= (10\ \text{kg})(1\ \text{m})^2 + 10\ \text{kg}\cdot\text{m}^2 = \underline{20\ \text{kg}\cdot\text{m}^2}$$

which is exactly the same value as that gotten for I_y in part (a).

(c) The distances in Figure 9-21c are $r_1 = r_2 = 1$ m and $r_3 = r_4 = 0$. Thus, the moment of inertia about the x axis is

$$I_x = \sum_i m_i r_i^2 = m_1 r_1^2 + m_2 r_2^2 + m_3 r_3^2 + m_4 r_4^2$$

$$= (2\ \text{kg})(1\ \text{m})^2 + (3\ \text{kg})(1\ \text{m})^2$$
$$\qquad\qquad + (2\ \text{kg})(0)^2 + (3\ \text{kg})(0)^2$$

$$= \underline{5\ \text{kg}\cdot\text{m}^2}$$

The distances in Figure 9-21d are $r_1 = 1$ m, $r_2 = \sqrt{5}$ m, $r_3 = 2$ m, and $r_4 = 0$. Thus, the moment of inertia about the z axis is

$$I_z = \sum_i m_i r_i^2 = m_1 r_1^2 + m_2 r_2^2 + m_3 r_3^2 + m_4 r_4^2$$

$$= (2\ \text{kg})(1\ \text{m})^2 + (3\ \text{kg})(\sqrt{5}\ \text{m})^2$$
$$\qquad\qquad + (2\ \text{kg})(2\ \text{m})^2 + (3\ \text{kg})(0)^2$$

$$= \underline{25\ \text{kg}\cdot\text{m}^2}$$

4. $I_A = \underline{1\ \text{kg}\cdot\text{m}^2}$, $I_x = \underline{3\ \text{kg}\cdot\text{m}^2}$, $I_y = \underline{3\ \text{kg}\cdot\text{m}^2}$, and $I_z = \underline{6\ \text{kg}\cdot\text{m}^2}$.

5. (a) The value for I_{cm} obtained from Table 9-1 is

$$I_{cm} = \tfrac{1}{12}ML^2$$

and the parallel-axis theorem is $I = Mh^2 + I_{cm}$, where h is the distance, labeled x in Figure 9-22 (on the next page), between the two axes.

Substituting into this formula gives

$$I = \underline{M\left(x^2 + \tfrac{1}{12}L^2\right)}$$

Figure 9-22

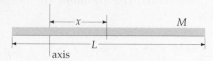

(b) For an axis perpendicular to the rod and passing through one end $x = L/2$. Substituting $L/2$ for x gives

$$I = M\left[\left(\frac{L}{2}\right)^2 + \frac{1}{12}L^2\right] = M\left(\frac{1}{4} + \frac{1}{12}\right)L^2 = \frac{1}{3}ML^2$$

which is the value listed in Table 9-1.

6. (a) $M\left(r^2 + \frac{1}{2}R^2\right)$ (b) $\frac{3}{4}MR^2$

7. From Table 9-1 we find the expression $\frac{1}{2}MR^2$ for the moment of inertia of a uniform circular disk about an axis through its center. Thus for the larger disk we have $I_{1,\text{cm}} = \frac{1}{2}M_1R^2$ and for the smaller disk $I_{2,\text{cm}} = \frac{1}{2}M_2(R/2)^2$, where M_2 is its mass. The mass per unit area of the two disks are the same and the area of a circle is proportional to the square of its radius, so the smaller disk has one fourth the mass of the larger disk. That is $M_2 = M_1/4$ so

$$I_{2,\text{cm}} = \frac{1}{2}\left(\frac{M_1}{4}\right)\left(\frac{R}{2}\right)^2 = \frac{1}{32}M_1R^2$$

To find the moment of inertia of the smaller disk about the axis through the center of the larger disk we use the parallel-axis theorem with the distance between the axes equal to $R/2$. This gives

$$I_2 = M_2h^2 + I_{2,\text{cm}} = \left(\frac{M_1}{4}\right)\left(\frac{R}{2}\right)^2 + \frac{1}{32}M_1R^2 = \frac{3}{32}M_1R^2$$

The moment of inertia of the two-disk composite is the sum of the two moments of inertia. That is

$$I = I_1 + I_2 = I_{1,\text{cm}} + I_2$$
$$= \frac{1}{2}M_1R^2 + \frac{3}{32}M_1R^2 = \frac{19}{32}M_1R^2$$

8. $I = \frac{1}{3}M_sL^2 + M_d\left(\frac{3}{4}L\right)^2 + \frac{1}{2}M_dR^2 = \underline{0.364 \text{ kg·m}^2}$

9. The moment of inertia about axis A is the sum of the moments of inertia of strips of width dx a distance x from A, shown in Figure 9-23.

Figure 9-23

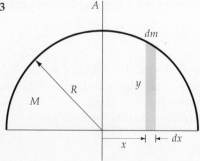

The differential mass dm of this strip is the product of the disk's mass per unit area σ and the differential area dA of the strip. Thus

$$dm = \sigma dA = \frac{M}{\pi R^2/2}y\,dx$$

The differential moment of inertia dI_A of the strip is the product of $x^2\,dm$, that is,

$$dI_A = x^2\,dm = \frac{2M}{\pi R^2}x^2y\,dx$$

The moment of inertia of the semicircle about axis A is the sum of the differential moments of inertia of all the strips that make up the disk. To sum differential quantities is to take an integral. Thus

$$I_A = \int_{-R}^{R} \frac{2M}{\pi R^2}x^2y\,dx$$

The integration variable is x whereas the integrand is a function of both x and y. In order to evaluate the integral we need the integrand to be a function of only x. The equation of the curved part of the semicircle is $y = \sqrt{R^2 - x^2}$, so to evaluate the integral we first substitute for y to get

$$I_A = \frac{2M}{\pi R^2}\int_{-R}^{R} x^2\sqrt{R^2 - x^2}\,dx = \frac{1}{4}MR^2$$

This is consistent with the result for a solid cylinder about its diameter through its center shown in Table 9-1, if we let $L \to 0$.

The moment of inertia of the semicircle about axis B is related to the moment of inertia about axis A by the parallel axis theorem. See Figure 9-24.

Figure 9-24

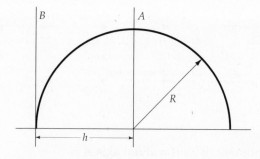

$I_{\text{cm}} = I_A$ because axis A, which is parallel to axis B, passes through the center of mass of the disk. Thus

$$I_B = Mh^2 + I_{\text{cm}} = MR^2 + I_A$$
$$= MR^2 + \frac{1}{4}MR^2 = \frac{5}{4}MR^2$$

The moment of inertia about axis C is determined from the moment of inertia of a semicircular strip of width dr a distance r from the axis, shown in Figure 9-25.

Figure 9-25

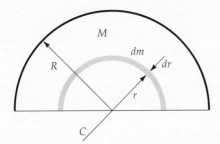

The differential mass dm of this strip is the product of the mass per unit area σ and the differential area dA of the strip. Thus

$$dm = \sigma dA = \frac{M}{\pi R^2/2}\,\pi r dr = \frac{2M}{R^2}\,r\,dr$$

where we have used $dA = \pi r dr$. This can be seen qualitatively by realizing that the strip's area is equal to that of a rectangle with length πr and width dr.

The differential moment of inertia dI_C of the strip is the product $r^2 dm$, that is,

$$dI_C = r^2 dm = r^2\,\frac{2M}{R^2}\,r dr = \frac{2M}{R^2}\,r^3\,dr$$

The moment of inertia of the semicircle about axis C is the sum of the differential moments of inertia of all the differential strips that make up the disk. To sum differential quantities is to take an integral. Thus

$$I_C = \frac{2M}{R^2}\int_0^R r^3\,dr = \frac{1}{2}\,MR^2$$

This is the same formula as is given for a whole disk (a solid cylinder) in Table 9-1. The mass distribution with respect to distance from the axis is the same for a half disk as for a whole disk so the formulas should be identical.

10. (a) $I_A = \frac{1}{6}\,Mb^2$

(b) $I_B = \frac{1}{6}\,Ma^2 = \frac{1}{24}\,Mb^2$

11. To analyze the motion of this system we apply Newton's second law to each block and Newton's second law for rotation to the pulley wheel. This means we draw a separate free-body diagram, shown in Figure 9-26, for the pulley wheel and for each block. Then we write down the laws in component form.

Figure 9-26

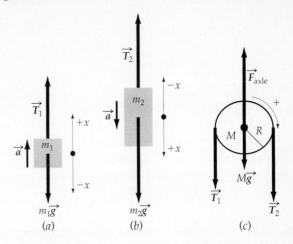

(a)   (b)   (c)

We apply Newton's second law to the block of mass m_1. This results in the equation

$$T_1 - m_1 g = m_1 a \tag{1}$$

We next apply Newton's second law to the block of mass m_2 and obtain

$$m_2 g - T_2 = m_2 a \tag{2}$$

And lastly we apply Newton's second law for rotations to the pulley disk. This gives

$$T_2 R - T_1 R = I\alpha \tag{3}$$

Equations 1, 2, and 3 have five unknowns T_1, T_2, I, a, and α, so before we can solve for the unknowns we must find additional equations. Because the string does not slip on the disk we have

$$a = R\alpha \tag{4}$$

and because the pulley wheel is a uniform disk we have

$$I = \tfrac{1}{2}MR^2 \tag{5}$$

Now we have five equations and five unknowns so we can proceed to solve for the desired unknowns a and α.

Substituting from Equation 4 into Equation 3 gives

$$T_2 R - T_1 R = I\frac{a}{R}$$

or

$$T_2 - T_1 = \frac{I}{R^2}\,a \tag{6}$$

Adding Equations 1 and 2 and rearranging gives

$$T_1 - T_2 = (m_1 - m_2)g + (m_1 + m_2)a \quad (7)$$

Adding Equations 6 and 7 gives

$$0 = (m_1 - m_2)g + (m_1 + m_2)a + \frac{1}{R^2}a$$

or

$$(m_2 - m_1)g = \left(m_1 + m_2 + \frac{I}{R^2}\right)a$$

Solving for the acceleration and then substituting for I from Equation 5 gives

$$a = \frac{m_2 - m_1}{m_1 + m_2 + \dfrac{I}{R^2}}g = \frac{m_2 - m_1}{m_1 + m_2 + \dfrac{1}{2}M}g$$

$$= \frac{(1.6 \text{ kg}) - (0.8 \text{ kg})}{(1.6 \text{ kg}) + (0.8 \text{ kg}) + 0.5\,(0.4 \text{ kg})}(9.81 \text{ m/s}^2)$$

$$= 3.02 \text{ m/s}^2$$

Solving for the angular acceleration using Equation 4 we get

$$\alpha = \frac{a}{R} = \frac{3.02 \text{ m/s}^2}{0.06 \text{ m}} = 50.3 \text{ rad/s}^2$$

Solving for T_1 and T_2 using Equations 1 and 2, respectively, we get

$$T_1 = m_1(g + a)$$
$$= (0.8 \text{ kg})(9.81 \text{ m/s}^2 + 3.02 \text{ m/s}^2) = \underline{10.3 \text{ N}}$$

and

$$T_2 = m_2(g - a)$$
$$= (1.6 \text{ kg})(9.81 \text{ m/s}^2 = 3.02 \text{ m/s}^2) = \underline{10.9 \text{ N}}$$

Follow-up: To verify the validity of our results we will substitute our calculated values into Equation 3.

$$T_2R - T_1R = I\alpha$$
$$(T_2 - T_1)R = \tfrac{1}{2}MR^2\alpha$$
$$(10.9 \text{ N} - 10.3 \text{ N})(0.06 \text{ m})$$
$$= \tfrac{1}{2}(0.4 \text{ kg})(0.06\text{m})^2(50.3 \text{ rad/s}^2)$$
$$3.6 \times 10^{-2} \text{ N·m} = 3.6 \times 10^{-2} \text{ kg·m}^2/\text{s}^2$$

Since $1 \text{ N} = 1 \text{ kg·m/s}^2$ we know our calculated values are consistent with Equation 3 (Newton's second law for rotation as applied to the pulley wheel).

12. (a) $\alpha = \underline{15.7 \text{ rad/s}^2 \text{ clockwise}}$
$a_1 = \underline{0.63 \text{ m/s}^2 \text{ down}}$
$a_2 = \underline{1.26 \text{ m/s}^2 \text{ up}}$
(b) $T_1 = \underline{16.7 \text{ N}}$ $T_2 = \underline{8.55 \text{ N}}$

13. The string does not slip so no energy is dissipated via friction and the rolling-without-slipping condition holds. It follows that there is no change in the mechanical energy of the system, so the final mechanical energy is equal to the initial mechanical energy.

$$K_{\text{trans f}} + K_{\text{rot f}} + U_{\text{g f}} = K_{\text{trans i}} + K_{\text{rot i}} + U_{\text{g i}}$$
$$\tfrac{1}{2}MV_{\text{cm f}}^2 + \tfrac{1}{2}I_{\text{cm}}\omega_f^2 + MgY_{\text{cm f}}$$
$$= \tfrac{1}{2}MV_{\text{cm i}}^2 + \tfrac{1}{2}I_{\text{cm}}\omega_i^2 + MgY_{\text{cm i}} \quad (1)$$

The wheel is released from rest so $V_{\text{cm i}} = \omega_i = 0$, and we will choose the zero of the y coordinate axis to be at the initial height so $Y_{\text{cm i}} = 0$, as shown in Figure 9-27.

Figure 9-27

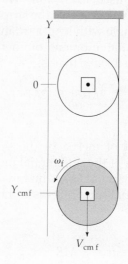

Substituting these values into Equation (1) gives

$$\tfrac{1}{2}MV_{\text{cm f}}^2 + \tfrac{1}{2}I_{\text{cm}}\omega_f^2 + MgY_{\text{cm f}} = 0 + 0 + 0 = 0$$

It is given that $V_{\text{cm f}} = 4.74 \text{ m/s}$, $Y_{\text{cm f}} = -2.00 \text{ m}$, and $R = 0.06 \text{ m}$. We are to solve for I_{cm}, and ω_f is an unknown. However, we eliminate ω_f by using the rolling-without-slipping condition $V_{\text{cm f}} = R\omega_f$. Substituting for ω_f gives

$$\tfrac{1}{2}MV_{\text{cm f}}^2 + \tfrac{1}{2}I_{\text{cm}}\frac{V_{\text{cm f}}^2}{R^2} + MgY_{\text{cm f}} = 0$$

Solving for I_{cm} we obtain

$$I_{cm} = MR^2\left(1 + \frac{2gY_{cm\,f}}{V_{cm\,f}^2}\right)$$

$$= -(2.00\text{ kg})(0.06\text{ m})^2$$

$$\left(1 + \frac{2(9.81\text{ m/s}^2)(-2.00\text{ m})}{(4.74\text{ m/s})^2}\right)$$

$$= \underline{5.37 \times 10^{-3}\text{ kg·m}^2}$$

Follow-up: The moment of inertia of a wheel like the one in this problem should be greater than that for a uniform solid disk of the same mass and radius and less than that for a thin cylindrical shell. That is, its moment of inertia should be given by $I_{cm} = \beta MR^2$, where $\frac{1}{2} < \beta < 1$. For our result

$$\beta = \frac{I_{cm}}{MR^2} = \frac{5.37 \times 10^{-3}\text{ kg·m}^2}{(2.00\text{ kg})(0.06\text{ m})^2} = 0.747$$

a value close to the center of the acceptable range.

14. $\dfrac{7v_0^2}{10g}$

15. The linear motion and the rotational motion, shown in Figure 9-28, are related. We analyze the linear motion via the relation $\Sigma F = ma_{cm}$, while we analyze the rotational motion via the relation $\Sigma \tau_{cm} = I_{cm}\alpha$. These parallel analyses are listed below. For convenience clockwise has been taken as positive.

Figure 9-28

Linear Motion	Rotational Motion
$\Sigma F_x = ma_{cm\,x}$	$\Sigma \tau_{cm} = I_{cm}\alpha$
$\mu_k mg = ma_{cm}$	$-\mu mgR = \frac{2}{5}mR^2\alpha$
$a_{cm} = \mu_k g$	$\alpha = -\dfrac{5\mu_k g}{2R}$
$v_{cm} = v_{0\,cm} + a_{cm}t$	$\omega = \omega_0 + \alpha t$
$v_{cm} = 0 + \mu_k gt$	$\omega = \omega_0 - \dfrac{5\mu_k g}{2R}t$

As time increases, the velocity of the center of mass increases and the angular velocity decreased.

The slipping will continue until $v_{cm} = R\omega$. To determine when this condition is reached we substitute into $v_{cm} = R\omega$ and solve for the time. That is

$$\mu_k gt = R\left(\omega_0 - \frac{5\mu_k g}{2R}t\right)$$

Solving for the time we get

$$t = \frac{2R\omega_0}{7\mu_k g}$$

$$= \frac{2(0.10\text{ m})\left(120\,\dfrac{\text{rev}}{\text{min}}\right)\left(\dfrac{2\pi\text{ rad}}{\text{rev}}\right)\left(\dfrac{1\text{ min}}{60\text{ s}}\right)}{7\,(0.80)(9.81\text{ m/s}^2)} = \underline{46\text{ ms}}$$

Substituting these values for t back into the expressions for v_{cm} and ω we find that by the time the skidding ceases, the ball is moving at

$$v_{cm} = 0.80\,(9.8\text{ m/s}^2)(0.046\text{ s}) = \underline{0.36\text{ m/s}}$$

and is rotating at

$$\omega = v_{cm}/R = \frac{0.36\text{ m/s}}{0.10\text{ m}} = 3.6\text{ rad/s (34 rev/min)}$$

During the first 46 ms the ball moves a distance of

$$\Delta x_{cm} = v_{avg}\,\Delta t = \frac{(v_0 + v)}{2}\,\Delta t$$

$$= \frac{(0.36\text{ m/s})}{2}(0.046\text{ s}) = \underline{0.0083\text{ m or 8.3 mm}}$$

while rotating through an angle of

$$\Delta\theta = \omega_{avg}\,\Delta t = \frac{(\omega_0 + \omega)}{2}\,\Delta t$$

$$= \frac{(12.6\text{ rad/s} + 3.6\text{ rad/s})}{2}(0.046\text{ s})$$

$$= \underline{0.37\text{ rad or 21.3°}}$$

Summary: During the first 46 ms the ball is spinning too fast to be rolling without slipping. During this interval it moves forward 8.3 mm and rotates through 21°, gaining a linear speed of 36 cm/s and reducing its angular speed from 120 rev/min to 34 rev/min. At $t = 46$ ms it acquires the rolling-without-slipping condition $v_{cm} = R\omega$ and continues to roll without slipping with a constant linear speed of 36 cm/s.

16. $\underline{49°}$

Chapter 10

Conservation of Angular Momentum

I. Key Ideas

In this chapter we'll continue to study the rotational motion of many-particle systems, and emphasize the vector aspects of such motion. The vector nature of angular velocity is presented in Section 1 of Chapter 9 of this Study Guide. Here we will introduce the vector nature of torque and angular momentum. Conservation of angular momentum is a fundamental law of nature.

10-1 *The Vector Nature of Rotation*   For a rotating object the angular velocity vector $\vec{\omega}$ is directed parallel to the rotation axis in the direction specified via a right-hand rule shown in Figure 9-2. (If you curl your fingers in the direction of rotation, then $\vec{\omega}$ is in the same direction as your thumb, using your right hand of course!)

The Cross Product   Torque $\vec{\tau}$ is mathematically best expressed as the **cross product** (or **vector product**) of $\vec{r}$ and $\vec{F}$. That is $\vec{\tau} = \vec{r} \times \vec{F}$, where $\vec{r}$ is the vector from the origin to the point of application of force $\vec{F}$.

The cross product of any two vectors $\vec{A}$ and $\vec{B}$ is perpendicular to both vectors. The mathematical convention for determining the direction of a cross product, illustrated in Figure 10-1, is called the *right-*

hand rule. The cross product is defined to be perpendicular to both vectors. However, the right-hand rule is needed to completely specify its direction. To apply this rule, place the fingers of your right hand in the direction of the first vector ($\vec{A}$) such that when you make a fist they curl toward the direction of the second vector ($\vec{B}$). Your thumb then shows the direction of $\vec{A} \times \vec{B}$.

If ϕ is the angle between the directions of the two vectors, and $\hat{n}$ is the unit vector in the direction given by the right-hand rule, then

Definition—Cross (or vector) product

$$\vec{C} = \vec{A} \times \vec{B} = (AB \sin \phi)\hat{n}$$

It follows that the cross product of any vector with itself is zero and that $\vec{A} \times \vec{B} = -\vec{B} \times \vec{A}$. Thus, for the unit vectors $\hat{i}$, $\hat{j}$, and $\hat{k}$ we have $\hat{i} \times \hat{i} = \hat{j} \times \hat{j} = \hat{k} \times \hat{k} = 0$ and

$$\hat{i} \times \hat{j} = \hat{k} \qquad \hat{j} \times \hat{i} = -\hat{k}$$
$$\hat{j} \times \hat{k} = \hat{i} \qquad \hat{k} \times \hat{j} = -\hat{i}$$
$$\hat{k} \times \hat{i} = \hat{j} \qquad \hat{i} \times \hat{k} = -\hat{j}$$

The signs of cross products such as $\hat{k} \times \hat{j}$ can be found by using the graphic shown in Figure 10-2. If the order of the factors is such that you travel in the direction of the arrows (counterclockwise) around the circle, then the sign of the product is positive. If you travel in the clockwise direction, then the sign of the product is negative. Check it out. For the product $\hat{j} \times \hat{k}$ you travel counterclockwise on the circle, so $\hat{j} \times \hat{k} = +\hat{i}$. However, for the product $\hat{k} \times \hat{j}$ you travel clockwise on the circle, so $\hat{k} \times \hat{j} = -\hat{i}$.

Figure 10-1

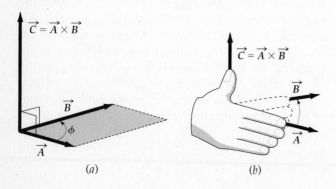

(a)   (b)

Figure 10-2

Using these rules for multiplying the unit vectors $\hat{i}$, $\hat{j}$, and $\hat{k}$ we can show

Cross product using $\hat{i}$, $\hat{j}$, and $\hat{k}$

$$\vec{C} = \vec{A} \times \vec{B}$$
$$= (A_y B_z - A_z B_y)\hat{i} + (A_z B_x - A_x B_z)\hat{j}$$
$$+ (A_x B_y - A_y B_x)\hat{k}$$

That is, $C_x = A_y B_z - A_z B_y$, $C_y = A_z B_x - A_x B_z$, and $C_z = A_x B_y - A_y B_x$. *Note:* In each of these three equations the subscript on the left side does not appear on the right side. Also, in each equation the unit vector corresponding to the first subscript equals the cross product of the unit vectors corresponding to the next two. For example, the first three subscripts in the equation $C_x = A_y B_z - A_z B_y$ are x, y, and z; the corresponding unit vectors are $\hat{i}$, $\hat{j}$, and $\hat{k}$, and $\hat{i} = \hat{j} \times \hat{k}$. By remembering this you can quickly calculate the cross product of two vectors.

10-2 *Angular Momentum* The angular momentum $\vec{L}_i$ of a moving particle is given by $\vec{r}_i \times \vec{p}_i$ where $\vec{r}_i$ is the particle's position vector and $\vec{p}_i$ is its linear momentum $m_i \vec{v}_i$. Like torque, angular momentum is defined relative to a point in space, rather then relative to an axis (line). The concept of angular momentum can be extended from that of a single particle to that of a system of particles where

Definition—Angular momentum of a particle

$$\vec{L}_i = \vec{r}_i \times \vec{p}_i$$

Angular momentum of a system

$$\vec{L} = \sum_i \vec{L}_i$$

Definition—Torque due to a force

$$\vec{\tau}_i = \vec{r}_i \times \vec{F}_i$$

Net torque on a system

$$\vec{\tau}_{net,ext} = \sum_i \vec{\tau}_i$$

It is often helpful to express the total angular momentum of a system as the sum of two terms: the orbital angular momentum $\vec{L}_{orbit}$, associated with the motion of the center of mass, and the spin angular momentum $\vec{L}_{spin}$, associated with any motion relative to the center of mass. That is,

Orbital and spin angular momentum

$$\vec{L} = \vec{L}_{orbit} + \vec{L}_{spin}$$
$$= \vec{r}_{cm} \times M\vec{V}_{cm} + \Sigma \vec{r}'_i \times m_i \vec{u}_i$$

where $\vec{L}_{orbit} = \vec{r}_{cm} \times M\vec{V}_{cm}$, $\vec{L}_{spin} = \Sigma \vec{r}'_i \times m_i \vec{u}_i$, $\vec{r}_{cm}$ is the center-of-mass position vector, and $\vec{r}'_i$ and $\vec{u}_i$ are the position and velocity of the ith particle relative to the center of mass. For example, the total angular momentum of the earth can be expressed as the sum of its orbital angular momentum, due to its orbital motion around the sun, plus its spin angular momentum, due to its spinning motion about an axis through its center of mass.

The impulse–momentum theorem relates the change in a particle's linear momentum to the net force acting on it. The rotational analog of this theorem is the angular version

Impulse–momentum theorem (angular version)

$$\Delta \vec{L} = \int_{t_i}^{t_f} \vec{\tau}_{net,ext}\, dt$$

Here torque and angular momentum are defined as vector quantities relative to a *point* rather than an axis. In the equations above, the $\Sigma \vec{\tau}$ terms refer to the sum of all torques, both internal and external. However since the internal torques sum to zero, this sum can be replaced by a sum over only the external torques. It follows from Newton's laws that

Newton's second law (angular version)

$$\vec{\tau}_{net,ext} = \frac{d\vec{L}}{dt}$$

For a symmetric rigid object that is rotating with angular velocity $\vec{\omega}$ about a fixed axis the angular momentum $\vec{L}$ is given by

Angular momentum for a symmetric rigid object about a fixed axis

$$\vec{L} = I\vec{\omega}$$

where I is the moment of inertia. Substituting this expression for angular momentum into the angular version of Newton's second law gives

Newton's second law for a rigid body about a fixed axis

$$\vec{\tau}_{\text{net,ext}} = I\vec{\alpha}$$

For an object rotating about a fixed axis, the kinetic energy K is given by $K = \frac{1}{2}I\omega^2$. Substituting for ω in this equation gives

Kinetic energy of a rigid object rotating about a fixed axis

$$K = \frac{L^2}{2I}$$

Gyroscopic motion illustrates some of the non-intuitive vector properties of rotational motion. A spinning gyroscope moves so the change in its angular momentum vector is in the same direction as the net external torque vector.

10-4   Conservation of Angular Momentum   If the net external torque $\vec{\tau}_{\text{net,ext}}$ remains zero, then the system's angular momentum $\vec{L}$ is constant. This follows from the angular version of Newton's second law and is called the *law of conservation of angular momentum*. Like the conservation of linear momentum this law is universally valid.

10-5   Quantization of Angular Momentum   One of the strange discoveries in physics that took place during the early twentieth century is that, like energy, angular momentum is quantized with the fundamental unit of angular momentum $\hbar$ (read h-bar) being Plank's constant h divided by 2π.

II. Numbers and Key Equations

Numbers

Fundamental unit of angular momentum

$$\hbar = \frac{h}{2\pi} = 1.05 \times 10^{-34} \text{ J·s}$$

Key Equations

Definition—Cross (or vector) product

$$\vec{C} = \vec{A} \times \vec{B} = (AB\sin\phi)\hat{n}$$

Cross product using $\hat{i}$, $\hat{j}$, and $\hat{k}$

$$\vec{C} = \vec{A} \times \vec{B}$$
$$= (A_yB_z - A_zB_y)\hat{i} + (A_zB_x - A_xB_z)\hat{j}$$
$$+ (A_xB_y - A_yB_x)\hat{k}$$

Definition—Angular momentum of a particle

$$\vec{L}_i = \vec{r}_i \times \vec{p}_i$$

Angular momentum of a system

$$\vec{L} = \sum_i \vec{L}_i$$

Definition—Torque due to a force

$$\vec{\tau}_i = \vec{r}_i \times \vec{F}_i$$

Net torque on a system

$$\vec{\tau}_{\text{net,ext}} = \sum_i \vec{\tau}_i$$

Orbital and spin angular momentum

$$\vec{L} = \vec{L}_{\text{orbit}} + \vec{L}_{\text{spin}}$$
$$= \vec{r}_{\text{cm}} \times M\vec{V}_{\text{cm}} + \Sigma\vec{r}'_i \times m_i\vec{u}_i$$

Impulse-momentum theorem (angular version)

$$\Delta\vec{L} = \int_{t_i}^{t_f} \vec{\tau}_{\text{net,ext}} \, dt$$

Newton's second law (angular version)

$$\vec{\tau}_{\text{net,ext}} = \frac{d\vec{L}}{dt}$$

Angular momentum for a symmetric rigid object about a fixed axis

$$\vec{L} = I\vec{\omega}$$

Newton's second law for a rigid body about a fixed axis

$$\vec{\tau}_{\text{net,ext}} = I\vec{\alpha}$$

Kinetic energy of a rigid object rotating about a fixed axis

$$K = \frac{L^2}{2I}$$

Conservation of angular momentum

If $\vec{\tau}_{\text{net,ext}} = 0$, then $\vec{L}_i = \vec{L}_f$

Quantization of orbital angular momentum of a particle

$$L = \sqrt{\ell(\ell + 1)}\hbar$$
$$\ell = 0, 1, 2, \ldots$$

III. Potential Pitfalls

Don't think the fact that angular velocity is a vector is too abstract to be useful. The direction of the vector specifies the direction of the rotation axis—often an important consideration.

Don't think that for an object rotating about a fixed axis, the angular velocity vector and angular momentum vector must be in the same direction. The angular momentum vector, unlike the angular velocity vector, depends on the mass distribution. Only for sufficiently symmetric mass distributions are the two in the same direction.

Don't think that conservation of angular momentum is useful only if the net torque on the system remains zero. For example, the z component of the angular momentum is conserved as long as the z component of the net torque remains zero, even if the other components of the net torque are nonzero.

IV. True and False and Responses

True and False

_____ 1. The vector product of two vectors is always perpendicular to both vectors.

_____ 2. A force directed toward a point exerts zero torque about that point.

_____ 3. The vector product is not commutative.

_____ 4. Both angular momentum vectors and torque vectors are defined relative to some particular origin or reference point.

_____ 5. The angular momentum vector is always parallel to the net torque on the body.

_____ 6. The angular momentum of a system of particles depends on the motion of the center of mass, but not on the motion of the particles relative to the center of mass.

_____ 7. A particle of mass m moving in a circle of radius R at a speed v has an angular momentum with a magnitude of mvR about *any* point.

_____ 8. If the total linear momentum of an arbitrary system of particles is conserved, the total angular momentum must also be conserved.

_____ 9. A gyroscope precesses, rather than falling over, because its spin counteracts the force of gravity.

_____ 10. A gyroscope supported at its center of mass will not precess under the influence of gravity.

_____ 11. Both the angular momentum and the angular velocity of a body that is rotating about a symmetry axis are parallel.

_____ 12. A rotating wheel is balanced if the axis of rotation passes through its center of mass.

Responses

1. True

2. True

3. True. $\vec{A} \times \vec{B} \neq \vec{B} \times \vec{A}$

4. True

5. False. The rate of change of the angular momentum vector is always parallel to the net torque.

6. False. This omits any angular momentum the system may have about the center of mass; that is, it omits the spin angular momentum.

7. False. Only about the center of the circle is the magnitude of the angular momentum equal to mvR.

8. False. A pair of non colinear (not acting along the same line) forces that are equal in magnitude but opposite in direction produce a net torque but not a net force. These forces will change the system's angular momentum, but not its linear momentum.

9. False. Spin isn't a force. For a spinning gyroscope supported at a point, it is the force of the support exerted on the gyroscope that prevents it from falling. The force of the support is equal in magnitude and opposite in direction to the gravitational force. However, they do not necessarily act along the same line unless the gyroscope is supported at a point along the vertical line through its center of mass. The gyroscope precesses as a result of the torque due to the gravitational force.

10. True. In that case, there is no gravitational torque.

11. True

12. False. This does not guarantee dynamic balance—only static balance.

V. Questions and Answers

Questions

1. If the polar ice caps melted tomorrow, what would be the effect on the earth's rotation rate?

2. A basketball thrown against a closed door causes it to rotate open. The door is stationary before the collision and rotating after it, so during the collision it clearly gains angular momentum. Where did this angular momentum come from? The ball didn't lose any, or did it?

3. The propeller of a light airplane is rotating clockwise, as seen from someone observing it from behind the plane. If the pilot pulls back the stick the flaps on the tail fins are raised. This results in the air pushing down on the plane's tail and, thus, causing the tail to lower and nose to raise. However, exerting a torque on an object with angular momentum causes it to precess. When the tail flaps are up which way, port or starboard, will the plane precess? (Port is to the left and starboard is to the right of a person facing the same direction as the plane.)

Answers

1. If all the water now frozen at the earth's poles were distributed over the earth's oceans, the earth's moment of inertia would increase. Its rate of rotation would have to decrease for its angular momentum to remain constant.

2. The ball does lose angular momentum during the collision. Its angular momentum is given by $\vec{L} = \vec{r} \times M\vec{V}_{cm}$, where $\vec{r}$ is the vector from the door's rotation axis to the ball, M is the ball's mass, and $\vec{V}_{cm}$ is the velocity of its center of mass. During the collision $\vec{V}_{cm}$ decreases so the angular momentum of the ball decreases. This decrease equals the increase in the door's angular momentum.

3. The tendency for the plane to turn sideways (yaw) is a gyroscopic effect. The propellers of single-engine aircraft of U.S. manufacture rotate clockwise as seen from the rear. In accord with the right-hand rule, the angular momentum vector $\vec{L}$ of the propeller points in the forward direction, parallel to the propeller shaft. The downward force of the air on the tail of the plane results in a vector torque about the center of mass in the starboard direction. This causes the plane to precess, slowly turning to starboard so that $\Delta\vec{L}$ is in the same direction as the net torque.
   Remark: To prevent this rotation the pilot uses the rudder to deflect the air to the port. The air then exerts a starboard force on the rudder which tends to turn the nose of the plane to port.

VI. Problems and Solutions

Problems

1. Consider the two vector quantities $\vec{A} = 2.4\hat{i} + 3.2\hat{j} + 3\hat{k}$ and $\vec{B} = 4.5\hat{i} + 6\hat{j}$. (a) Find their vector product $\vec{C} = \vec{A} \times \vec{B}$ directly from the components of $\vec{A}$ and $\vec{B}$. (b) Use your part (a) result to verify that $|\vec{A} \times \vec{B}| = AB \sin \phi$.

How to Solve It
• In part (b) use the scalar (or dot) product to find the angle.

2. Consider the two vector quantities $\vec{A} = 2.4\hat{i} + 3.2\hat{j} - 3\hat{k}$ and $\vec{B} = 3\hat{i} + 4\hat{j}$. Find the angle between $\vec{A} \times \vec{B}$ and $\vec{C} = 3\hat{i} + 4\hat{j}$.

How to Solve It
• Take the dot product of $\vec{A} \times \vec{B}$ and $\vec{C}$.

3. A particle released from rest falls freely under the influence of only gravity. Pick a coordinate system with an origin not on the line of fall and show explicitly that $\Sigma\vec{\tau} = d\vec{L}/dt$ holds for this motion.

How to Solve It
• Express the position, force, and linear momentum in terms of m, g, v, t, and the unit vectors $\hat{i}$, $\hat{j}$, and $\hat{k}$. Substitute into $\Sigma\vec{\tau} = d\vec{L}/dt$ and turn the math crank.

4. A particle, projected with an initial horizontal velocity, falls under the influence of only gravity. Pick a coordinate system with an origin not on the line of fall and show explicitly that $\Sigma\vec{\tau} = d\vec{L}/dt$ holds for this motion.

How to Solve It
• The only force is the weight which equals $m\vec{g}$.

• Draw the position vector $\vec{r}$ from the orgin to the particle, write out expressions for the torque vector $\vec{\tau}$ and the angular momentem vector $\vec{L}$, and take the time derivitive of the angular momentum vector.

• If you express the angular momentum vector in component form using unit vectors you may find it easier.

5. A 2.00-kg rabbit is sitting on a small, stationary merry-go-round 70 cm from the axis of rotation. The merry-go-round has a moment of inertia about the axis of 3.00 kg·m². The rabbit becomes startled and executes a horizontal jump in a direction perpendicular with the radial direction. This leaves the merry-go-round with an angular velocity of 4.00 rad/s. How fast is the rabbit moving relative to the ground just after her feet leave the merry-go-round? Assume that air resistance and friction in the merry-go-round's bearings are negligible.

How to Solve It
• The net external torque on the rabbit–merry-go-round system about the axis is zero. It follows that the system's angular momentum is conserved.

• The pre-jump angular momentum of the system is zero. Equate the magnitudes of the post-jump angular momenta of the rabbit and the merry-go-round.

6. A merry-go-round consists of a circular platform mounted on a frictionless vertical axis. It has a 2.40-m radius and a 210-kg·m/s² moment of inertia and is

rotating with an angular velocity of 2.40 rad/s with a 38.0-kg boy standing on it next to its edge. The boy then runs around its perimeter, running at a constant speed. As he comes up to speed the platform's angular speed decreases to 1.91 rad/s. How fast is the boy moving relative to the platform?

How to Solve It
- There are no external torques on the boy–platform system so its angular momentum remains constant. Find the initial angular momentum of the boy–platform system.

- When the boy is running obtain an expression for the angular momentum of the system and set it equal to the initial angular momentum. Solve for the speed of the boy relative to the ground.

- Calculate the speed of the edge of the platform. Knowing it and the boy's speed relative to the ground, find his speed relative to the platform.

7. Consider the sun as a uniform sphere of radius 1.39×10^9 m and mass 1.99×10^{30} kg, rotating on its axis with a period of 26 days. If, at the end of the sun's life, all the mass collapses inward to form a neutron star of radius 16 km, what will its period be?

How to Solve It
- The net external torque on the system is zero, thus angular momentum is conserved. Equate the values of the angular momentum of the system before and after the collapse.

- Use the relation between the angular speed and the period.

8. A student standing on a frictionless, horizontal platform with her arms extended out from her sides is completing one rotation every 2.40 s. She raises her arms straight up over her head and her rate of rotation increases to once every 0.80 s. Determine the percentage by which the moment of inertia decreased when she raised her arms.

How to Solve It
- The net external torque about a vertical axis is zero, thus angular momentum about this axis is conserved. Equate the values of the angular momentum of the system before and after she raises her arms.

9. In Figure 10-3, an 80-kg man runs up and jumps onto the bottom of a stationary Ferris wheel whose moment of inertia is 1300 kg·m². The radius of the wheel is 2.40 m and friction on its axle is negligible. After the man jumps on, the wheel rotates freely, swinging him up above the ground. With what minimum initial speed v_0 must he run if it swings him over the top?

Figure 10-3

How to Solve It
- The man–Ferris wheel collision is inelastic, so the total mechanical energy of this system is not conserved.

- During the collision, the angular momentum of the man–Ferris-wheel system is conserved. Equate the values of the angular momentum of the system immediately prior to and immediately following the collision.

- Following the collision, mechanical energy is conserved. Equate the final mechanical energy with the mechanical energy immediately following the collision.

10. Tarzan, the ape man, swings down from a tree branch to kick a bad guy in the head. In your calculations, model Tarzan as a uniform thin rod 1.90 m long with a mass of 91 kg. When he starts swinging, his body makes an angle of 40° with the vertical, as shown in Figure 10-4. At what speed are Tarzan's feet moving when they whack the bad guy in the head, which is at the bottom of the swing?

Figure 10-4

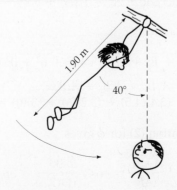

How to Solve It

• Because no work is done by forces other than gravity, Tarzan's total mechanical energy (potential plus kinetic) is conserved.

• Set the initial mechanical energy equal to the mechanical energy at the bottom of the swing the instant before Tarzan's feet hit the bad guy's head. Remember, Tarzan's gravitational potential energy depends on the height of his center of mass, and his kinetic energy depends on his moment of inertia and his angular speed. Determine his angular speed at the bottom of the swing.

• Determine the speed of Tarzan's feet just prior to impact.

Solutions

1. (*a*). The vector product $\vec{C} = \vec{A} \times \vec{B}$ computed directly from components is

$$\vec{C} = \vec{A} \times \vec{B}$$
$$= (A_y B_z - A_z B_y)\hat{i} + (A_z B_x - A_x B_z)\hat{j}$$
$$\qquad + (A_x B_y - A_y B_x)\hat{k}$$
$$= (3.2 \cdot 0 - 3 \cdot 6)\hat{i} + (3 \cdot 4.5 - 2.4 \cdot 0)\hat{j}$$
$$\qquad + (2.4 \cdot 6 - 3.2 \cdot 4.5)\hat{k}$$
$$= -18\hat{i} + 13.5\hat{j}$$

(*b*). We are asked to verify $|\vec{A} \times \vec{B}| = AB \sin \phi$ using the result from part (*a*). We calculate $|\vec{A} \times \vec{B}|$ using the Pythagorean theorem. This gives

$$|\vec{A} \times \vec{B}| = |-18\hat{i} + 13.5\hat{j}|$$
$$= \sqrt{(-18)^2 + 13.5^2} = 22.5 \qquad (1)$$

Our task is to show that $AB \sin \phi$ also equals 22.5. To accomplish this we find the angle ϕ between $\vec{A}$ and $\vec{B}$ using the dot product relation:

$$\vec{A} \cdot \vec{B} = AB \cos \phi \qquad (2)$$

To solve this for the angle ϕ we first find A, B, and $\vec{A} \cdot \vec{B}$ from the components. That is

$$A = \sqrt{A_x^2 + A_y^2 + A_z^2} = \sqrt{2.4^2 + 3.2^2 + 3^2} = 5$$

and

$$B = \sqrt{B_x^2 + B_y^2 + B_z^2} = \sqrt{4.5^2 + 6^2 + 0^2} = 7.5$$
$$\vec{A} \cdot \vec{B} = A_x B_x + A_y B_y + A_z B_z$$
$$= (2.4)(4.5) + (3.2)(6) + (3)(0) = 30$$

Solving Equation (2) for ϕ gives

$$\phi = \cos^{-1} \frac{\vec{A} \cdot \vec{B}}{AB} = \cos^{-1} \frac{30}{(5)(7.5)} = 36.9°$$

Substituting our calculated values for A, B, and ϕ we obtain

$$AB \sin \phi = (5)(7.5)\sin 36.9° = 22.5$$

which is in agreement with our calculation of $|\vec{A} \times \vec{B}|$ in Equation (1).

2. 90°

3. The torque $\vec{\tau}$ and the angular momentum $\vec{L}$ are given by

$$\vec{\tau} = \vec{r} \times \vec{F} \qquad \text{and} \qquad \vec{L} = \vec{r} \times \vec{p}$$

where $\vec{r} = x\hat{i} + y\hat{j}$, $\vec{p} = -mv\hat{j}$, and $\vec{F} = -mg\hat{j}$, as shown in Figure 10-5.

Figure 10-5

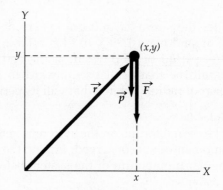

The angular momentum is then

$$\vec{L} = \vec{r} \times \vec{p} = (x\hat{i} + y\hat{j}) \times (-mv\hat{j}) = -xmv\hat{k}$$

Differentiating both sides of this equation with respect to time yields

$$\frac{d}{dt}\vec{L} = \frac{d}{dt}(-xmv\hat{k}) = -xm\frac{dv}{dt}\hat{k} = -xmg\hat{k}$$

Because the particle falls straight down, x is constant, and because it is, free fall $dv/dt = g$. We now compute the torque about the origin caused by the sole force of gravity acting on the particle. From the definition of torque we have

$$\vec{\tau} = \vec{r} \times \vec{F} = (x\hat{i} + y\hat{j}) \times (-mg\hat{j}) = -xmg\hat{k}$$

which is identical with the result obtained above for $d\vec{L}/dt$. Thus we have shown that $\Sigma\vec{\tau} = d\vec{L}/dt$.

4. Try modeling your solution after the solution to Problem 4, except for this problem the linear momentum is expressed as $\vec{p} = mv_x\hat{i} + mv_y\hat{j}$.

5. In this problem there are no external torques about the axis so the angular momentum is conserved. Initially the system is at rest, so its angular momentum is zero. After the rabbit jumps (see Figure 10-6) both the rabbit and the merry-go-round do have angular momentum, one is clockwise and the other counterclockwise. The magnitude of the angular momentum of the rabbit is

$$L_{\text{rabbit}} = |\vec{r} \times m\vec{v}| = rmv \sin \phi$$
$$= mv\,(r \sin \phi) = mvr_\perp$$

where $r_\perp$ is the distance between the axis and the rabbit's line of motion, $\vec{r}$ is its position relative to the axis, and m and $\vec{v}$ are its mass and velocity.

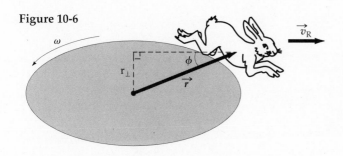

Figure 10-6

The magnitude of the angular momentum of the merry-go-round is $I\omega$ where I is its moment of inertia and ω its angular speed. Setting the angular momenta equal to each other and solving for the speed v gives

$$r_\perp mv = I\omega$$
$$v = \frac{I\omega}{r_\perp m} = \frac{(3.00 \text{ kg·m}^2)(4.00 \text{ rad/s})}{(0.70 \text{ m})(2.00 \text{ kg})} = \underline{8.57 \text{ m/s}}$$

6. $\underline{2.31 \text{ m/s}}$

7. There are no significant external torques acting on the sun, therefore its angular momentum about its rotation axis is conserved. Thus

$$L = I_1\omega_1 = I_2\omega_2$$

The angular speed ω of a rotating body equals $2\pi/T$, where T is the period (it rotates 2π radians each period). Expressing the angular speeds in terms of the period gives

$$I_1\frac{2\pi}{T_1} = I_2\frac{2\pi}{T_2}$$

The moment of inertia of a uniform sphere of mass M and radius R is $2/5\ MR^2$. Solving for the final period of the sun gives

$$T_2 = \frac{I_2}{I_1}T_1 = \frac{\frac{2}{5}MR_2^2}{\frac{2}{5}MR_1^2}T_1 = \left(\frac{R_2}{R_1}\right)^2 T_1$$
$$= \left(\frac{1.6 \times 10^4 \text{ m}}{1.39 \times 10^9 \text{ m}}\right)^2 26 \text{ days}\left(\frac{24 \text{ h}}{\text{day}}\right)\left(\frac{3600 \text{ s}}{\text{h}}\right)$$
$$= \underline{0.30 \text{ ms}}$$

(*Note:* We assumed that the mass of the sun did not change.)

8. $\underline{67\%}$

9. During the collision, the angular momentum of the man–Ferris-wheel system is conserved. The wheel is initially at rest so its angular momentum is zero. When the man is running, the perpendicular distance $r_\perp$ between the line of his motion and the rotation axis equals the radius R of the wheel. Immediately following the collision the man–wheel system rotates as a single rigid body with an angular velocity ω_1. The moment of inertia I of the man–wheel system is that of the wheel plus that of the man. Thus, the angular momentum of this man–wheel system is $(I_w + mR^2)\omega_1$, where I_w is the moment of inertia of the wheel alone, m is the mass of the man, and ω_1 is the angular speed immediately after the collision—before the wheel has had time to rotate much at all. Equating the pre-collision and post-collision angular momenta we have

$$L_0 = L_1$$
$$Rmv_0 = I\omega_1 \qquad (1)$$

After the collision, as the wheel rotates while raising the man, the mechanical energy (kinetic plus potential) of the system is conserved. We will now equate the mechanical energy immediately after the collision with the mechanical energy when George is at the top of the wheel, which has come to rest.

Following the collision George and the Ferris wheel rotate as a single object. The kinetic energy of a rotating object is $K = \frac{1}{2}I\omega^2$. As the wheel rotates, its center of mass remains at rest so its gravitational potential energy remains constant. However, George's gravitational potential energy is given by the formula $U_g = mgh$. Equating the mechanical energy immediately following the collision with the mechanical energy when George has swung up a height h gives

$$K_1 + U_{g\,1} = K_2 + U_{g\,2}$$
$$\tfrac{1}{2}I\omega_1^2 + mgh_1 = \tfrac{1}{2}I\omega_2^2 + mgh_2$$
$$\frac{1}{2}I\left(\frac{Rmv_0}{I}\right)_1^2 + 0 = 0 + mgh$$
$$\frac{(Rmv_0)^2}{2I} = mgh$$

where, using Equation (1), Rmv_0/I has been substituted for ω_1.

To determine the minimum initial speed v_0 needed for George to swing up to the top of the wheel, we substitute $I_w + mR^2$ for I, set $h = 2R$, and solve for v_0. This gives

$$v_0^2 = 4gR\left(\frac{I_w}{mR^2} + 1\right)$$

$$= 4(9.81 \text{ m·s}^2)(2.4 \text{ m})\left(\frac{1300 \text{ kg·m}^2}{(80 \text{ kg})(2.4 \text{ m})^2} + 1\right)$$

so

$$v_0 = \underline{19.0 \text{ m/s}}$$

If George could actually run this fast, he could run 100 m in 5.3 s. Quite a feat.

10. $\underline{3.6 \text{ m/s}}$

Chapter 11

Gravity

I. Key Ideas

11-1 *Kepler's Laws* Around 1600, Johannes Kepler inferred from many observations three empirical laws about planetary motion:

Law 1. Planetary orbits are ellipses with the sun at one focus.

Law 2. Each planet moves such that an imaginary line joining it and the sun sweeps out equal areas in equal times.

Law 3. The square of the period of each planet is proportional to the cube of its mean distance from the sun.

11-2 *Newton's Law of Gravity* Isaac Newton showed that a gravitational force law together with his own three laws of motion could be used to derive Kepler's empirically established laws. **Newton's law of gravity** states that every object in the universe attracts every other object with a force that is directly proportional to the product of the masses of the two objects and inversely proportional to the square of the distance between them:

Newton's law of gravity

$$\vec{F}_{1,2} = -\frac{Gm_1m_2}{r_{1,2}^2}\,\hat{r}_{1,2}$$

where $\vec{F}_{1,2}$ is the force exerted by object 1 on object 2, $r_{1,2}$ is the magnitude of the relative position vector $\vec{r}_{1,2}$ from object 1 to object 2, $\hat{r}_{1,2}$ is the unit vector in the direction of $\vec{r}_{1,2}$ (that is, $\hat{r}_{1,2} = \vec{r}_{1,2}/r_{1,2}$), and G is the universal gravitational constant

Universal gravitational constant

$$G = 6.67 \times 10^{-11}\ \text{N·m}^2/\text{kg}^2$$

Newton showed that Kepler's second law, the law of equal areas, is a direct consequence of the

conservation of angular momentum and holds for any central force. He also showed that planets with circular orbits have a period given by

Kepler's third law

$$T^2 = \frac{4\pi^2}{GM_S}r^3$$

where r is the radius of the planet's orbit and M_S is the mass of the sun.

Newton's law of gravity describes the gravitational force between particles. Although he had to invent calculus to do it, Newton was able to prove that the force of gravity exerted by any spherically symmetric object on a point mass on or outside its surface is the same as if all the mass of the object were concentrated at its center. Thus, the gravitational force exerted on a particle of mass m a distance r from the earth's center is

Gravitational force exerted by the earth

$$F = \frac{GM_E m}{r^2} \qquad r \geq R_E$$

where M_E is the earth's mass and R_E is its radius. It follows that the gravitational field $g(r)$ of the earth is

Gravitational field of the earth

$$g(r) = \frac{F}{m} = \frac{GM_E}{r^2} \qquad r \geq R_E$$

To verify this equation, Newton compared the free-fall acceleration of objects falling here at the earth's surface, a distance of one earth radius from the earth's center, with the acceleration of the moon in its nearly circular orbit at a distance of approximately 60 earth radii from the earth's center.

Measurement of the Universal Gravitational Constant   The universal gravitational constant G can be experimentally determined by measuring the force of gravitational attraction between two objects of known mass. This measurement is difficult because the force is extremely small for objects of laboratory (rather than planetary) size. Knowledge of G is of great scientific value because it enables the experimental determination of the mass of the earth and other objects in the solar system. The mass of the earth is found to be 5.98×10^{24} kg.

Gravitational and Inertial Mass   Newton's law of gravity makes it clear that the *mass* of an object is a measure of two distinct properties: (1) that property responsible for the gravitational force the object exerts on other objects, referred to as its **gravitational mass**, and (2) that property responsible for the object's resistance to acceleration, referred to as its **inertial mass**. The equivalence of these two properties is by no means obvious, but it has been confirmed experimentally to very high precision. Therefore, we do not ordinarily distinguish between the two, and just use the term mass. This **principle of equivalence** is a very important concept in Einstein's theory of general relativity.

11-3   *Gravitational Potential Energy*   Gravity is a conservative force, so it can be described in terms of potential energy. The **gravitational potential energy** of a particle at a location is defined as the negative of the work done by the force of gravity on the particle as it moves to that location from a reference location where its gravitational potential energy is zero. The gravitational potential energy $U(r)$ of a particle in the earth's gravitational field at a distance r from the center of the earth is

Gravitational potential energy ($U = 0$ at $r = R_E$)

$$U(r) = \frac{GM_E m}{R_E} - \frac{GM_E m}{r} \qquad r \geq R_E$$

if the reference location, where its potential energy is chosen to be zero, is at the earth's surface. (This potential energy is, of course, a property not of the particle but of the particle–earth system.)

Consider an object of mass m leaving the earth's surface with some initial speed. At the earth's surface, where $r = R_E$, the potential energy of the object is zero. As the object rises, it gains potential energy and loses kinetic energy. As the object approaches an infinite distance from the earth, its potential energy approaches $GM_E m / R_E$ which we will call U_{max}. If the object leaves the earth's surface at sufficient speed so that its kinetic energy exceeds U_{max}, the earth's grav-

ity will never be able to stop it and pull it back to earth. The object will still have some kinetic energy when it is an arbitrarily great distance away. When the object leaves the earth's surface with a kinetic energy equal to U_{max}, its speed is called the **escape speed** of the earth v_e. The escape speed, calculated by equating the initial kinetic energy and U_{max}, is

Escape speed

$$v_e = \sqrt{\frac{2GM_E}{R_E}} = \sqrt{2gR_E}$$

which is about 11 km/s (or 7 mi/s).

Classification of Orbits by Energy   In the equation for gravitational potential energy, the surface of the earth is the reference location where the potential energy was chosen to be zero. However, the reference location is arbitrary. We can choose it to be anywhere outside the earth. If we choose it to be an infinite distance from the earth, the equation for the gravitational potential energy becomes

Gravitational potential energy with $U = 0$ at infinite separation

$$U(r) = -\frac{GM_E m}{r} \qquad r \geq R_E$$

(The reason for making this choice is, of course, that it results in a simpler expression for the potential energy.) With this choice, the potential energy is negative for any finite separation and approaches zero as the separation approaches infinity. Any orbit in which the energy of the orbiting object is not sufficient for the object to escape is called a *bound orbit*. The total mechanical energy of an object is negative if the object is in a bound orbit (an ellipse) and zero or positive if the object is in an unbound orbit (a parabola or hyperbola).

11-4   *The Gravitational Field of a Spherical Shell and a Solid Sphere*   The gravitational field of a thin uniform spherical shell of mass M and radius R at a distance r from the center of the shell is

Gravitational field of a thin uniform spherical shell

$$\vec{g}(r) = -\frac{GM}{r^2} \hat{r} \qquad r > R$$

$$\vec{g}(r) = 0 \qquad r < R$$

A uniform solid sphere can be thought of as a collection of concentric spherical shells. The gravitational field of a solid sphere of radius R and mass M a distance r from its center is

Gravitational field of a uniform solid sphere

$$\vec{g}(r) = -\frac{GM}{r^2}\,\hat{r} \qquad\qquad r \geq R$$

$$\vec{g}(r) = -\frac{G}{r^2}\left(\frac{Mr^3}{R^3}\right)\hat{r} = -\frac{GMr}{R^3}\,\hat{r} \qquad r \leq R$$

where (Mr^3/R^3) is the mass inside a sphere of radius r.

II. Numbers and Key Equations

Numbers

Universal gravitational constant

$$G = 6.67 \times 10^{-11}\ \text{N·m}^2/\text{kg}^2$$

Radius of the earth

$$R_E = 6.37 \times 10^6\ \text{m}$$

Radius of the earth's orbit (mean earth–sun distance)

$$r_E = 1.50 \times 10^{11}\ \text{m} = 1\ \text{AU (astronomical unit)}$$

Mass of the earth

$$M_E = 5.98 \times 10^{24}\ \text{kg}$$

Mass of the sun

$$M_S = 1.99 \times 10^{30}\ \text{kg}$$

Length of a year

$$1\ \text{yr} = 3.16 \times 10^7\ \text{s}$$

Key Equations

Newton's law of gravity

$$\vec{F}_{1,2} = -\frac{Gm_1 m_2}{r_{1,2}^2}\,\hat{r}_{1,2}$$

Kepler's third law

$$T^2 = \frac{4\pi^2}{GM_S}\,r^3$$

Gravitational field of the earth

$$g(r) = \frac{F}{m} = \frac{GM_E}{r^2} \qquad r \geq R_E$$

Escape speed

$$v_e = \sqrt{\frac{2GM_E}{R_E}} = \sqrt{2gR_E}$$

Gravitational potential energy ($U \rightarrow 0$ as $r \rightarrow \infty$)

$$U(r) = -\frac{GM_E m}{r} \qquad r \geq R_E$$

Gravitational field of a thin uniform spherical shell

$$\vec{g}(r) = -\frac{GM}{r^2}\,\hat{r} \qquad r > R$$

$$\vec{g}(r) = 0 \qquad\qquad r < R$$

Gravitational field of a uniform solid sphere

$$\vec{g}(r) = -\frac{GM}{r^2}\,\hat{r} \qquad\qquad r \geq R$$

$$\vec{g}(r) = -\frac{G}{r^2}\left(\frac{Mr^3}{R^3}\right)\hat{r} = -\frac{GMr}{R}\,\hat{r} \qquad r \leq R$$

where $M(r^3/R^3)$ is the mass inside a sphere of radius r.

III. Potential Pitfalls

Every formula for gravitational potential energy (or for any potential energy, for that matter) *assumes* a particular zero location. Keep clearly in mind where this location is when working a particular problem. The expression *mgh* for gravitational potential energy means the potential energy is zero at $h = 0$.

The force of gravity exerted by a spherically symmetric object (a collection of uniform concentric spherical shells) on an object located outside the spherical object is the same as if the mass of the spherically symmetric object were concentrated at its geometric center. This is true *only* for objects where the mass is distributed with spherical symmetry.

In problems where the distances between planets and moons and such appear, the distances are measured between their *centers*.

A body moving freely in the earth's gravity is not necessarily moving toward the earth! However, such bodies, even those in circular orbits, are always *accelerating* toward the earth. This is true no matter what direction they are moving.

The earth's escape speed is 11 km/s or 7 mi/s. Remember that this number refers only to escape from the earth's surface. A body starting from a higher altitude requires a lower initial velocity to escape.

Every planet, moon, or star has its own escape speed, determined by its mass and radius.

Be careful of signs when working problems that deal with the energy of orbital motion. Use the simplest form of the equation for gravitational potential energy, $-GM_E m/r$, for which the potential energy is always negative and approaches zero as the distance between the object and the earth approaches infinity. For objects in bound orbits, the total mechanical energy is negative, whereas for objects in unbound orbits, the total mechanical energy is never negative.

For the sake of illustration, we often do problems involving the earth's gravity in which we neglect air drag. Don't worry if the results differ widely from reality. Without air drag, an object falling from a very large distance would strike the earth's surface at a speed approaching the earth's escape speed. However, due to air drag the actual impact speed is an order of magnitude or so smaller than the escape speed. (The object may also burn up in the atmosphere and never reach the earth's surface.)

IV. True and False and Responses

True and False

_____ 1. Kepler's second law, the law of equal areas, implies that the force of gravity varies inversely with the square of the distance.

_____ 2. Kepler's second law, the law of equal areas, follows from the fact that an orbiting planet's angular momentum about the sun is constant.

_____ 3. According to Kepler's third law, the period of a planet's orbital motion varies as the $\frac{3}{2}$ power of its orbital radius.

_____ 4. For the special case of a circular orbit, the inverse-square force law can be derived from Kepler's laws using Newton's laws of motion.

_____ 5. A confirmation of Newton's law of gravity is the fact that the acceleration of the moon in its orbit is the same as that of objects falling freely near the surface of the earth.

_____ 6. Cavendish's experiment provides a direct measurement of the mass of the earth, from which the gravitational constant G can be inferred.

_____ 7. The gravitational field at a point is defined as the gravitational force on a unit mass at that point.

_____ 8. The gravitational field is a scalar quantity.

_____ 9. The gravitational field due to a uniform spherical shell of matter is zero throughout the region $r < R_{inner}$, where R_{inner} is the inner radius of the shell.

_____ 10. The gravitational field due to a spherically symmetric distribution of matter is identical to that of a point mass only at distances from its center that are *very large* compared to its radius.

Responses

1. False. The law of equal areas follows directly from the conservation of angular momentum. It holds for any and all central forces, not just those that vary inversely with the square of the distance.

2. True

3. True; (period)$^2 \propto$ (radius)^3.

4. True

5. False. The ratio of the moon's acceleration to the acceleration of objects falling freely at the surface of the earth is R_E^2/r^2, where R_E is the radius of the earth and r is the radius of the moon's orbit. (This is the confirmation of Newton's law of gravity.)

6. False. It's the other way around. Cavendish's experiment provides a direct measurement of the gravitational constant G from which the mass of the earth can be inferred.

7. True

8. False. It is the force per unit mass, a vector.

9. True

10. False. The gravitational field due to a spherical mass distribution is identical to that of a point mass at any distance from its center that is *larger than* its radius. (For a nonspherical mass distribution, the distribution's gravitational field approaches that of a point mass as the distance from its center becomes large compared to its "radius." The larger this distance, the less impact the details of the actual mass distribution have on the gravitational field.)

V. Questions and Answers

Questions

1. Some communications satellites remain stationary over one point on the earth. How is this accomplished?

2. Must the satellite orbit described in Question 1 be circular?

3. The two satellites of the planet "Krypton" have near-circular orbits with diameters in the ratio 1.7 to 1. What is the ratio of their periods?

4. Estimate the force of gravity exerted upon you by a mountain 2-km high. (You'll have to make some simplifying assumptions about the shape of the mountain, its density, and so forth. See Figure 13-1 on page 375 of the text for the densities of various materials.)

5. The earth's orbit isn't a perfect circle; the earth is a little closer to the sun in January than it is in July. How can you tell this from the apparent motion of objects in the heavens?

6. The drag force of the atmosphere on an orbiting satellite has a tendency to make the satellite's orbit more nearly circular. Why?

7. What does it mean to say that an astronaut in a satellite orbiting the earth is "weightless?"

Answers

1. If a satellite is in a circular equatorial orbit with a period equal to that of the earth's rotation, it will appear to be stationary relative to a single spot on the ground. (Such an orbit is called a *geosynchronous orbit*.) The altitude of the required orbit, which can be found using Kepler's third law ($T^2/r^3 = $ constant) using the moon's orbital data and the radius of the earth, works out to be about 35,800 km or 22,400 mi.

2. Yes, the orbit must be circular if the satellite is to appear truly stationary. If the satellite were in an elliptical orbit with a period equal to that of the earth's rotation, its angular speed would be faster than the earth's rotation rate part of the time and slower at other times—in accord with Kepler's law of equal areas.

3. By Kepler's third law, the period T of an orbit is proportional to $R^{3/2}$, where R is the radius of the orbit. Thus, the ratio of the periods is $(1.7)^{3/2}$ to 1 or 2.22 to 1.

4. The model I used for the mountain had the mass of a cone 2 km high and 10 km in diameter at the base. I assumed that I was standing at the base of the cone and that the gravity of the mountain acted as though from a point one-third of the way up its axis. These thoroughly arbitrary assumptions gave me a force of 2×10^{-4} N, which is about 2×10^{-7} times my weight. You may get something quite different, depending on the assumptions you make, but it will be small relative to your weight. Gravity is a weak force unless the mass of one of the objects is astronomically large!

5. From Kepler's second law we know the line joining the earth and the sun sweeps out equal areas in equal times. In January, when this line is shortest, it sweeps out angle the fastest. It follows that viewed in January from the earth, the sun's apparent motion

against the background of the "fixed" stars is greater than it is in July. This can be observed by comparing the sidereal day and the solar day. (The sidereal day is the time for each "fixed" star to reach its zenith on consecutive days whereas the solar day is the time for the sun to reach its zenith on consecutive days. The zenith is the highest point reached.)

6. There is more residual atmosphere at lower altitudes. Thus, most of the effect of atmospheric drag on the satellite occurs when it is near perigee (its point of closest approach to the earth). Every time it comes around to perigee, it loses a little speed. As a result, it doesn't climb as far away from the Earth during the next orbit, so its path becomes a little more nearly circular each time.

7. Properly speaking, the astronaut isn't weightless since weight is the force that gravity exerts on him or her. The earth's gravity hasn't gone away. However, both the astronaut and scale are in free fall, so the scale exerts no net force against the astronaut and his or her apparent weight is zero.

VI. Problems and Solutions

Problems

1. An astronaut has landed on a 7940-km-diameter planet in another solar system. The astronaut drops a stone from the port of his landing craft and observes that it takes 1.78 s to fall the 4.5-m distance to the planet's surface. Using this data she calculates the mass of the planet. What value does she obtain?

How to Solve It
• From the distance it falls and the fall time, calculate the local free-fall acceleration.

• Knowing the planet's radius and the free-fall acceleration at its surface, use Newton's law of gravity to calculate its mass.

2. The moons of Mars are Deimos and Phobos ("Terror" and "Fear"). Deimos's orbital radius is 14,600 mi with a period of 30 h and Phobos's orbital period is 7.7 h. (*a*) What is the radius of the orbit of Phobos? (*b*) What is the mass of Mars?

How to Solve It
• Since you know the radius of one orbit and orbital periods, you can get the radius of Phobos's orbit using Kepler's third law ($T^2/r^3 = $ constant).

• To calculate the mass of Mars use Kepler's third law again.

• Don't forget that you have to convert units before calculating the mass of Mars.

3. The average distance of Saturn from the sun is 1.40×10^9 km. Assuming a circular orbit, find its orbital speed in meters per second.

How to Solve It

• If Saturn's orbit is circular, then its orbital speed is constant.

• Using Saturn's orbital radius and earth's orbital radius and period, calculate Saturn's orbital period using Kepler's third law (T^2/r^3 = constant).

• Determine the distance Saturn travels in one orbital period. Then calculate its average speed.

4. The planet Jupiter takes 11.9 years to orbit the sun. What is the radius of Jupiter's orbit?

How to Solve It

• The earth takes one year to orbit the sun.

• Use Kepler's third law (T^2/r^3 = constant) to relate the orbital radii and periods of the earth and Jupiter.

5. A space probe is launched from the earth with sufficient speed so that after escaping from the planet its speed, relative to the earth, is equal to the escape speed from the earth's surface. The probe is launched so that after escaping the earth it is still one astronomical unit from the sun, moving in the same direction as the earth is. Does it have sufficient energy to escape from the sun?

How to Solve It

• Find an expression for the escape speed from the earth's surface.

• If the probe is moving in the direction of the earth's orbital motion, the speed of the probe relative to the sun equals the speed of the probe relative to the earth plus the orbital speed of the earth. Find an expression for the earth's orbital speed by applying Newton's second law ($\Sigma \vec{F} = m\vec{a}$) to it.

• Determine the minimum speed needed at a distance of 1 AU for the probe to escape the solar system. This can be done using the conservation of mechanical energy.

6. What is the escape speed from the surface of the moon? Take the mass of the moon to be 7.40×10^{22} kg and its radius to be 1.74×10^6 m.

How to Solve It

• When an object is at escape speed, it has just enough mechanical energy to get very far away from the attracting body (here, the moon).

• The escape speed is the minimum speed that makes the object's initial mechanical energy nonnegative.

7. If the resistance of the atmosphere is neglected, (a) at what speed would an object have to be projected directly upward from the earth's surface in order to reach an altitude of 200 km? (b) Once it reaches 200 km, how much more energy would the 800-kg object have to be given in order to be put into a circular orbit at this altitude?

How to Solve It

• The air resistance is negligible so the satellite's mechanical energy is conserved.

• In part (a), the object must start upward from the ground with enough kinetic energy to supply the additional potential energy it has at an altitude of 200 km.

• In part (b), it must be given additional kinetic energy for it to attain orbital speed at an altitude of 200 km. Use Newton's second law ($F = ma$) to find the orbital speed.

8. An 1800-kg Earth satellite is to be placed in a circular orbit at an altitude of 150 km above the earth's surface. (a) How much work must be done on the satellite to accomplish this? Neglect the resistance of the atmosphere. (b) How much of the work goes into increasing the potential energy of the satellite and how much into increasing its kinetic energy?

How to Solve It

• The air resistance is negligible so the work that must be done on the satellite is equal to the change in its total mechanical energy.

• The change in the satellite's potential energy is due to its increased height, just as in the previous problem.

• The change in its kinetic energy corresponds to its orbital speed.

• *Hint:* Do part (b) first

9. A research satellite is in an elliptical orbit around the earth. Its closest approach to the earth's surface (perigee) is 100 km, and its greatest distance (apogee) is 1600 km. Find its speed at each of these two points.

How to Solve It

• The only force acting on the satellite is the gravitational force exerted by the earth. This force is directed toward the earth's center, so we know that the torque exerted on the satellite about the earth's center equals zero. It follows that the satellite's angular momentum about the earth's center is conserved. Equate expressions for the magnitude of the satellite's angular momentum at apogee and at perigee to obtain an equation relating the satellite's speeds at those points.

- Because the gravitational force is conservative, we know that the satellite's mechanical energy is conserved. Equate expressions for the satellite's mechanical energy at apogee and at perigee to obtain a second equation relating its speeds at those points.

- Solve the two simultaneous equations for the desired speeds.

10. A satellite is in an elliptical orbit around the earth. At its closest approach to the earth (perigee) it is 112 km above the earth's surface and is moving at a speed of 8032 m/s. When it is at apogee, how far above the earth's surface is it?

How to Solve It
- The only force acting on the satellite is the gravitational force exerted by the earth. This force is directed toward the earth's center, so we know that the torque exerted on the satellite about the earth's center equals zero. It follows that the satellite's angular momentum about the earth's center is conserved. Equate expressions for the magnitude of the satellite's angular momentum at apogee and at perigee to obtain an equation relating the satellite's speeds at those points. (see solution for Problem 9).

- Because the gravitational force is conservative, we know that the satellite's mechanical energy is conserved. Equate expressions for the satellite's mechanical energy at apogee and at perigee to obtain a second equation relating its speeds at those points. (see solution for Problem 9).

- Eliminate the satellite's speed at apogee from the two equations, and solve for the earth–satellite distance at apogee.

Solutions

1. The distance d of the free fall from rest of an object that is influenced only by gravity is given as a function of time by

$$d = \tfrac{1}{2} a_f t^2 \qquad (1)$$

where a_f is the free-fall acceleration. Applying Newton's second law ($\Sigma F = ma$) to the stone gives

$$\frac{GMm}{R^2} = ma_f$$

or

$$a_f = \frac{GM}{R^2}$$

Substituting this expression for a_f into Equation 1 we obtain

$$d = \frac{1}{2} \frac{GM}{R^2} t^2$$

or

$$M = \frac{2dR^2}{Gt^2} = \frac{2(4.5 \text{ m})(3.97 \times 10^6 \text{ m})^2}{(6.67 \times 10^{-11} \text{ N·m}^2/\text{kg}^2)(1.78 \text{ s})^2}$$

$$= \underline{6.71 \times 10^{23} \text{ kg}}$$

2. (a) $\underline{5900 \text{ mi}}$ (b) $\underline{6.59 \times 10^{23} \text{ kg}}$

3. Using Kepler's third law, we can compare the orbits of Saturn and earth:

$$\frac{T_S^2}{R_S^3} = \frac{T_E^2}{R_E^3}$$

or

$$T_S = \left(\frac{R_S}{R_E}\right)^{3/2} T_E = \left(\frac{1.40 \times 10^9 \text{ km}}{1.50 \times 10^8 \text{ km}}\right)^{3/2} (1 \text{ y})$$

$$= \underline{28.5 \text{ y}}$$

Saturn's orbital speed equals the circumference of the orbit divided by the period. Thus,

$$v = \frac{2\pi R_S}{T_S} = \frac{2\pi(1.40 \times 10^{12} \text{ m})}{(28.5 \text{ y})(3.16 \times 10^7 \text{ s/y})}$$

$$= \underline{9.76 \times 10^3 \text{ m/s}}$$

4. $\underline{7.82 \times 10^{11} \text{ m (5.21 AU)}}$

5. The speed v_p of the probe with respect to the sun equals the earth's orbital speed v_E plus the escape speed $v_{E,e}$ of the earth from its surface

$$v_p = v_E + v_{E,e} \qquad (1)$$

To find an expression for the escape speed from the earth's surface we apply conservation of mechanical energy to a probe launched at the surface with this speed. This gives

$$U_{g,f} + K_f = U_{g,i} + K_i$$

$$0 + 0 = -\frac{GM_E m_p}{R_E} + \frac{1}{2} m_p v_{E,e}^2$$

so

$$v_{E,e} = \sqrt{\frac{2GM_E}{R_E}} \qquad (2)$$

By applying Newton's second law ($F = ma$) to the earth we obtain an expression relating the earth's orbital speed v_E and the radius r_E of the earth's orbit

$$F = ma$$

$$\frac{GM_S M_E}{r_E^2} = M_E \frac{v_E^2}{r_E}$$

so

$$v_E = \sqrt{\frac{GM_S}{r_E}} \tag{3}$$

Substituting our expressions for v_E and $v_{E,e}$ into Equation 1 gives

$$v_p = \sqrt{\frac{GM_S}{r_E}} + \sqrt{\frac{2GM_E}{R_E}}$$

To see if this exceeds the escape speed from the sun we next determine the minimum speed needed, at a distance equal to the radius of the earth's orbit, for the probe to escape the solar system. By comparing these two speeds we will find out whether or not the probe will eventually escape from the solar system.

The formula (Equation 2) for the escape speed from a planet (or star) is developed by equating the mechanical energy of a probe at the planet's surface with the probe's potential energy at an infinite distance from the planet. Because there is nothing special about starting from the surface, this formula will give the escape speed starting at any distance above the surface. Thus, the formula for escape speed from the sun $v_{S,e}$ starting at a distance from the sun equal to the earth's orbital radius r_E is

$$v_{S,e} = \sqrt{\frac{2GM_S}{r_E}} \tag{4}$$

The ratio of the probe speed to the escape speed from the sun is

$$\frac{v_p}{v_{S,e}} = \frac{\sqrt{\dfrac{GM_S}{r_E}} + \sqrt{\dfrac{2GM_E}{R_E}}}{\sqrt{\dfrac{2GM_S}{r_E}}}$$

$$= \sqrt{\frac{M_E r_E}{M_S R_E}} + \frac{1}{\sqrt{2}}$$

$$= \sqrt{\frac{(5.98 \times 10^{24}\,\text{kg})(1.50 \times 10^{11}\,\text{m})}{(1.99 \times 10^{30}\,\text{kg})(6.37 \times 10^6\,\text{m})}} + \frac{1}{\sqrt{2}}$$

$$= \underline{0.973}$$

Because this ratio is less than one, we can conclude that the probe will almost not escape from the sun. *Note:* The orbital speed of the earth, calculated from Equation 3, is 29.7 km/s (18.6 mi/s) and, starting at the earth's orbit, the escape speed from the sun, calculated from Equation 4, is 42.1 km/s (26.1 mi/s).

6. 2.38×10^3 m/s

7. (a) The potential energy of an object of mass m above the earth's surface and distance r from its center is

$$U = -\frac{GM_E m}{r}$$

where M_E is the mass of the earth. Because we are neglecting air resistance, the initial energy of the object equals its final energy:

$$U_{g,f} + K_f = U_{g,i} + K_i$$

$$-\frac{GM_E m}{R_E + y} + \frac{1}{2}m(0)^2 = -\frac{GM_E m}{R_E} + \frac{1}{2}mv_i^2$$

so

$$v_i^2 = \frac{2GM_E}{R_E} - \frac{2GM_E}{R_E + y} = \frac{2GM_E y}{R_E(R_E + y)}$$

$$= \frac{2(6.67 \times 10^{-11}\,\text{N·m}^2/\text{kg}^2)(5.98 \times 10^{24}\,\text{kg})(2.00 \times 10^5\,\text{m})}{(6.37 \times 10^6\,\text{m})(6.37 \times 10^6\,\text{m} + 2.00 \times 10^5\,\text{m})}$$

$$= 3.81 \times 10^6\,\text{m}^2/\text{s}^2$$

so

$$v_i = 1.95 \times 10^3\,\text{m/s (or 1.95 km/s)}$$

(b) A satellite in circular orbit has a speed v such that the centripetal force on it equals the gravitational force of the earth:

$$\frac{GM_E m}{r^2} = \frac{mv^2}{r}$$

$$v^2 = \frac{GM_E}{r}$$

Thus, the required additional energy is

$$K = \frac{1}{2}mv^2 = \frac{1}{2}m\frac{GM_E}{r} = \frac{GM_E m}{2(R_E + y)}$$

$$= \frac{(6.67 \times 10^{-11}\,\text{N·m}^2/\text{kg}^2)(5.98 \times 10^{24}\,\text{kg})(800\,\text{kg})}{2(6.37 \times 10^6\,\text{m} + 2.00 \times 10^5\,\text{m})}$$

$$= \underline{2.43 \times 10^{10}\,\text{J}}$$

8. (*a*) 5.77×10^{10} J (*b*) The kinetic energy is increased by $\underline{5.51 \times 10^{10}\,\text{J}}$ and the potential energy is increased by $\underline{0.26 \times 10^{10}\,\text{J}}$

9. Figure 11-1 shows the motion of the satellite. The angular momentum of the satellite about the earth's center is conserved because the gravitational force exerts no torque about that point. For a particle of mass m moving with velocity v, the angular momentum L of the particle about a point is

$$\vec{L} = \vec{r} \times m\vec{v}$$

Figure 11-1

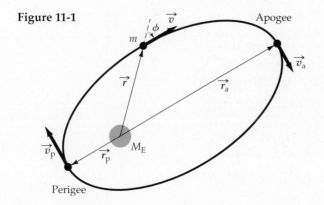

where $\vec{r}$ is the vector from the point to the particle. At apogee and perigee, $\vec{r}$ and $\vec{v}$ are perpendicular, so the magnitudes of the angular momenta at these points, L_a and L_p, are

$$L_a = r_a m v_a$$

and

$$L_p = r_p m v_p$$

where r_a and r_p are the distances from the earth's center to the satellite when it is at apogee and perigee and v_a and v_p are the speeds of the satellite at those points. These angular momenta are equal:

$$L_a = L_p$$

so we obtain

$$r_a v_a = r_p v_p \qquad (1)$$

The only force acting on the satellite is the conservative force of the earth's gravity. Therefore mechanical energy is conserved. That is,

$$E_a = E_p$$

$$U_a + K_a = U_p + K_p$$

$$-\frac{GM_E m}{r_a} + \frac{1}{2}mv_a^2 = -\frac{GM_E m}{r_p} + \frac{1}{2}mv_p^2 \qquad (2)$$

$$-\frac{2GM_E}{r_a} + v_a^2 = -\frac{2GM_E}{r_p} + v_p^2$$

With Equations 1 and 2 we have two simultaneous equations and two unknowns, v_a and v_p. We now solve for v_a by obtaining an expression for v_p from Equation 1 and substituting it into Equation 2. Solving Equation 1 for v_p, we obtain

$$v_p = \frac{r_a}{r_p}v_a \qquad (3)$$

Substituting this expression for v_p into Equation 2 we obtain

$$-\frac{2GM_E}{r_a} + v_a^2 = -\frac{2GM_E}{r_p} + \left(\frac{r_a}{r_p}\right)^2 v_a^2$$

Solving this for v_a gives

$$\left(\frac{r_a^2}{r_p^2} - 1\right)v_a^2 = \frac{2GM_E}{r_p} - \frac{2GM_E}{r_a}$$

$$\frac{r_a^2 - r_p^2}{r_p^2}v_a^2 = \frac{2GM_E(r_a - r_p)}{r_p r_a}$$

$$\frac{(r_a - r_p)(r_a + r_p)}{r_p^2}v_a^2 = \frac{2GM_E(r_a - r_p)}{r_p r_a}$$

$$v_a^2 = \frac{2GM_E r_p}{r_a(r_a + r_p)}$$

The values of r_a and r_p are obtained by adding 1600 km and 100 km, respectively, to the radius of the earth R_E. Thus,

$$v_a^2 =$$

$$\frac{2(6.67 \times 10^{-11}\,\text{N·m}^2/\text{kg}^2)(5.98 \times 10^{24}\,\text{kg})(6.47 \times 10^6\,\text{m})}{(7.97 \times 10^6\,\text{m})(6.47 \times 10^6\,\text{m} + 7.97 \times 10^6\,\text{m})}$$

or

$$v_a = \underline{6.70 \times 10^3\,\text{m/s}}$$

Substituting this value into Equation 3, we obtain

$$v_p = \frac{r_a}{r_p} v_a$$

$$= \frac{7.97 \times 10^6 \text{ m}}{6.47 \times 10^6 \text{ m}} (6.70 \times 10^3 \text{ m/s})$$

$$= \underline{8.25 \times 10^3 \text{ m/s}}$$

10. $r_a = 7.14 \times 10^6$ m

so

$$h_a = r_a - R_E = \underline{7.71 \times 10^5 \text{ m (or 771 km)}}$$

Chapter 12

Static Equilibrium and Elasticity

I. Key Ideas

If an object is stationary and remains stationary, it is said to be in **static equilibrium**. In this chapter we study the conditions necessary for static equilibrium to exist. The concept of center of gravity will be introduced, and the parameters affecting stability will be examined.

12-1   Conditions for Equilibrium   For an object to remain in static equilibrium, the acceleration of its center of mass must remain zero, which means that the net external force acting on the object must remain zero. However, the object can rotate even when the net external force remains zero. For an object to be in static equilibrium, the net external torque on the object about any point must also be zero. Thus the two necessary conditions for a body to be in static equilibrium are

Conditions for equilibrium

$$\Sigma \vec{F}_{ext} = 0$$

$$\Sigma \vec{\tau}_{ext} = 0 \quad \text{(about any point)}$$

12-2   The Center of Gravity   In a gravitational field, a gravitational force is exerted on each constituent particle of an object. The vector sum of this collection of forces is considered a single force, the object's weight. In order to calculate the net torque exerted by the object's weight, the weight is best thought of as acting at a single point, the **center of gravity**. The center of gravity is defined so that the torque produced by the resultant weight force about any point equals the sum of the torques produced by the many individual gravitational forces about the same point. The x coordinate X_{cg} of the center of gravity of the object is related to the x coordinates of the particles that make up the object according to the equation

Center of gravity

$$WX_{cg} = \sum_i w_i x_i = \sum_i m_i g x_i$$

where w_i is the weight of the ith particle, x_i is the x coordinate of the position of the ith particle, and $W = \Sigma w_i$. The center of mass is defined as $MX_{cm} = \Sigma m_i x_i$. If the gravitation field $\vec{g}$ is uniform then the above equation can be simplified to $WX_{cg} = g\Sigma m_i x_i = gMX_{cm}$. Since $W = Mg$, we have

$$X_{cg} = X_{cm}$$

That is, in a uniform gravitational field the center of gravity coincides with the center of mass.

When calculating the torque about a point due to the force of gravity acting on an object we consider the weight as a single force that is applied at the object's center of gravity.

12-3   Some Examples of Static Equilibrium   See Problems 1, 3, 5, and 7 and their solutions at the end of this chapter.

12-4   Couples   A **couple** consists of a pair of oppositely directed forces of equal magnitude acting along different lines of action. Although the torque exerted by a single force depends on the axis about which the torque is calculated, the torque exerted by a couple is, perhaps surprisingly, independent of the point about which it is calculated. The magnitude of the torque exerted by a couple is

Torque exerted by a couple

$$\tau = FD$$

where F is the magnitude of one of the forces and D is the perpendicular distance between the lines of action of the two forces.

12-5 Static Equilibrium in an Accelerated Frame
If the forces on an object are analyzed in its rest frame (a frame of reference in which that object is at rest), and if the rest frame is an inertial reference frame, then the conditions for static equilibrium stated in Section 12-1 apply. Interestingly, if the rest frame is a noninertial reference frame, then the forces can be analyzed by a modified version of the conditions stated in Section 12-1. These modified conditions are:

1. $\Sigma \vec{F} = m\vec{a}_{cm}$

where $\vec{a}_{cm}$ is the acceleration of the center of mass in a noninertial reference frame.

2. $\Sigma \vec{\tau}_{cm} = 0$

The sum of the torques *about the center of mass* must be zero. Condition 2 restricts us to computing torques only about axes passing through the center of mass.

12-6 Stability of Rotational Equilibrium An object is in **stable rotational equilibrium** if small angular displacements of the object away from equilibrium result in a torque (or torques) that tends to return the object to its equilibrium position. For example, the pendulum of a tall clock when it is at rest hanging vertically is in stable rotational equilibrium. If rotated slightly from the vertical and released, the torque due to the gravitational force will tend to rotate it back to the vertical.

An object is in **unstable rotational equilibrium** if small angular displacements of the object away from equilibrium result in torques that tend to rotate the object further away from the equilibrium position. An example of an object in unstable rotational equilibrium is a pencil balanced on its point. If rotated slightly from the vertical and released, the torque due to the gravitational force will tend to rotate it further from the vertical.

An object is in **neutral rotational equilibrium** if both the net force and the net torque remain zero following small linear and/or angular displacements of the object from an equilibrium position. A meter stick suspended by a nail though a small hole drilled through its center of mass and driven into a wall is an example. The torque due to the gravitational force is zero, so if the stick is rotated slightly from an equilibrium position and released, there is no torque to either return it to or move it farther from its initial position.

When you rotate something slightly, if the potential energy increases, remains the same, or decreases, then the rotational equilibrium is stable, neutral, or unstable, respectively.

12-7 Indeterminate Problems The conditions for static equilibrium consist of two vector equations, $\Sigma \vec{F} = 0$ and $\Sigma \vec{\tau} = 0$. By taking x, y, and z components of each of these we end up with six component equations: $\Sigma F_x = 0$, $\Sigma F_y = 0$, $\Sigma F_z = 0$, $\Sigma \tau_x = 0$, $\Sigma \tau_y = 0$, and $\Sigma \tau_z = 0$. However, with these six equations we can solve for, at most, six unknowns. Any problem that has more unknowns than available (independent) equations cannot be solved and is said to be indeterminate. However, for static equilibrium problems that concern a deformed object, additional equations can often be obtained by relating the deformation to the forces acting on it. Such equations are introduced in the following section.

12-8 Stress and Strain Most solids behave elastically. If stretched or deformed by external forces, they tend to return to their original size and shape when the deforming stress is removed. For a given material, this is true up to some **elastic limit;** exceeding this limit will cause permanent deformation.

If an applied force tends to stretch an object, the force per unit area (perpendicular to the force) is called the **tensile stress:**

Tensile stress

$$\text{Stress} = \frac{F}{A}$$

The resulting fractional change in the object's length is the **tensile strain:**

Tensile strain

$$\text{Strain} = \frac{\Delta L}{L}$$

Up to the proportional limit—a point somewhat lower than the elastic limit for a substance—stress is proportional to strain. The constant of proportionality, which is the ratio of stress to strain, is called **Young's modulus** Y:

Young's modulus

$$Y = \frac{\text{stress}}{\text{strain}} = \frac{F/A}{\Delta L/L}$$

The maximum tensile stress a material can withstand is called its **tensile strength**. Young's modulus for both the tensile stress and the tensile strength of a given material may or may not be the same as Young's modulus for the compressive stress and the compressive strength of that material.

Figure 12-1

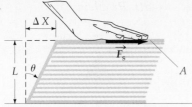

A sideways deformation resulting from forces that act parallel to the surface of a material is called a **shear strain** (see Figure 12-1). Within the proportional limit of a material, **shear stress** (shear force per unit area) is proportional to **shear strain**. The constant of proportionality is the **shear modulus** M_s:

Shear modulus

$$M_s = \frac{\text{shear stress}}{\text{shear strain}} = \frac{F_s/A}{\Delta X/L} = \frac{F_s/A}{\tan\theta}$$

II. Numbers and Key Equations

Numbers

There are no new numbers for this chapter.

Key Equations

Conditions for equilibrium

$$\Sigma \vec{F}_{\text{ext}} = 0$$

$$\Sigma \vec{\tau}_{\text{ext}} = 0 \qquad \text{(about any point)}$$

Center of gravity

$$WX_{\text{cg}} = \sum_i w_i x_i = \sum_i m_i g x_i$$

Torque exerted by a couple

$$\tau = FD$$

III. Potential Pitfalls

Don't just draw force vectors anywhere. When drawing free-body diagrams, the line of action of a force must pass through the point of application of the force. The effective point of application of the gravitational force acting on an object is the object's center of gravity.

For an object to be in equilibrium, the net external torque must be zero about *any* axis.

For a particular problem, we are free to select the axis about which we calculate torques. Some axis choices are better than others because they result in simpler equations that are more easily solved. A useful rule of thumb is to choose an axis that passes through the point of application of the force that is the least specified. A force is completely specified when its point of application and both its magnitude and direction (or its components) are known.

Don't let signs confuse you. To determine whether the sign of a torque $\vec{\tau}$ associated with a force $\vec{F}$ is positive (counterclockwise) or negative (clockwise), imagine that the object is constrained to rotate about the selected axis and that $\vec{F}$ is the only force acting on it. The torque due to $\vec{F}$ is positive if the object would rotate in the positive direction (counterclockwise) and negative if it would rotate in the negative direction. An alternative way to determine the sign of the torque τ is to determine the direction of $\vec{r} \times \vec{F}$, where $\vec{r}$ is the vector from the axis to the point of application of the force. (The direction of a cross product is obtained by applying the right-hand rule.) If the direction of $\vec{r} \times \vec{F}$ is up out of the paper, then $\vec{\tau}$ is positive (counterclockwise), and if the direction is into the paper, then it is negative.

Don't think that the elastic moduli are always constants. They are constants only for stresses that are less than the proportional limits of a material. They vary considerably for stresses approaching the elastic limit or the tensile or compressive strength of the material.

IV. True and False and Responses

True and False

_____ **1.** When an object is in static equilibrium, the net external force acting on it must equal zero.

_____ **2.** When an object is in static equilibrium, the net external torque about *any* axis must equal zero.

_____ **3.** The lever arm of a force about a point P always equals the distance from P to the point of application of the force.

_____ **4.** The torque exerted by a couple about a point P is independent of the location of P, unless P is located between the lines of action of the two forces.

_____ **5.** Strain has dimensions of length and the SI unit for it is the meter.

_____ **6.** Young's modulus and the shear modulus have the same dimensions.

Responses

1. True

2. True

3. Not necessarily. The lever arm equals the perpendicular distance from P to the line of action of the force.

4. False. The torque exerted by a couple about a point P is always independent of the location of P.

5. False. Strain is dimensionless.

6. True. They have dimensions of $\dfrac{[M]}{[L][T]^2}$ and SI units of N/m².

V. Questions and Answers

Questions

1. Must the center of gravity of an object be located inside the material of the object?

2. If exactly three external forces act on an object in equilibrium and if two of their lines of action intersect, must the third line of action intersect at the same point?

Answers

1. No. For example, the center of gravity of a bowl is outside the material of the bowl.

2. Yes. If a point is on the line of action of a force, then the force's torque about that point must equal zero. Conversely, if the torque of a force about a point is zero, then the point must lie on the force's line of action. (Convince yourself that these assertions are true. Make some drawings, etc.) The intersection point P of the lines of action of two of the forces lies on both lines, so the torques of both forces about P each equal zero. Since the sum of all three torques about P must equal zero, and the torques of two of the forces are both zero, the torque of the third force must also equal zero. It follows that the line of action of the third force must also pass through P.

VI. Problems and Solutions

Problems

1. A uniform, 0.1-kg meterstick is suspended by a pivot at the 50-cm mark. As shown in Figure 12-2, the system is balanced when a 0.3-kg mass is suspended from the 20-cm mark and an unknown mass m is suspended from the 70-cm mark. Determine the mass m and the force exerted on the stick by the pivot.

Figure 12-2

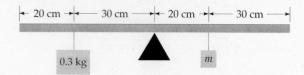

How to Solve It

• Sketch a free-body diagram of the stick. There are four forces acting on it: the force of gravity, and the contact forces exerted by the pivot and the two suspended masses.

• Because the suspended masses are each in equilibrium, each exerts a force equal to its weight on the stick. The weight of the stick is applied at the stick's center of gravity. Because the stick is uniform, the center of gravity of the stick is at the 50-cm mark.

• Select an axis that passes through the point of application of the force that is the least specified. In this case, the least specified force is the force exerted by the pivot. (We know neither its magnitude nor its direction.)

• Because the stick is in equilibrium, the net torque about the selected axis is zero. Write out an equation setting the sum of the torques to zero. If you have any difficulties determining the sign of a torque, read the next-to-last paragraph of the Potential Pitfalls section of this chapter. Solve this equation for the mass m.

• Because the stick is in equilibrium, the net force acting on the stick is zero. Write out an equation setting the vector sum of the forces acting on the stick to zero. Solve this equation for the force exerted by the pivot.

2. A uniform 0.3-kg meterstick is supported by a hinge and a vertical string. A 0.6-kg mass is supported by the stick, as shown in Figure 12-3. Determine the tension in the string and the force exerted by the hinge on the stick.

Figure 12-3

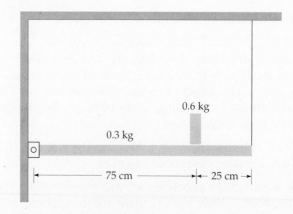

How to Solve It

- Sketch a free-body diagram of the stick. There are four forces acting on it, the force of gravity and the contact forces exerted by the pivot, the string, and the 0.6-kg mass.

- Because the 0.6-kg mass is in equilibrium, it exerts a force on the stick equal to its weight. The weight of the stick is applied at the stick's center of gravity. Because the stick is uniform, its center of gravity is 50 cm from either end.

- Select an axis that passes through the point of application of the force that is the least specified. Because we know neither its magnitude nor its direction the least specified force is that exerted by the hinge pin.

- Because the stick is in equilibrium, the net torque about the selected axis is zero. Write out an equation setting the sum of the torques to zero. If you have any difficulties determining the sign of a torque, read the next-to-last paragraph of the Potential Pitfalls section of this chapter. Solve this equation for the tension in the string.

- Because the stick is in equilibrium, the net force acting on it is zero. Write out an equation setting the vector sum of the forces acting on the stick to zero. Solve this equation for the force exerted by the hinge.

3. A seesaw consists of a uniform plank, 4 m long, with a mass of 10 kg. As shown in Figure 12-4, a 40-kg boy sits 0.5 m from the short end of the seesaw and a girl of mass m_g sits 0.5 m from the other end. The plank is balanced when the pivot is located 25 cm from the center of the plank. Determine the mass of the girl.

Figure 12-4

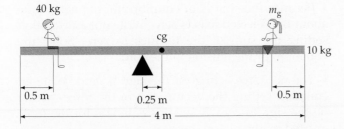

How to Solve It

- Sketch a free-body diagram of the plank. There are four forces acting on it: the force of gravity and the contact forces of the pivot and the two children.

- Because the children are each in equilibrium, each child pushes on the plank with a force equal to the weight of that child. The weight of the plank is ap-

plied at the plank's center of gravity. Because the plank is uniform, its center of gravity is at its geometric center.

- Select an axis that passes through the point of application of the force that is the least specified. Because we know neither its magnitude nor its direction the least specified force is that exerted by the pivot.

- Because the stick is in equilibrium, the net torque about the selected axis is zero. Write out an equation setting the sum of the torques to zero. If you have any difficulties determining the sign of a torque, read the next-to-last paragraph of the Potential Pitfalls section of this chapter.

4. A uniform 30-kg plank, 10 ft long, is supported at points 2 ft from each end as shown in Figure 12-5. A man walks along the plank toward the right. When he reaches a point 1 ft from the right end of the plank, the plank starts to tilt. Determine the mass of the man.

Figure 12-5

How to Solve It

- Sketch a free-body diagram of the plank. There are four forces acting on the plank: the force of gravity, the contact forces exerted by the two supports, and the force exerted by the man.

- When the man reaches the point 1 ft from the right end of the plank and the plank starts to tilt, the force exerted by the left support on the plank is zero. Select an axis that passes through the point of application of the force that is the least specified. Because we know neither its magnitude nor its direction, the least specified force is that exerted by the right support.

- As the man approaches the point 1 ft from the end of the plank, the plank remains in equilibrium. Thus the net torque about the selected axis is zero. Write out an equation setting the sum of the torques to zero. If you have any difficulties determining the sense (clockwise or counterclockwise) of a torque, read the next-to-last paragraph of the Potential Pitfalls section of this chapter.

5. A uniform 20-kg strut supporting a 30-kg object is suspended by a string and a hinge as shown in Figure 12-6. Determine the tension in the string and the force exerted on the strut by the hinge.

Figure 12-6

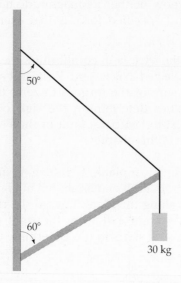

Figure 12-7

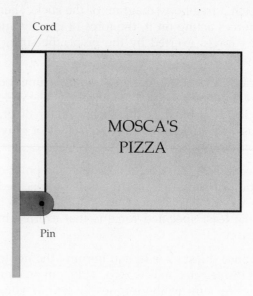

the upper left corner. Determine the tension in the cord and the force exerted by the pin on the sign.

How to Solve It
• Draw a free-body diagram of the sign. There are three forces acting on the sign: the force of gravity, the force of the pin, and the force of the cord.

• Select an axis that passes through the point of application of the force that is the least specified. In this case, the least specified force is the force exerted by the pin. We know neither its magnitude nor its direction.

• Show the lever arm for each force on your diagram. Recall that the lever arm is the perpendicular distance from the axis to the line of action of the force.

• Because the sign is in equilibrium, the net torque about the selected axis is zero. Write out an equation setting the sum of the torques to zero. Solve the equation for the tension in the cord.

• The net force acting on the sign is zero. Write an equation setting the vector sum of the forces acting on the sign to zero. Solve this equation for the force exerted by the pin on the sign.

7. A thin 30-cm rod of nonuniform composition is supported in a horizontal position by a string that passes over two frictionless pulleys as shown in Figure 12-8. The vertical segments of the string are attached to the rod at points 1 cm and 2 cm from the ends of the rod. Determine the location of the rod's center of gravity.

How to Solve It
• Draw a free-body diagram of the strut. There are four forces acting on the strut: the force of gravity, the force of the hinge, the force of the upper string, and the force of the lower string.

• Select an axis that passes through the point of application of the force that is the least specified. In this case, the least specified force is the force exerted by the hinge. We know neither its magnitude nor its direction.

• Indicate the lever arm for each force on your diagram. Recall that the lever arm is the perpendicular distance from the axis to the line of action of the force.

• Because the strut is in equilibrium, the net torque about the selected axis is zero. Write out an equation setting the sum of the torques to zero. Solve the equation for the tension in the cord.

• Because it is in equilibrium, the net force acting on the strut is zero. Write an equation setting the vector sum of the forces acting on the strut to zero. Solve this equation for the force exerted by the hinge on the strut.

6. A sign is constructed from a uniform 15-kg sheet of plywood 80 cm high and 120 cm wide. As shown in Figure 12-7, the sign is suspended via a pin through its lower left corner and a horizontal cord attached to

Figure 12-8

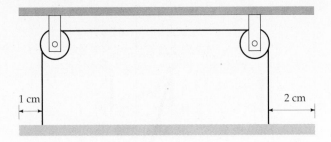

1 cm   2 cm

Figure 12-9

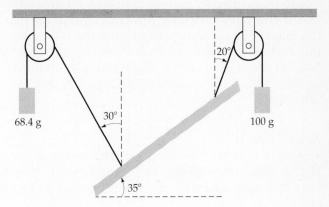

68.4 g   30°   100 g   20°   35°

How to Solve It
• Draw a free-body diagram of the rod. Choose a point on the rod an arbitrary distance x from its left end as the center of gravity. There are three forces acting on the rod: the force of gravity and the forces exerted on the rod by the opposite ends of the string.

• We select an axis that passes through the point of application of the force that is the least specified. There are three forces acting on the rod, the two tension forces and the weight, with the weight being the least specified. That is because, unlike the tension forces, its point of application is not known. You should select an axis through the point of application of the weight force.

• Indicate the lever arm for each force on your diagram. Recall that the lever arm is the perpendicular distance from the axis to the line of action of the force.

• The stick is in static equilibrium so the net force acting on it equals zero. Also, because the pulleys are frictionless we know that the tension T is the same in the two vertical string segments. Use this result (that the tensions are equal) to determine the relation between the weight of the stick and the tension in a string.

• The stick is in static equilibrium, so the net torque acting on the stick about any axis is zero. Set the sum of the torques about the selected axis to zero and solve for the position of the center of gravity.

8. A nonuniform meterstick is suspended from two light strings attached to the 10- and 90-cm marks. Masses of 68.4 g and 100 g are suspended from the strings as shown in Figure 12-9. The friction in the pulleys is negligible. Determine both the mass of the meterstick and the location of its center of gravity.

How to Solve it
• Draw a free-body diagram of the stick. Choose a point on the stick an arbitrary distance x from the left end of the stick as the center of gravity. There are three forces acting on the stick: the force of gravity and the forces exerted on the stick by the two strings.

• Because the stick is in equilibrium, the net force acting on the stick is zero. Write an equation setting the vector sum of the forces acting on the stick to zero. Solve this equation for the weight of the stick and then use the relation $w = mg$ to determine the mass.

• The magnitudes, directions, and points of application of the forces exerted by the two strings are all known. Also, we know the magnitude and direction of the gravitational force, but not its point of application (the center of gravity). Following our rule-of-thumb, we select the axis through the point of application of this force since it is the least specified.

• To calculate the torque due to each of the two tension forces, use the formula $|\vec{\tau}| = |\vec{r} \times \vec{F}| = rF \sin \phi$, where $\vec{r}$ is the vector from the axis to the point of application of the force and ϕ is the angle between the directions of $\vec{r}$ and $\vec{F}$. For the tension force applied at the 10-cm mark, $r = x - 0.10$ m and $\phi = 95°$. For the tension force applied at the 90-cm mark find the corresponding expressions for r and ϕ. (Doing this will require that you construct a diagram in which the angles and lengths are carefully drawn and labeled.)

• Determine which of these torques is positive (counterclockwise) and which is negative. Set the sum of the torques equal to zero and solve for x.

9. A piece of tungsten wire 0.8 m long and a 0.5-m length of steel wire, each of diameter 1.0 mm, are joined end to end, making a total length of 1.3 m. If a mass of 22 kg is suspended from the ceiling by this vertical wire how much does the wire stretch? Neglect the weight of the wire. (Young's modulus is 2.0×10^{11} N/m² for steel; 3.6×10^{11} N/m² for tungsten.)

How to Solve It
• Draw a sketch of the situation. The tensile stress is the same throughout both wires since they have the same diameter and are connected end to end.

• Obtain expressions for the strain in each wire using the appropriate Young's modulus.

• Add the two extensions for each wire to obtain the net extension.

10. If a mass of 40 kg is hung from a 2-m copper wire of diameter 1.5 mm, by how much does the wire stretch? (Young's modulus for copper is 1.1×10^{11} N/m².)

How to Solve It
• Obtain a formula relating the stress and the strain.

• Solve for the extension of the wire.

Solutions

1. Figure 12-10 is a free-body diagram of the meterstick, where $\vec{w}_1$, $\vec{w}$, and $\vec{w}_{ms}$ are the weights of the 0.3-kg mass, the unknown mass, and the meterstick and $\vec{F}$ is the force exerted on the stick by the pivot. (Actually $\vec{w}_1$, the weight of the 3-kg mass, acts on the 3-kg mass and not on the meterstick. There are two forces acting on the 3-kg mass, the force exerted by the string and $\vec{w}_1$, the force exerted by gravity. Because the 3-kg mass is in equilibrium we know these forces are equal and opposite. Because the force exerted by the meterstick on the string attached to the 3-kg mass and the force exerted by the string on the meterstick form an action–reaction pair, we know that these forces must be equal in magnitude and oppositely directed. Thus we conclude that the force exerted by the string attached to the 3-kg mass on the meterstick is equal to $\vec{w}_1$.) Neither the magnitude nor the direction of $\vec{F}$ is known, but we do know that its line of action passes through the axis.

Setting the sum of the torques to zero yields

$$\sum_i \tau_i = w_1 x_1 - w x_2 + w_{ms}(0) + F(0) = 0$$

Figure 12-10

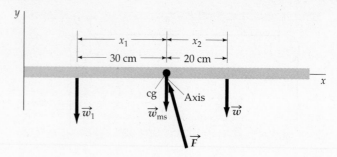

where x_1 and x_2 are the lever arms of the forces $\vec{w}_1$ and $\vec{w}$, with counterclockwise taken as the positive direction. Writing the weights in terms of the masses and mg gives

$$m_1 g x_1 - m g x_2 = 0$$

Solving for the mass m gives

$$m = m_1 \frac{x_1}{x_2} = (0.3 \text{ kg}) \frac{0.3 \text{ m}}{0.2 \text{ m}} = \underline{0.45 \text{ kg}}$$

To find the force $\vec{F}$ we set the sum of the forces acting on the meterstick to zero. Thus

$$\vec{F} + \vec{w}_1 + \vec{w}_{ms} + \vec{w} = 0$$
$$\vec{F} = -(\vec{w}_1 + \vec{w}_{ms} + \vec{w})$$
$$\vec{F} = (m_1 + m_{ms} + m)g\hat{j}$$
$$= (0.3 \text{ kg} + 0.1 \text{ kg} + 0.45 \text{ kg}) \, 9.81 \text{ m/s}^2 \hat{j}$$
$$= \underline{8.34 \text{ N}\hat{j}}$$

2. $T = \underline{5.89 \text{ N}}$, $\vec{F} = \underline{2.94 \text{ N, upward}}$

3. Figure 12-11 is a free-body diagram of the plank, where $\vec{w}_b$, $\vec{w}_g$, and $\vec{w}_p$ are the weights of the boy, the girl, and the plank, and $\vec{F}$ is the force exerted by the pivot. Neither the magnitude nor the direction of $\vec{F}$ is known. The axis is at the point of application of $\vec{F}$.

Figure 12-11

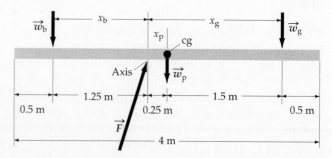

Setting the sum of the torques to zero yields

$$\sum_i \tau_i = w_b x_b - w_p x_p - w_g x_g + F(0) = 0$$

where x_b, x_p, and x_g are the lever arms of the forces $\vec{w}_b$, $\vec{w}_p$, and $\vec{w}_g$. (The lever arm of $\vec{F}$ equals zero.) The torques exerted by $\vec{w}_p$ and $\vec{w}_g$ are clockwise (negative) while the torque exerted by $\vec{w}_b$ is counterclockwise (positive). Writing the weights in terms of the masses and the acceleration due to gravity g gives

$$m_b g x_b - m_p g x_p - m_g g x_g = 0$$

Solving for the mass of the girl gives

$$m_g = \frac{m_b x_b - m_p x_p}{x_g}$$
$$= \frac{(40 \text{ kg})(1.25 \text{ m}) - (10 \text{ kg})(0.25 \text{ m})}{1.75 \text{ m}}$$
$$= 27.1 \text{ kg}$$

4. <u>90 kg</u>

5. Figure 12-12 is a free-body diagram of the strut. The least specified force is the force exerted by the hinge on the strut, so we will choose the point of application of that force as the axis. The strut makes an angle of 60° with the y axis and thus makes an angle of 30° with the x axis. Because the sum of the angles in a triangle equals 180° we know the angle between the strut and the string supporting the strut is 70°. Thus, $\ell = L \sin 70°$, where L is the length of the strut and ℓ is the lever arm of the force $\vec{T}$. The lever arms for $\vec{w}_s$ and $\vec{w}$ are $(L/2) \cos 30°$ and $L \cos 30°$, respectively.

Figure 12-12

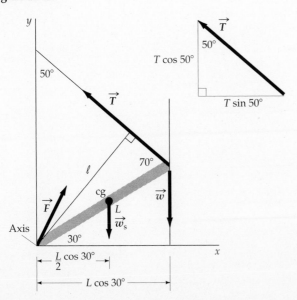

Because the strut is in equilibrium, the sum of the torques about the axis through the hinge equals zero:

$$\sum_i \tau_i = F(0) - w_s \frac{L}{2} \cos 30°$$
$$- wL \cos 30° + TL \sin 70° = 0$$

Expressing the weights in terms of the masses and the acceleration due to gravity g yields

$$m_s g \frac{L}{2} \cos 30° + mgL \cos 30° = TL \sin 70°$$

Solving for the tension gives

$$T = \frac{(m_s + 2m)g(\cos 30°)}{2 \sin 70°}$$
$$= \frac{(80 \text{ kg})(9.81 \text{ m/s}^2)(\cos 30°)}{2 \sin 70°} = \underline{362 \text{ N}}$$

The net external force acting on the strut equals zero. Thus

$$\vec{F} + \vec{w}_1 + \vec{w}_2 + \vec{T} = 0$$
$$\vec{F} = -(\vec{w}_1 + \vec{w}_2 + \vec{T})$$

From the illustration to the right of the free-body diagram we can see that

$$\vec{T} = -T \sin 50° \, \hat{i} + T \cos 50° \, \hat{j}$$

Using this expression for $\vec{T}$ and expressing the weights as the products of the masses and the acceleration due to gravity g we have

$$\vec{F} = -[(m_s + m)\vec{g} + \vec{T}]$$
$$= T \sin 50° \, \hat{i} + [(m_s + m)g - T \cos 50°]\hat{j}$$
$$= (362 \text{ N})(\sin 50°)\hat{i} + [(50 \text{ kg})(9.81 \text{ m/s}^2) - (362 \text{ N})(\cos 50°)]\hat{j}$$
$$= \underline{277 \text{ N}\hat{i} + 258 \text{ N}\hat{j}}$$

6. $T = \underline{110 \text{ N}}$, $\vec{F} = \underline{110 \text{ N}\hat{i} + 147 \text{ N}\hat{j}}$

7. Figure 12-13 is a free-body diagram of the rod.

Figure 12-13

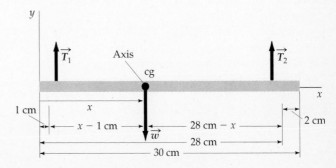

The rod is in equilibrium so the net torque about an axis through the center of gravity is zero. The lever arm about this axis for the torque exerted by $\vec{T}_1$ is $x - 1$ cm, where x is the distance of the center of gravity from the left end, and the lever arm for the torque exerted by $\vec{T}_2$ is 28 cm $- x$. The sum of the torques about the axis is zero. Thus,

$$\sum_i \tau_i = -T_1(x - 1.0 \text{ cm}) + w(0)$$
$$+ T_2(28 \text{ cm} - x) = 0 \qquad (1)$$

Because the string is massless and the pulleys are frictionless the tension is the same at all points in the string. Thus

$$T_1 = T_2 = T$$

Substituting T for both T_1 and T_2 in Equation 1 and solving for x yields

$$x = \underline{14.5 \text{ cm}}$$

(The center of gravity is 14.5 cm from the stick's left end.)

8. The mass of the stick is 153.2 g, and the center of gravity is at the 46.6 cm mark.

9. As shown in Figure 12-14, the total length L of the compound wire equals the length L_s of the steel segment plus the length L_t of the tungsten segment. It follows that the extension ΔL equals the extension ΔL_s plus the extension ΔL_t. Because the tension F and the cross-sectional area $A = \pi d^2/4$ is the same in each segment, the stress (F/A) is the same in each

Figure 12-14

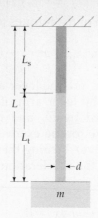

segment. The two strains ($\Delta L/L$) are related to the stress by the equations

$$Y_s = \frac{\text{stress}}{\text{strain}} = \frac{F/A}{\Delta L_s/L_s}$$

$$Y_t = \frac{\text{stress}}{\text{strain}} = \frac{F/A}{\Delta L_t/L_t}$$

The tension F equals mg, the weight of the 22-kg mass. Thus the total extension is

$$\Delta L = \Delta L_s + \Delta L_t = \frac{F}{A}\frac{L_s}{Y_s} + \frac{F}{A}\frac{L_t}{Y_t}$$

$$= \frac{F}{A}\left(\frac{L_s}{Y_s} + \frac{L_t}{Y_t}\right) = \frac{4mg}{\pi d^2}\left(\frac{L_s}{Y_s} + \frac{L_t}{Y_t}\right)$$

$$= \frac{4(22 \text{ kg})(9.81 \text{ m/s}^2)}{\pi(1.0 \times 10^{-3} \text{ m})^2}$$

$$\times \left(\frac{0.5 \text{ m}}{2 \times 10^{11} \text{ N/m}^2} + \frac{0.8 \text{ m}}{3.6 \times 10^{11} \text{ N/m}^2}\right)$$

$$= 1.30 \times 10^{-3} \text{ m} = \underline{1.30 \text{ mm}}$$

10. $\underline{4.03 \text{ mm}}$

Chapter 13

Fluids

I. Key Ideas

Matter in bulk can be either solid or fluid. Solids are more or less rigid and tend to maintain a fixed shape. Fluids (liquids and gases) flow more or less freely and assume the shape of their container.

13-1 Density The **density** ρ of a substance is its mass per unit volume:

Density

$$\rho = \frac{m}{V}$$

One cubic centimeter of water at 4°C has a mass of 1 gram (this was the original definition of the gram), so the density of water is 1 g/cm³. An object with a density lower than that of water floats in water; an object of greater density sinks. The ratio of the density of a substance to that of water is the **specific gravity** of the substance. Solids and liquids have densities that are roughly independent of external conditions, but the density of a gas is strongly affected by its pressure and temperature. Thus, the temperature and pressure must be specified when stating the density of a gas. Standard conditions are atmospheric pressure at sea level and a temperature of 0°C.

Weight density is the product of the mass density ρ and the acceleration due to gravity g; it is the weight per unit volume of a substance. It is most often used in the U.S. customary system of units. The weight density of water is 62.4 lb/ft³.

13-2 Pressure in a Fluid The distinction between fluids and solids is that fluids are unable to support any shear stress and so conform to the shape of their container. The force per unit area exerted by a fluid is called the **pressure** P:

Pressure

$$P = \frac{F}{A}$$

An increase in the pressure on a material tends to compress it in all directions at once. The ratio of the increase in pressure ΔP to the resulting fractional decrease in the volume $-\Delta V/V$ of the material is called the **bulk modulus** B:

Bulk modulus

$$B = -\frac{\Delta P}{\Delta V/V}$$

The **compressibility** k of the material is the reciprocal of the bulk modulus:

Compressibility

$$k = \frac{1}{B}$$

The pressure at any point in a fluid is the same in all directions, so it is a scalar quantity. Also, it increases with increasing depth because of the weight of the fluid above. For a static fluid the pressure is the same at all points at the same depth (in any horizontal plane). Any change in pressure is transmitted undiminished throughout the fluid. This is a statement of **Pascal's principle**. Many pressure gauges measure the difference between the "absolute" pressure P and the atmospheric pressure P_{at}. This difference between absolute pressure and atmospheric pressure is called the gauge pressure P_{gauge}:

Gauge pressure

$$P_{\text{gauge}} = P - P_{\text{at}}$$

Liquids are highly incompressible. As a result, the pressure in a liquid increases linearly with depth:

Pressure in a static liquid

$$P = P_0 + \rho g h$$

where P_0 is the pressure at the top of the liquid and h is the depth below the top. Gases, on the other hand, are highly compressible, so they have densities that are approximately proportional to the pressure. Therefore the pressure of the atmosphere decreases more or less exponentially with increasing altitude—that is, the pressure decreases by a constant fraction for a given increase in height.

13-3 Buoyancy and Archimedes' Principle

There is an upward, **buoyant force** on an object submerged in a fluid that is equal to the weight of the fluid displaced by the object. An object will float if it displaces a quantity of fluid equal in weight to the weight of the object.

Surface Tension and Capillarity The surface of a liquid is under tension due to the mutual attraction of the molecules of the liquid. Thus it exerts a force that resists any stretching or breaking. The force per unit length, denoted by γ, is the **coefficient of surface tension** of the liquid. Surface tension is the reason that small droplets of liquid tend to assume a spherical shape.

The behavior of a fluid at its surface when it is in contact with another material depends on the relative strength of (1) the *cohesive forces* between the molecules of the fluid and (2) the *adhesive forces* between the molecules of the fluid and the molecules of the second material. Thus, water "wets" glass because the glass–water adhesive forces are much stronger than the cohesive forces within the water.

If the liquid is one that "wets" the material of the tube, adhesive forces will cause a liquid in a small tube to rise above the level of the surrounding liquid. This is called **capillary action**. The height h to which a liquid with a density ρ and a coefficient of surface tension γ will rise in a vertical tube of radius r is

Capillary action

$$h = \frac{2\gamma \cos \theta_c}{\rho r g}$$

where θ_c, the contact angle between the liquid surface and the vertical direction, has a value deter-

mined by the type of liquid and the material of which the tube is made.

13-4 Fluids in Motion and Bernoulli's Equation

The general flow of a fluid can be very complicated, so we restrict our study here to *steady, nonturbulent flow*. If the fluid is incompressible (a liquid), then the **volume flow rate** I_V (volume per unit time) in a tube or pipe is the same throughout its length. That is,

Continuity equation for incompressible flow

$$I_V = Av = \text{constant}$$

where A is the cross-sectional area of the pipe and v is the speed of the fluid. The mass flow rate equals ρI_V (mass per unit volume × volume per unit time). For the steady flow of both compressible and incompressible fluids, the mass flow rate is the same throughout the length of the tube. By applying the work–energy theorem and neglecting any dissipation of energy, we obtain **Bernoulli's equation**, the fundamental dynamic equation for the steady, nonviscous flow of an incompressible fluid:

Bernoulli's equation

$$P + \rho g h + \tfrac{1}{2}\rho v^2 = \text{constant}$$

A pressure gradient is required to accelerate a nonviscous fluid; the acceleration is in the direction of decreasing pressure. Thus, in accordance with Newton's second law, a fluid speeds up when it flows into a region of lower pressure and slows down when it flows into a higher pressure region. (These remarks apply to horizontal flow, for which the height h is a constant; we speak here of acceleration due to a pressure gradient, not that due to gravity.) The pressure drop associated with the increase in the speed of a fluid is known as the **Venturi effect**.

Viscous Flow Nonviscous flow is an idealization. When real fluids flow, an internal shear stress is produced that opposes the flow. This shear stress increases with the speed of the flow. This property of fluid flow is known as **viscosity**. Because of viscosity, a pressure gradient is required to cause a fluid to flow through a horizontal pipe at constant speed. The pressure drop (which occurs along the direction of flow) is proportional to the flow rate:

Poiseuille's law

$$\Delta P = \frac{8\eta L}{\pi r^4} I_v$$

where L and r are the length and radius of the pipe and η, the **coefficient of viscosity**, is a measure of a fluid's resistance to flow.

Poiseuille's law and Bernoulli's equation apply only to steady, laminar (nonturbulent) flow. However, laminar flow ceases and turbulence sets in when the velocity of a fluid flowing through a tube becomes sufficiently great. The **Reynolds number** N_R is a dimensionless parameter that characterizes the degree of turbulence of the flow of a fluid. It is defined by

Reynolds number

$$N_R = \frac{2r\rho v}{\eta}$$

where r is the radius of the tube. The transition from laminar to turbulent flow occurs as the Reynolds number increases from 2000 to 3000.

II. Numbers and Key Equations

Numbers

1 pascal (Pa) $= 1 \, N/m^2 = 1 \, kg/m{\cdot}s^2$

1 atmosphere (atm) $= 760 \, mmHg$

$\qquad\qquad\qquad = 1.013 \times 10^5 \, Pa$

1 bar $= 10^3$ millibars (mbar) $= 10^5 \, Pa$

1 torr $= 1 \, mmHg = 133.3 \, Pa$

1 poise $= 0.1 \, Pa{\cdot}s$

Table 13-1 Approximate Values of the Bulk Modulus B of Various Materials

Material	B, GN/m^2
Aluminum	70
Brass	61
Copper	140
Iron	100
Lead	7.7
Mercury	27
Steel	160
Tungsten	200
Water	2.0

Key Equations

Density

$$\rho = \frac{m}{V}$$

Pressure

$$P = \frac{F}{A}$$

Bulk modulus

$$B = -\frac{\Delta P}{\Delta V/V}$$

Compressibility

$$k = \frac{1}{B}$$

Gauge pressure

$$P_{gauge} = P - P_{at}$$

Pressure in a static liquid

$$P = P_0 + \rho g h$$

Capillary action

$$h = \frac{2\gamma \cos \theta_c}{\rho r g}$$

Continuity equation for incompressible flow

$$I_V = Av = \text{constant}$$

Bernoulli's equation

$$P + \rho g h + \tfrac{1}{2}\rho v^2 = \text{constant}$$

Poiseuille's law

$$\Delta P = \frac{8\eta L}{\pi r^4} I_v$$

Reynolds number

$$N_R = \frac{2r\rho v}{\eta}$$

III. Potential Pitfalls

It's easy to confuse density (mass density) with either weight density or specific gravity. Density is the mass per unit volume of a substance; multiplying it by g gives the weight per unit volume or weight density. Contrary to the way it may sound, specific gravity is not defined in terms of weight. It is the ratio of the density of a substance to the density of water. It is thus a dimensionless number that is approximately numerically identical to the density expressed in grams per cubic centimeter.

Don't forget that the decrease of pressure with height (or the increase with depth) is linear only for a fluid that is incompressible and therefore has a constant density. The decrease of atmospheric pressure with height is not at all linear. This is because air is compressible so its density decreases with increasing height.

When solving problems that deal with the pressure at some depth in a fluid, don't forget that ρgh is the pressure difference over a vertical distance h. It's not necessarily the pressure at either the top or bottom.

The equilibrium condition for an object floating in a fluid is that the weight of the object is equal to the weight of the volume of fluid that the submerged portion of the object displaces. The volume of the fluid displaced is, of course, less than or equal to the volume of the object.

Just because an object sinks in a liquid, don't think that there is no buoyant force acting on it. The buoyant force is simply less than the object's weight.

Bernoulli's equation and the continuity equation (Av = constant) are applicable only for the flow of incompressible fluids, so they are of limited use for studying gas flow. The Bernoulli equation applies only for nonviscous laminar flow and Poiseuille's law applies only for laminar flow.

You can really mess up calculations with Bernoulli's equation if you are not careful to use consistent units for all the quantities. It's best to put everything into SI units.

When doing numerical problems be sure to write out all units and determine the units of your answer. Use the conversions 1 Pa = 1 N/m², 10 poise = 1 Pa·s = 1 N·s/m², and the like.

IV. True and False and Responses

True and False

_____ 1. Fluids flow—changing their shape—freely, but they maintain a nearly constant volume.
_____ 2. Solids tend to be rigid (to maintain their shape and volume).
_____ 3. The bulk modulus and Young's modulus have the same dimensions.
_____ 4. The bulk modulus is the negative of the compressibility.
_____ 5. The density of a substance is the weight per unit volume of a substance.

_____ 6. The density of a substance can be expressed as the product of its specific gravity and the density of water.
_____ 7. An elastic material is one that is easily deformed.
_____ 8. The ratio of the increase in the pressure on a material to the fractional decrease in its volume is called its compressibility.
_____ 9. The pressure in a static fluid is the same at every point throughout the fluid.
_____ 10. The buoyant force on an object completely submerged in a fluid depends on the density of the fluid and the volume of the object but not on the shape or composition of the object.
_____ 11. A body that floats in a liquid does so at a depth at which it displaces an amount of fluid equal in weight to its own weight.
_____ 12. Among units of pressure, 1 bar is equal to 1 atmosphere.
_____ 13. The surface tension in a liquid arises from forces of attraction between the liquid and the walls of the container holding it.
_____ 14. The free surface of water in a narrow glass tube is concave because the cohesive forces are stronger than the adhesive forces.
_____ 15. Buoyancy arises from the increase in pressure with depth in a fluid.
_____ 16. Shearing forces cannot exist in a fluid.
_____ 17. Bernoulli's equation applies to any flow of an incompressible fluid.
_____ 18. Bernoulli's equation gives incorrect results when viscosity is significant.
_____ 19. Bernoulli's equation is derived by requiring that the total mechanical energy of an element of the fluid remains constant.

Responses

1. False. The statement is true for a liquid, but the term fluid also refers to gases.

2. True

3. True

4. False. The bulk modulus is the reciprocal of the compressibility.

5. False. Density is the mass per unit volume.

6. True. This is the definition of specific gravity.

7. False. Easily deformed materials aren't necessarily elastic. What "elastic" means is that deformations of the material are reversible. Elastic materials can be stiff or spongy.

8. False. The statement describes the material's bulk modulus, which is the reciprocal of its compressibility.

9. False. The pressure varies with depth because of the weight of the fluid lying above. However, any change in pressure at one point in a fluid results in an equal change at all points in it.

10. True

11. True

12. Not exactly; 1 atm = 1.013 bar.

13. False. Surface tension arises from attractive forces between molecules of the liquid.

14. False. It is concave because the adhesive forces between the glass wall and the water are stronger than the cohesive forces within the water.

15. True

16. False. A fluid cannot sustain a static shearing stress, but shearing stresses (called viscous forces) do exist in a real flowing fluid. They do not exist in an *ideal* fluid.

17. False. It applies only to the steady, nonturbulent flow of a nonviscous, incompressible fluid.

18. True

19. False. It is derived by applying the work–energy theorem to account for the change in mechanical energy.

V. Questions and Answers

Questions

1. Assuming that you float in fresh water with about 5% of your body above water, estimate the volume of your body.

2. The "fog" produced by evaporating "dry ice" (frozen CO_2) stays at ground level and spreads along the ground. Why is this?

3. Water bugs can walk on the surface of a pond or lake. Could a monster water bug (in a horror movie) 20 times larger in all its dimensions pull the same trick? Why or why not?

4. Figure 13-1 shows a small vessel upended in water,

Figure 13-1

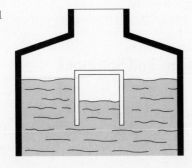

which is contained in a larger vessel, so as to trap some air inside. This gadget is known as a Cartesian diver. If the air pressure above the water is increased (for example, by pressing on a flexible lid on the larger container), the small vessel sinks to the bottom. Why does it sink? How can it be made to rise again?

5. Once, in the course of an experiment, I needed to know the volume within a large, irregular, thin-walled aluminum container. The volume was of the order of 1 m³. Someone suggested the obvious, which was to fill the container with water from vessels of known volume. The only drawback was that a cubic meter of water weighs about a ton and would have deformed the thin aluminum walls of the container. Can you suggest a way to solve this problem?

6. A colleague of mine likes to do the following demonstration. He pours some red-dyed water into a glass U-tube that is open at both ends to show that the fluid "seeks its own level" (that is, it rises to the same height on both sides of the U-tube). He then pours in some more red liquid, and the fluid in the U-tube comes to rest as shown in Figure 13-2. What's the trick here?

Figure 13-2

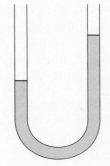

7. If the height to which capillary action would raise water in a tube is greater than the tube's actual height, will the water rise and flow out the top of the tube like a fountain?

8. An ice cube floats in a glass of water. As the ice melts, does the water level rise or fall? Explain.

9. Aircraft carriers steam into the wind at an angle so that the planes take off headed directly into the wind. Why? In what direction should they head (into or with the wind) when the planes return? Assume that relative to the flight deck the planes travel in the same direction during recovery as during launch.

10. The fluid pressure in a pipe decreases as you go downstream, even when the pipe is level and the fluid is incompressible. Why?

11. No pump that works by suction can raise water in a pipe higher than about 34 feet. Why not?

12. You can't breathe underwater by drawing air through a tube stretching from the water's surface (a "snorkel") to a depth of much more than a foot. Why not?

Answers

1. You can answer this question only if you know your mass. Suppose your mass is 70 kg. For you to float, the buoyant force of the water must equal your weight. Thus, by Archimedes' principle, you must displace 70 kg of water. A 70-kg mass of water has a volume of 70 L or 0.070 m³. The submerged 95% of your body must therefore have a volume of 0.070 m³, so the total volume of your body must be 0.074 m³ (= 0.070 m³/0.95).

2. The "fog" that you see is water vapor condensing out of the air locally due to the presence of the cold CO_2 vapor. Although the CO_2 vapor itself is transparent, you can see where it is by means of the condensed water vapor droplets. The CO_2 vapor stays at ground level because it is substantially more dense than air.

3. It seems to me I saw an old black-and-white second feature a long time ago in which this was shown. In any case, it couldn't really happen. A scaled-up water bug 20 times larger in every direction would weigh 20³ (or 8000) times as much as one of normal size. However, the supporting force due to surface tension would increase proportionally to the circumference of the bug's foot, so it would be only 20 times greater. This force would be far too small to hold the bug up.

4. The increased pressure on the water surface is transmitted to the trapped air inside the small vessel; this is in accordance with Pascal's principle. The trapped air is compressed and therefore displaces less water, so in accordance with Archimedes' principle the buoyant force holding the small vessel up is reduced. Increase the pressure enough, and the small vessel sinks. Decrease it enough, and the trapped air expands enough to raise the sunken vessel again.

5. We floated the container in a swimming pool and filled it with water from a calibrated vessel. The water pressure inside and outside the container was then about the same, so there was little or no stress on its fragile walls.

6. The fluid he added the second time wasn't water but some liquid that is less dense than water and doesn't mix readily with it. Thus, a taller column of the second fluid was required to produce the same pressure in the liquid at the bottom of the U-tube.

7. No. The only reason the water rises is because of the adhesive force attracting it to the vertical wall of the tube pulling it upward. Thus, it can only rise as high as the vertical wall of the tube.

8. When the ice is floating, the part of it that is underwater is displacing a volume of water that is equal in weight to the whole weight of the ice. This is exactly the volume of the water produced by the melted ice, so the water level goes neither up nor down but stays the same. (Note that water shrinks on melting.)

9. To handle both launch and recovery most easily, you want a plane's speed, relative to the flight deck, to be as slow as safety permits. Adequate lift for takeoff requires a certain air speed. By taking off with the flight deck moving into the wind, the plane acquires the required air speed while minimizing the speed of the plane relative to the deck. During touchdown, the plane also requires a certain air speed (not necessarily the same as the air speed required for takeoff). To minimize the speed of the plane relative to the flight deck, the carrier heads into the wind during recovery as during launch.

10. The pressure decreases because the fluid has viscosity. If a steady flow is to be maintained, there must be a pressure difference to maintain the flow of the fluid against this friction.

11. A suction pump raises water by creating a pressure in the pipe that is lower than atmospheric pressure; it is the pressure of the atmosphere that raises the water. Since the absolute pressure at the pump cannot be less than zero, the largest pressure available to raise the water is equal to atmospheric pressure, which corresponds to a height of about 34 ft of water.

12. To draw air into your lungs, your diaphragm must lower the pressure in your lungs to less than atmospheric pressure. At a depth of a foot, the water outside your body is already at a pressure around 3 kPa higher than atmospheric. Because 4 to 7 kPa is the maximum external pressure difference your diaphragm can handle, you can't suck air through a tube at a depth of much more than a foot.

VI. Problems and Solutions
Problems

1. The density of water is 1.000 g/cm³ at 4°C. Suppose that a 500-mL flask is filled exactly full of water at a temperature of 60°C, where the density of water is (say) 0.980 g/cm³. When the flask of water is cooled to 4°C, how much more water must be added to fill the flask again? Neglect the thermal expansion or contraction of the flask itself.

How to Solve It
• How many grams of water are in the flask at 60°C?

• How many cubic centimeters does this much water occupy at 4°C?

2. The mass of a small analytic flask is 15.2 g. When the flask is filled with water, the mass of the flask and water total 119.0 g; when it is filled with an unknown fluid, the total mass is 96.7 g. What is the specific gravity of the unknown fluid?

How to Solve It
• The total mass of the filled vessel is the mass of the fluid in the flask plus the mass of the flask itself.

• Thus, you know the mass of the same (unknown) volume of water and of the unknown fluid.

• Specific gravity is defined as the ratio of the density of a substance to that of water. Since equal volumes are involved here, the ratio of the mass of the unknown fluid to that of water equals the specific gravity of the unknown fluid.

3. A lead brick measuring $5.00 \times 10.0 \times 20.0$ cm is dropped into a swimming pool 3.5 m deep. By how much does the volume of the brick change due to the pressure of the water?

How to Solve It
• What is the pressure at the bottom of 3.5 m of water? Be careful: should you include the pressure of the atmosphere in this calculation?

• The bulk modulus of lead is given in Table 13-1 on page 133 of this Study Guide.

4. Blood pressure is normally measured on a patient's arm at approximately the level of the heart. If it were measured instead on the leg of a standing patient, how significantly would the reading be affected? Normal blood pressure is in the range of 70 to 140 torr, and the specific gravity of normal blood is 1.06.

How to Solve It
• The question is simply, how much does pressure change with height in blood?

• You'll have to pick some approximate value for the height between arm and leg—say, one meter.

• Calculating the pressure difference over this height won't give you precisely the correct pressure in the leg because the blood is actually flowing and

it is viscous. It will give you an idea of how significant the difference is, however.

5. Some colleagues and I once had to carry out an experiment on a lake. To do this we built a raft out of Styrofoam, 1 ft (0.305 m) thick and 8 ft (2.44 m) square. Take the specific gravity of Styrofoam to be 0.035. (*a*) How deep in the water did the unloaded raft float? (*b*) With three men (whose average mass was 88 kg each) and 120 kg of experimental equipment on the raft, how deep did it float?

How to Solve It
• When it is floating in equilibrium, the weight of the raft plus the weight of the load is equal to the weight of the displaced water.

• The volume of water displaced is the area of the raft multiplied by the depth to which it sinks.

6. The density of air under "ordinary" (not standard) conditions is about 1.2 kg/m³, whereas that of helium is 0.17 kg/m³. What must the radius of a spherical helium-filled balloon be if it is to lift a total load of 350 kg, not including the helium itself?

How to Solve It
• The balloon will rise if its mass (including the helium and the load) is less than that of the air it displaces.

• Express the weight of a sphere of air in terms of its density and the radius of the sphere.

• Equate the weight of the displaced air to the weight of the helium plus the weight of the load.

7. In Figure 13-3, oil with a specific gravity of 0.68 is floating on top of some water in a container. A wooden object is floating at the fluid boundary such that one third of its volume lies below the boundary. What is the density of the wood?

Figure 13-3

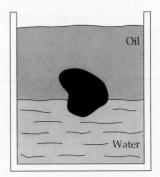

How to Solve It
• The wooden object will float such that the weight of the fluid it displaces is equal to its weight.

• The fluid it displaces is one-third water and two-thirds oil, by volume.

• What is the total mass of the displaced fluids? This must equal the mass of the wooden object.

8. A water bug stands on the surface of a pond on six legs and does not sink. The surface tension of water is about 0.073 N/m. Assuming that the diameter of the end of each leg is 0.107 mm, what is the maximum possible mass of the bug?

How to Solve It
• The circumference of one of the bug's feet multiplied by the coefficient of surface tension of water gives the maximum support force that the water can provide for each foot.

• The bug has six feet.

9. A vertical capillary tube, with an inside diameter 0.8 mm, extends 15 cm below the surface of a body of water. What is the minimum (gauge) pressure required to blow air down through the tube and out its bottom? The surface tension of water is 0.073 N/m and the glass–water contact angle is 0°.

How to Solve It
• Draw a picture of the tube and the water and consider the forces acting on the column of water in the tube. Obtain an expression relating the pressure of the air in the tube, the coefficient of surface tension, the density of water, the length of the water column, and the pressure at the bottom of the tube.

• Obtain a second expression for the pressure at a depth equal to the depth of the bottom of the tube. This expression should relate this pressure with the atmospheric pressure of the air above the body of water, the density of water, and the depth at which the bottom of the tube is submerged.

• Equate the two expressions for the pressure at the bottom of the tube and solve for the air pressure in the tube when the length of the water column in the tube equals zero.

10. Water is flowing smoothly at 15 ft/s in a horizontal pipe with a 2-in inside diameter at an absolute pressure of 40 lb/in². Neglect the viscosity of water. At a certain point the diameter of the pipe necks down to 1.05 in. (*a*) How fast does the water flow in the narrow section? (*b*) What is its pressure there (in lb/in²)?

How to Solve It
• Before you do anything else, convert the flow velocity, pressure, and pipe diameters to SI units. It's hopeless to try to do a complicated problem without putting all quantities into a consistent system of units.

• Use the continuity equation to find the water's speed in the narrow section.

• Since the pipe is horizontal, there are only two terms (the pressure and velocity terms) in Bernoulli's equation.

• Use Bernoulli's equation to find the pressure difference required to cause this change in speed.

11. A town's water tank is supported above ground on posts. Its diameter is 18 m, and the water in the tank is 7.5 m deep. The top is open to the atmosphere. If a hole is punched in the bottom of the tank and the water flows out in a stream 1 cm in diameter, how long does it take for the water level to drop by 10 cm? Neglect the viscosity of the water.

How to Solve It
• Use Bernoulli's equation and the equation of continuity to relate the flow at the top of the tank with the flow out of the 1-cm hole.

• You know enough to calculate all three terms in Bernoulli's equation at the top surface of the water.

• Neglect the speed of descent of the water at the top of the tank.

• The volume that leaked out equals the volume flow rate times the time.

12. At a certain point in a sloping pipe 2.5 cm in diameter, water is flowing at 10 m/s. From there the pipe gradually descends vertically by 7 m, and over the same distance its diameter increases to 3.5 cm. What is the pressure difference between the upper and lower points?

How to Solve It
• Using the continuity equation, you can determine how fast the water is flowing where the pipe diameter is 3.5 cm.

• Use Bernoulli's equation to calculate the pressure difference between the upper and lower points.

13. Water is supplied to an outlet from a pumping station 5 km away. From the pumping station to the outlet there is a net vertical rise of 19 m. Take the coefficient of viscosity of water to be 0.01 poise. The pipe leading from the pumping station to the outlet is 1 cm in diameter, and the gauge pressure in the pipe at the point where it exits the pumping station is 520 kPa. At what volume flow rate does water flow from the outlet?

How to Solve It

• Bernoulli's equation will not work here since the water is viscous.

• There will be a pressure drop due to viscous flow in the pipe and a drop due to the 19-m vertical rise.

• You know the total pressure drop from the pumping station to the outlet and can calculate the drop due to the vertical rise. The difference is the pressure drop due to viscous drag in the pipe.

• Assume the flow is laminar and use Poiseuille's law to find the volume flow rate at the outlet.

14. Blood flows in the finer capillaries of the body at a rate of around 1 mm/s. If blood flows at this rate through a capillary 8 μm in diameter, what is the pressure difference required to move blood at this rate through a capillary 1.8 mm long? Assume that the coefficient of viscosity of blood is 4.0 mPa·s.

How to Solve It

• Calculate the volume flow rate in the capillary.

• If you get all the units consistent, this is just a plug into Poiseuille's law.

• It may interest you to put the result in torr to compare it to ordinary blood pressure values of 70 to 140 torr.

15. Water ($\eta = 0.01$ poise) is pumped into one end of a level 50-ft-long 1-in-diameter pipe at a gauge pressure of 40 lb/in². (a) If the flow is nonturbulent, how much water can be pumped through the pipe in 1 min if its other end is open to the atmosphere? (b) Calculate the Reynolds number for this flow to see whether or not your assumption was justified.

How to Solve It

• Use Poiseuille's law to get the volume flow rate in the pipe.

• From the volume flow rate, calculate the velocity at which the water is flowing.

• From the velocity and the other data given, calculate the Reynolds number.

16. In Problem 13, Poiseuille's law, which holds only for nonturbulent flow, predicts that water ($\eta = 0.01$ poise) flows through the 1-cm-diameter pipe at a rate of 1.64×10^{-5} m³/s. Determine the Reynolds number for this flowing water, and state whether or not the assumption of nonturbulent flow is appropriate.

How to Solve It

• Calculate the speed of the water from the volume flow rate.

• Determine the Reynolds number. What does this number tell you about the probable turbulence of the flow?

Solutions

1. At 60°C the density of water is 0.980 g/cm³. The mass of water that fills the flask at that temperature is thus

$$m = \rho V = (0.980 \text{ g/cm}^3)(500 \text{ cm}^3) = 490 \text{ g}$$

At 4°C the density of water is 1.000 g/cm³, so at that temperature the 490 g of water occupies 490 cm³. Thus <u>10 cm³</u> must be added to fill the flask again.

2. <u>0.785</u>

3. The brick's original dimensions were presumably measured at atmospheric pressure. Hence it is the increase in pressure that compresses the brick. The bulk modulus is the negative of the ratio of the change in pressure to the fractional change in volume. Thus,

$$B = -\frac{\Delta P}{\Delta V / V}$$

so

$$\Delta V = -\frac{V \Delta P}{B} = -\frac{V \rho g \Delta h}{B}$$

$$= -\frac{(0.001 \text{ m}^3)(1000 \text{ kg/m}^3)(9.81 \text{ m/s}^2)(3.5 \text{ m})}{7.7 \times 10^9 \text{ N/m}^2}$$

$$= -4.46 \times 10^{-9} \text{ m}^3 \text{ (or } -4.46 \text{ mm}^3)$$

where we have used

$$V = (0.05 \text{ m})(0.1 \text{ m})(0.2 \text{ m}) = 0.001 \text{ m}^3$$

This is a fractional decrease in volume of

$$-\frac{\Delta V}{V} = \frac{4.46 \times 10^{-9} \text{ m}}{0.001 \text{ m}^3}$$

$$= 4.46 \times 10^{-6} \text{ (or about 0.0004 percent)}$$

which is minuscule and not likely to be noticeable.

4. The excess pressure at a depth of 1 m in blood would be 10.6 kPa or 78 torr, so the difference is significant.

5. (a) To float, the raft must displace a weight of water equal to the weight of the raft and its cargo. The volume V of displaced water equals the product Ah where A is the area of the downward-facing surface and h is the depth at which this surface floats. The mass m_w of the displaced water equals the product of the displaced volume and the density ρ of water. The mass m_S of the Styrofoam slab equals the product of the volume of the slab, the density of water, and the specific gravity (s.g.) of Styrofoam. Therefore

$$m_w g = m_S g$$

$$Ah\rho g = At\rho(\text{s.g.})g$$

where t is the thickness of the slab. Solving for h we have

$$h = t(\text{s.g.}) = (1 \text{ ft})(0.035) = \underline{0.035 \text{ ft (or 1.07 cm)}}$$

(b) With the three men and equipment on the raft we have

$$m_w g = (m_S + 3m_m + m_{eq})g$$

$$Ah\rho g = [At\rho(\text{s.g.}) + 3m_m + m_{eq}]g$$

where m_m is the average mass of a man and m_{eq} is the mass of the equipment. Solving for h we have

$$h = t(\text{s.g.}) + \frac{3m_m + m_{eq}}{A\rho}$$

$$= (0.305 \text{ m})(0.035) + \frac{3(88 \text{ kg}) + 120 \text{ kg}}{(2.44 \text{ m})^2(1000 \text{ kg/m}^3)}$$

$$= \underline{0.0752 \text{ m (or 7.52 cm)}}$$

It sat virtually on top of the water.

6. The radius must be $\underline{4.32 \text{ m}}$.

7. Let V be the volume of the wooden object. Then the total mass m of fluid displaced is

$$m = \rho_{water}V_{water} + \rho_{oil}V_{oil}$$

$$= \rho_{water}\frac{V}{3} + (\text{s.g.})\rho_{water}\frac{2V}{3}$$

$$= [1 + 2(\text{s.g.})]\frac{\rho_{water}V}{3}$$

To solve for the density of the wood we divide though the equation by V:

$$\rho_{wood} = \frac{m}{V} = [1 + 2(\text{s.g.})]\frac{\rho_{water}}{3}$$

$$= \frac{1 + 2(0.68)}{3}(1000 \text{ kg/m}^3) = \underline{787 \text{ kg/m}^3}$$

8. $\underline{0.015 \text{ g}}$

9. Figure 13-4 shows that a vertical tube with an inside radius r and a cross-sectional area $A = \pi r^2$ is partially full of water with density ρ and coefficient of surface tension γ. Consider the forces acting on the column of water in the tube.

Figure 13-4

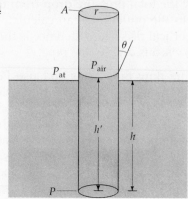

The air in the tube pushes down on this column with a force equal to $P_{air}A$, the water immediately below the tube pushes up on it with a force equal to PA, where P is the pressure in the liquid at the bottom of the tube, and gravity pulls down on the column with a force equal to $\rho gh'A$, where h' is the height of the water column. Also, the wall of the cylinder pulls up on the column with a net force equal to $\gamma 2\pi r \cos\theta$, where $2\pi r$ is the length of the wall-surface boundary, γ is the coefficient of surface tension (the force per unit length), and θ is the angle the surface-tension forces make with the vertical. The column is in static equilibrium so the sum of the downward forces acting on it equals the sum of the upward forces. That is,

$$P_{air}A + \rho gh'A = PA + \gamma 2\pi r \cos\theta$$

The pressure P at the bottom of the tube is $P_{at} + \rho gh$. Substituting we obtain

$$P_{air}A + \rho gh'A = (P_{at}A + \rho ghA) + \gamma 2\pi r \cos\theta$$

In our case, we want to know $P_{gauge} = P_{air} - P_{at}$ when $h' = 0$. Also, $A = \pi r^2$. Solving for $P_{air} - P_{at}$ when $h' = 0$ yields

$$P_{gauge} = P_{air} - P_{at} = \rho gh + \frac{\gamma 2\pi r \cos\theta}{\pi r^2}$$

$$= \rho gh + \frac{2\gamma \cos\theta}{r}$$

$$= (1000 \text{ kg/m}^3)(9.81 \text{ m/s}^2)(0.15 \text{ m})$$
$$+ \frac{2(0.073 \text{ N/m})(\cos 0)}{0.4 \times 10^{-3} \text{ m}}$$

$$= \underline{1.84 \text{ kPa}}$$

10. (*a*) 54.4 ft/s   (*b*) 21.6 lb/in²

11. The equation of continuity for an incompressible fluid states that the volume flow rate $I_V = Av$ is the same at different locations in the fluid. Because we are assuming a steady, smooth flow, and because we are neglecting viscous effects, we may use the Bernoulli and continuity equations. Thus,

$$P_1 + \rho g h_1 + \tfrac{1}{2}\rho v_1^2 = P_2 + \rho g h_2 + \tfrac{1}{2}\rho v_2^2$$

and

$$A_1 v_1 = A_2 v_2$$

where 1 and 2 refer to locations at the top of the water and at the 1-cm hole, respectively. Because the top of the tank and the hole are open to the atmosphere, $P_1 = P_2 = P_{at}$. The height $h_1 = h = 7.5$ m and $h_2 = 0$. Thus,

$$P_{at} + \rho g h + \tfrac{1}{2}\rho v_1^2 = P_{at} + \rho g(0) + \tfrac{1}{2}\rho v_2^2$$

By subtracting P_{at} from both sides and then multiplying through by $2/\rho$ we obtain

$$2gh + v_1^2 = v_2^2$$

Solving the equation $A_1 v_1 = A_2 v_2$ for v_1 and then substituting for it we obtain

$$2gh + \left(\frac{A_2}{A_1}\right)^2 v_2^2 = v_2^2$$

Solving for v_2^2 yields

$$v_2^2 = 2gh\left[1 - \left(\frac{A_2}{A_1}\right)^2\right]^{-1} \approx 2gh$$

where we have neglected $(A_2/A_1)^2$ in comparison with 1. Using the values for the diameters that are given in the problem statement we find $(A_2/A_1)^2 \approx 10^{-13}$. Thus, the volume flow rate is

$$I_V = A_2 v_2 = A_2 \sqrt{2gh}$$

The volume of water that leaks out when the level drops by $d = 10$ cm is $A_1 d$. Since the volume flow rate equals the volume that leaks out divided by the time, we have

$$I_V = \frac{A_1 d}{t}$$

or

$$t = \frac{A_1 d}{I_V} = \frac{A_1 d}{A_2 \sqrt{2gh}} = \frac{d}{\sqrt{2gh}}\left(\frac{D_1}{D_2}\right)^2$$

$$= \frac{0.1 \text{ m}}{\sqrt{2(9.81 \text{ m/s}^2)(7.5 \text{ m})}}\left(\frac{18 \text{ m}}{0.01 \text{ m}}\right)^2$$

$$= 2.67 \times 10^4 \text{ s (or 7.42 h)}$$

where we have used the result that the ratio of the areas of two circles is equal to the square of the ratio of the diameters.

12. 106 kPa

13. There will be a pressure drop due to viscous flow in the pipe (given by Poiseuille's law) plus a pressure drop due to the 19-m vertical rise. Since the outlet is open to the atmosphere, the total pressure difference is the 520 kPa given. The pressure drop due to the vertical rise is

$$P = \rho g \Delta y = (1000 \text{ kg/m}^3)(9.81 \text{ m/s}^2)(19 \text{ m})$$

$$= 1.86 \times 10^5 \text{ Pa} = 186 \text{ kPa}$$

The remaining pressure drop of 520 kPa − 186 kPa = 334 kPa must be due to viscosity in the pipe. From Poiseuille's law,

$$\Delta P = \frac{8\eta L}{\pi r^4} I_V$$

so

$$I_V = \frac{\pi r^4}{8\eta L}\Delta P$$

$$= \frac{\pi(0.005 \text{ m})^4}{8(0.001 \text{ Pa·s})(5 \times 10^3 \text{ m})}(3.34 \times 10^5 \text{ Pa})$$

$$= 1.64 \times 10^{-5} \text{ m}^3/\text{s (or 16 cm}^3/\text{s)}$$

14. 3.6 kPa (or 27 torr)

15. (*a*) We get the flow rate from Poiseuille's law. The pressure drop is 40 lb/in² = 276 kPa, the pipe's diameter is 1 in = 0.0254 m, and its length is 50 ft = 15.2 m. Thus,

$$I_V = \frac{\pi r^4}{8\eta L}\Delta P$$

$$= \frac{\pi(1.27 \times 10^{-2} \text{ m})^4}{8(0.001 \text{ Pa·s})(15.2 \text{ m})}(2.76 \times 10^5 \text{ Pa})$$

$$= 0.185 \text{ m}^3/\text{s} = 11.1 \text{ m}^3/\text{min} = 2930 \text{ gal/min!}$$

(b) The velocity of the water leaving the pipe outlet is

$$v = \frac{I_V}{A} = \frac{0.185 \text{ m}^3/\text{s}}{\pi(1.27 \times 10^{-2} \text{ m})^2} = 365 \text{ m/s}$$

The Reynolds number is therefore

$$N_R = \frac{2r\rho v}{\eta}$$

$$= \frac{2(1.27 \times 10^{-2} \text{ m})(1000 \text{ kg/m}^3)(365 \text{ m/s})}{0.001 \text{ Pa·s}}$$

$$= \underline{9.3 \times 10^6}$$

Turbulent flow occurs when the Reynolds number exceeds 3000. By assuming that the flow is laminar and using Poiseuille's law, we have obtained a Reynolds number = 9,000,000, so the assumption is wrong. In turbulent flow both the actual flow rate and the actual Reynolds number will be considerably less.

16. $N_R = \underline{2090}$ Flow is laminar (nonturbulent) for Reynolds numbers less than about 2000 and turbulent for Reynolds numbers greater than 3000. A Reynolds number of 2090 is near the lower boundary of the transition region from laminar to turbulent flow. Thus, the use of Poiseuille's law is probably, but not necessarily, justified.

Chapter 14

Oscillations

I. Key Ideas

Oscillatory, periodic motions are very important in nature. Oscillation occurs when a system is disturbed from a stable equilibrium; a restoring force causes oscillation around the equilibrium position. Wave motion is a closely related phenomenon.

14-1 Simple Harmonic Motion Simple harmonic motion is the oscillatory motion that occurs under a restoring force that is proportional to the displacement of the system from equilibrium. That is,

Linear restoring force (Hooke's Law)

$$F_x = -kx$$

This is nearly always the case provided that the displacement from equilibrium is small enough. The resulting motion has acceleration that is proportional to the position.

Acceleration proportional to displacement

$$a_x = -\omega^2 x$$

The motion is a sinusoidal oscillation; that is, a graph of the displacement versus time has the form of a sine function. A typical example would be a mass oscillating on the end of a spring.

Period and Frequency The frequency f of an oscillating system is the number of cyles completed per unit of time. The period T of the motion is the time required for one cycle of the motion. These are related by the equation

Period

$$T = \frac{1}{f}$$

The period is related to the force constant k of the spring and the mass m of the oscillating object by the equation

Period for particle on a linear spring

$$T = 2\pi\sqrt{\frac{m}{k}}$$

For simple harmonic motion, the period is independent of amplitude.

Circular Motion and Simple Harmonic Motion There is a very close connection between simple harmonic motion and circular motion with constant speed. The projection on one coordinate axis of a point undergoing uniform circular motion is a simple harmonic motion. The frequency is related to the angular velocity ω of the circular motion by the equation

Frequency–angular velocity

$$f = \frac{\omega}{2\pi}$$

For simple harmonic motion the position and velocity of the oscillating body are expressed as functions of time by the relations

Position in simple harmonic motion

$$x = A \cos(\omega t + \delta)$$

Velocity in simple harmonic motion

$$v = \frac{dx}{dt} = -\omega A \sin(\omega t + \delta)$$

where A is the amplitude of the motion and δ is the phase constant. The argument of the trigonometric functions $(\omega t + \delta)$ is the phase of the motion.

14-2 Energy in Simple Harmonic Motion
An object undergoing simple harmonic motion has a constant total energy, but its potential energy and its kinetic energy vary with time. For a linear restoring force (Hooke's law) the potential energy is $\frac{1}{2}kx^2$. At the turning points, the kinetic energy is zero and the potential energy has its maximum value. As the system passes through its equilibrium point, the reverse is true. That is, $E_{\text{total}} = U_{\text{max}} = U + K = K_{\text{max}}$ or

Conservation of energy

$$E_{\text{total}} = \tfrac{1}{2}kA^2 = \tfrac{1}{2}kx^2 + \tfrac{1}{2}mv^2 = \tfrac{1}{2}mv_{\text{max}}^2$$

14-3 Some Oscillating Systems
A familiar oscillating system is a *physical pendulum*—an object supported so it freely swings back and forth under the influence of gravity. The motion is not, in general, simple harmonic motion, but it approaches simple harmonic motion for small amplitudes. The period of the pendulum is independent of its mass, but depends on its moment of inertia I, the distance D between its rotation axis and the center of mass, and g, the gravitational field strength. The period T for small-amplitude oscillations is

Period of a physical pendulum

$$T = 2\pi \sqrt{\frac{I}{MgD}}$$

At first glance at this formula, it appears that the period depends upon the mass M. However, this is not the case. Since the moment of inertia is proportional to the mass, the ratio I/M does not depend on M and neither does T.

For a *simple pendulum* (a point mass attached to the end of a massless string) $I = ML^2$ and $D = L$, where L is the length of the string. Substituting these expressions for I and D into the formula for the period gives

Period of a simple pendulum

$$T = 2\pi \sqrt{\frac{L}{g}}$$

14-4 Damped Oscillations
In real oscillations, the motion is not conservative; it is always damped by frictional forces. As a result, the energy and amplitude of the motion decrease with time. If damping forces are small, the energy decreases exponen-

tially, that is, by a constant fraction in a given time interval. That is,

Position as a function of time for a damped oscillator

$$x = A_0 e^{-(b/2m)t} \cos(\omega' t + \delta)$$

where

Frequency of a damped oscillator

$$\omega' = \omega_0 \sqrt{1 - \left(\frac{b}{2m\omega_0}\right)^2}$$

ω_0 is the natural (undamped) frequency of the oscillator, and b is the damping constant. (The damping force is $-bv$, a linear drag force.)

The quality factor Q of a damped oscillation expresses the amount of damping; a high Q means the oscillation takes a long time to die out.

Q factor for a damped oscillator

$$Q = \omega_0 \tau = 2\pi \frac{E}{|\Delta E|}$$

where ω_0 is the oscillator's natural frequency, $\tau = 2m/b$, and $E/|E|$ is the ratio of the energy at the begining of a cycle to the energy dissipated during the cycle. Expressed in terms of the damping constant b, $Q = b/8m\omega_0$. If the damping is strong enough, there is no oscillation; the displacement just dies out without crossing the equilibrium position.

14-5 Driven Oscillations
Here we consider the motion of an oscillator that is driven by a repetitive (periodic) driving force. Consideration is restricted to the steady-state motion that results when the energy per cycle transferred by the driving force equals that dissipated by the frictional forces. This is simple harmonic motion at the frequency of the driving force. This is expressed

Position as a function of time for a driven damped oscillator

$$x = A \cos(\omega t - \delta)$$

where

Amplitude of a driven damped oscillator

$$A = \frac{F_0}{\sqrt{m^2(\omega_0^2 - \omega^2)^2 + b^2\omega^2}}$$

and

Phase constant of a driven damped oscillator

$$\delta = \tan^{-1} \frac{b\omega}{m(\omega_0^2 - \omega^2)}$$

If the frequency of the driving force is equal or almost equal to the oscillator's natural frequency, the amplitude of the driven motion will be large. This enthusiastic response to the driving force is called resonance. If the oscillator has high Q, the resonance response is strong, but only over a narrow frequency range.

II. Numbers and Key Equations

Numbers

1 hertz (Hz) = 1 cycle/s

Key Equations

Linear restoring force (Hooke's Law)

$$F_x = -kx$$

Acceleration proportional to displacement

$$a_x = -\omega^2 x$$

Frequency

$$f = \frac{\omega}{2\pi}$$

Period

$$T = \frac{1}{f}$$

Period for particle on a linear spring

$$T = 2\pi \sqrt{\frac{m}{k}}$$

Period of a physical pendulum

$$T = 2\pi \sqrt{\frac{I}{MgD}}$$

Period of a simple pendulum

$$T = 2\pi \sqrt{\frac{L}{g}}$$

Position in simple harmonic motion

$$x = A \cos(\omega t + \delta)$$

Velocity in simple harmonic motion

$$v = -\omega A \sin(\omega t + \delta)$$

Conservation of energy

$$E_{\text{total}} = \tfrac{1}{2}kA^2 = \tfrac{1}{2}kx^2 + \tfrac{1}{2}mv^2 = \tfrac{1}{2}mv_{\text{max}}^2$$

Q factor for a damped oscillator

$$Q = \omega_0 \tau = 2\pi \frac{E}{|\Delta E|}$$

Position as a function of time for a damped oscillator

$$x = A_0 e^{-(b/2m)t} \cos(\omega' t + \delta)$$

Frequency of a damped oscillator

$$\omega' = \omega_0 \sqrt{1 - \left(\frac{b}{2m\omega_0}\right)^2} = \omega_0 \sqrt{1 - \frac{1}{4Q^2}}$$

Q factor for a driven damped oscillator

$$Q = \frac{\omega_0}{\Delta\omega} = \frac{f_0}{\Delta f}$$

Position as a function of time for a driven damped oscillator

$$x = A \cos(\omega t - \delta)$$

Amplitude of a driven damped oscillator

$$A = \frac{F_0}{\sqrt{m^2(\omega_0^2 - \omega^2)^2 + b^2\omega^2}}$$

Phase constant of a driven damped oscillator

$$\delta = \tan^{-1} \frac{b\omega}{m(\omega_0^2 - \omega^2)}$$

III. Potential Pitfalls

Don't be careless and use any of the constant-acceleration formulas for the motion of a harmonic oscillator! The force is proportional to the displacement from equilibrium, so the acceleration is definitely not constant.

In the formulas involving sine and cosine functions, the argument of the trig function (that is, the quantity whose sine or cosine is taken) must be a dimensionless number; if thought of as an angle, it must be an angle in radians.

In almost all cases, it's best to put the origin of the x axis at the equilibrium position of the oscillating particle since that is the zero point of the displacement of, and thus of the force on, the oscillating particle.

Remember that the formula for the period of a simple pendulum applies only to the simple pendulum. If the actual system is something other than a particle on the end of a massless string, then the period will depend on how the mass is distributed in space.

The formulas we write for simple harmonic motion apply to any motion under a restoring force that is directly proportional to the displacement from an equilibrium position. The "spring constant k" used in the formulas is whatever quantity occupies that position in the force equation, $F = -kx$, whether or not there are springs involved.

The potential and kinetic energy of a harmonic oscillator both vary with time; it is only the total energy that is constant. The total energy may be all kinetic, all potential, or anything in between at different points in the cycle.

The motion of a mass on a spring hanging vertically is simple harmonic motion even through a second force—gravity—is involved. The weight simply changes the equilibrium point about which the mass oscillates.

IV. True and False and Responses

True and False

_____ 1. Periodic motion is any motion that repeats itself cyclically.

_____ 2. Simple harmonic motion is periodic motion in which the position of a particle varies sinusoidally with the time.

_____ 3. The time it takes for one full cycle of simple harmonic motion to be completed is called the *phase constant* of the motion.

_____ 4. The frequency of simple harmonic motion is the number of completed cycles per unit time.

_____ 5. At any instant, the acceleration of a moving particle undergoing simple harmonic motion is directed opposite to its displacement from equilibrium.

_____ 6. When the speed of a particle undergoing simple harmonic motion is a maximum, the magnitude of its acceleration is a minimum.

_____ 7. If a particle is undergoing simple harmonic motion due to the action of a force $F_x = -kx$, its kinetic energy is $\frac{1}{2}kA^2$, where A is the amplitude of its oscillation.

_____ 8. The total mechanical energy of a particle undergoing simple harmonic motion is constant.

_____ 9. An object hanging on the end of a spring oscillates vertically with a period that is independent of its mass.

_____ 10. An object swinging on the end of a string as a simple pendulum oscillates with a period that is independent of its mass.

_____ 11. The motion of a simple pendulum is necessarily a simple harmonic motion.

_____ 12. The motion of real oscillatory systems can, at best, be only approximately described as simple harmonic motion.

_____ 13. The motion of a weakly damped harmonic oscillator is very nearly simple harmonic motion with a decreasing amplitude.

_____ 14. For a damping force that is proportional to the velocity, the energy of a damped oscillator diminishes linearly with time.

_____ 15. The period of a damped oscillator is always shorter than it would be if there were no damping.

_____ 16. Critical damping is the condition in which the displaced particle returns to equilibrium most rapidly, without oscillating.

_____ 17. For $b = 0$, a driven harmonic oscillator has a maximum amplitude given by F_0/m when the frequency of the driving force is equal to the natural frequency of the oscillator.

_____ 18. The fact that a driven harmonic oscillator absorbs maximum power from the driving force when the frequency of the driving force is nearly equal to the natural frequency of the oscillator is referred to as *resonance*.

_____ 19. The width (in frequency) of the resonance curve increases as the damping constant b increases.

Responses

1. True

2. True

3. False. This time is the *period* of the motion.

4. True

5. True

6. True. In fact, if v is a maximum or minimum, $a = dv/dt$ is zero.

7. False. This is its total mechanical energy; its kinetic energy isn't constant.

8. True

9. False; $T = 2\pi/\omega = 2\pi\sqrt{m/k}$.

10. True; $T = 2\pi\sqrt{L/g}$.

11. False. It is simple harmonic motion only if the amplitude is small.

12. True. There is some dissipation of energy, and therefore some damping of the motion, in any real oscillating system.

13. True

14. False. It diminishes exponentially.

15. False. The frequency is always a little less, so the period is a little longer. The difference is pretty small unless the damping is near critical.

16. True

17. False. With $b = 0$ the amplitude at resonance would go to infinity. Of course, the proportional limit of the spring will be exceeded, or something will break, before that happens.

18. True

19. True

V. Questions and Answers

Questions

1. A particle is undergoing simple harmonic motion. How far does it move in one full period?

2. A mass oscillates on the end of a certain spring at a frequency f. The spring is cut in half, and the same mass is set oscillating on one of the pieces. What is its new frequency of oscillation?

3. The mass of the string is usually neglected in treating the motion of a simple pendulum. If the mass of the string is not completely negligible, how is the motion of the pendulum affected?

4. A mass on the end of a string, which is hung over a nail, is set swinging as a simple pendulum of length L (see Figure 14-1). While it is swinging, the string is paid out until the free-swinging length is $2L$. What happens to the pendulum's frequency? Can we use the conservation of energy to find out what happens to its amplitude? Neglect friction at the nail.

Figure 14-1

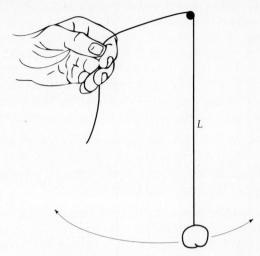

5. If a mass-and-spring system is set oscillating, why does its motion eventually stop?

6. A mass-and-spring system is undergoing simple harmonic motion with an amplitude A. If the mass is decreased by half but the amplitude is unchanged, how does the total energy of the oscillation change? How does the total energy change when the original mass is set oscillating with half the initial amplitude?

7. Would it have made any difference in our discussion in this chapter if we had defined *simple harmonic motion* as motion that obeys

$$x = A \sin (\omega t + \phi)$$

rather than defining it in terms of the cosine?

Answers

1. The particle's net displacement in one period is, of course, zero. The total distance it covers, however, is from $x = 0$ (its initial position) to $x = A$, from there to $x = -A$, and from $x = -A$ back to its initial position. It therefore covers a total distance of $4A$.

2. What we have to find first is the force constant of half a spring. The same tension stretches one-half of the original spring half as much as it would the whole. Therefore, if the force constant of the whole spring is k, that of one of the halves is $2k$. Since the frequency is proportional to the square root of k, cutting the spring in half increases the frequency by $\sqrt{2}$.

3. If the string has mass, a small fraction of the total mass that is swinging is closer to the pivot than the pendulum length L, so the effective length is a little less than L. Consequently, the motion of the pendulum has a slightly shorter period than it would if the string were massless.

4. The period of a simple pendulum is proportional to $\sqrt{L}$, so the period increases by a factor of $\sqrt{2}$ and the frequency decreases to $1/\sqrt{2}$ its original value. The energy of the swinging pendulum is not a constant because (negative) work is done on it by the hand that pays out the string. Since we don't know how to evaluate this work, we can't apply the conservation of energy to this problem. (We know the displacement of the hand, but the tension in the string is variable.)

5. The motion dies out because real oscillations are damped. The drag of the air and some of the internal forces in the material of the spring dissipate the energy of the oscillating mass.

6. When the mass is halved, the frequency of the oscillation increases, but the total energy, $\frac{1}{2}kA^2$, is unaffected. When the amplitude is halved, the total energy decreases to $\frac{1}{4}$ its original value.

7. It would have made no difference whatever. Defining *simple harmonic motion* in terms of the sine does change the phase constant by 90°:

$$x = A \sin(\omega t + \phi)$$
$$= A \cos(\omega t + \phi - \pi/2)$$
$$= A \cos(\omega t + \delta)$$

where $\delta = \phi - \pi/2$.

VI. Problems and Solutions

Problems

1. A 0.10-kg mass is suspended from a spring of negligible mass with a spring constant of 40 N/m. The mass oscillates vertically with an amplitude of 0.06 m. (*a*) Determine the angular frequency of the motion. Express the height y of the mass above the equilibrium position as a function of time so that initially (at $t = 0$) the mass is (*b*) at its highest point; (*c*) 0.03 m above the equilibrium position and moving downward.

How to Solve It
• The angular frequency ω is equal to the square root of the spring constant k divided by the mass m.

• To find the position as a function of time requires that the amplitude A and phase constant δ be determined from the initial conditions for the position and velocity. In part (*b*), the initial position equals 0.06 m and the initial velocity equals zero; while in part (*c*), the initial position is 0.03 m and the initial velocity is negative. In each part, simultaneously solve the position and velocity equations at time $t = 0$ for the amplitude and phase constant.

2. The motion of a mass oscillating on the end of a spring is graphed in Figure 14-2. What are (*a*) the amplitude, (*b*) the period, and (*c*) the frequency of its motion? (*d*) If the mass is 600 g, what is the force constant of the spring?

How to Solve It
• The maximum displacement is the amplitude.

• The period is the time required for the motion to complete one cycle. During this time the phase increases by 2π. The frequency is the reciprocal of the period.

• The force constant of the spring can be calculated from the known mass and period.

3. A certain mass hanging in equilibrium on the end of a vertical spring stretches the spring by 2 cm. (*a*) If the mass is then pulled down 5 cm farther and released, at what frequency does it oscillate? (*b*) At what speed is it moving as it passes through its equilibrium position?

How to Solve It
• The period and frequency of the oscillation depend on the ratio k/m. If you know that ratio, you don't need to know k or m individually.

• You can calculate k/m from the distance the spring is stretched by the mass hanging in equilibrium.

Figure 14-2

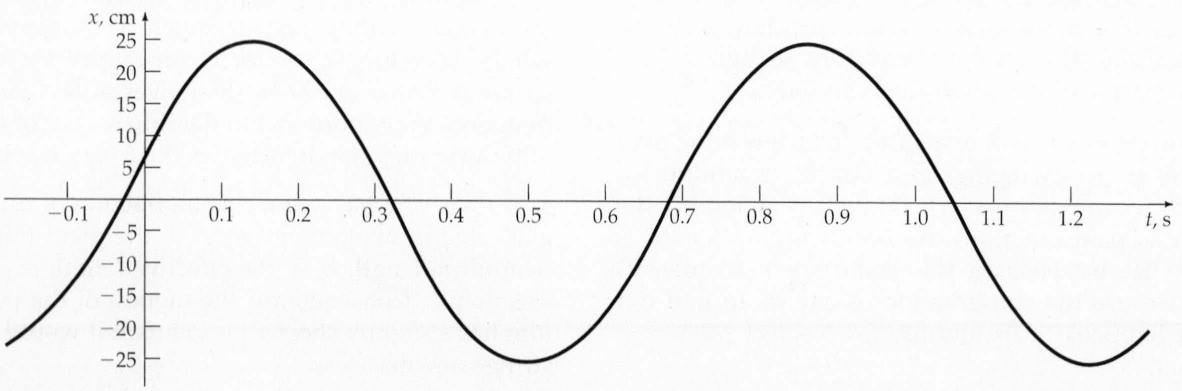

• As it passes through equilibrium, the mass has its maximum speed.

4. A 5-kg block on the end of a spring oscillates at a frequency of 5 Hz with an amplitude of 5 cm. (*a*) What is the force constant of the spring? (*b*) What is the period of the block's motion? (*c*) What is the maximum speed at which the block moves?

How to Solve It
• The force constant can be determined from the mass of the block and the frequency of its oscillation.

• The period is the reciprocal of the frequency.

• The maximum speed of the oscillating block can be determined from its amplitude and frequency.

5. You are riding on a 10-m diameter Ferris wheel that is rotating at a constant rate of 1.00 rev every 10.0 s. (*a*) Determine your angular speed in rad/s. Express your height y above the rotation axis as a function of time t so that initially (at $t = 0$) you are (*b*) 3.00 m above the rotation axis and moving downward, (*c*) at the top of the wheel, and (*d*) 2.00 m below the rotation axis and moving upward.

How to Solve It
• The angular frequency ω is equal to the frequency (in rev/s) times a conversion factor.

• To express the height above the rotation axis as a function of time requires that the phase constant δ be determined from the initial conditions. This is best done by drawing the wheel and indicating both the initial position and the position at time t for an arbitrary initial position. Use trigonometry to relate the initial height to the initial angular position, as well as the height at time t to angular position at time t.

6. A phonograph record 30 cm in diameter revolves once every 1.80 s. (*a*) What is the linear speed of a point on its rim? (*b*) What is its angular velocity ω? (*c*) Write an equation for the y component of the point's position as a function of time, given that it is at its maximum value on the y axis when $t = 0$.

How to Solve It

• You know the diameter of the record. How far does a point on its rim move in 1.80 s?

• The angular velocity of rotation is the linear speed divided by the radius of the disk.

• The amplitude of the oscillation is the radius of the record.

7. An object of mass 1 kg oscillates on a spring with an amplitude of 12 cm. Its maximum acceleration is 5.0 m/s². Find its total energy.

How to Solve It
• Relate the total energy with the spring constant and the amplitude.

• Use Newton's second law to relate the spring constant with the displacement, the mass, and the acceleration. Using this result, relate the maximum acceleration with the amplitude to obtain an expression for the spring constant in terms of the mass, the maximum acceleration, and the amplitude.

• Solve for the energy.

8. A 400-g mass oscillates on the end of a spring with an amplitude of 2.5 cm. Its total energy is 2.0 J. What is the frequency of the oscillation?

How to Solve It
• Since you know the amplitude of the oscillation, you can use the total energy given to find an expression for the force constant of the spring.

• Use this expression and the mass of the oscillating object to find the frequency of the motion.

9. Two simple pendulums, each of length 1.8 m, with masses of 1.4 kg and 2.0 kg, hang side by side such that, at rest, their round steel bobs just touch (see Figure 14-3). The lighter pendulum on the left is pulled back 6 cm and released. It swings down and strikes the other, which is initially at rest. If the collision is elastic, to what distance does each pendulum rebound after the collision?

Figure 14-3

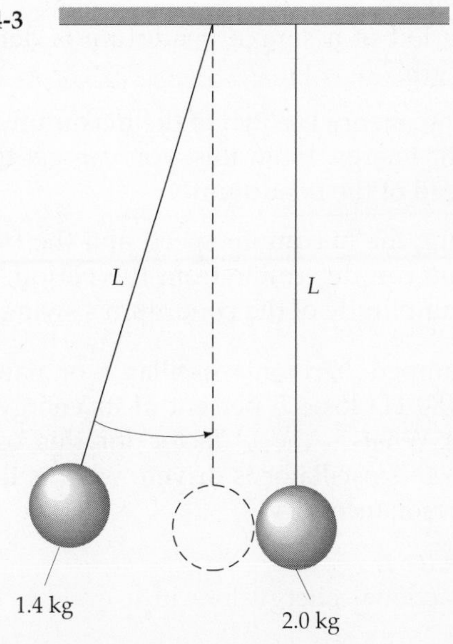

1.4 kg

2.0 kg

How to Solve It

• Relate the velocity of a pendulum to the amplitude of its motion and the angular frequency. The angular frequency can, in turn, be expressed in terms of the length of the pendulum and the acceleration due to gravity g. Obtain an expression relating the pendulum's maximum speed with its length, its amplitude, and g.

• Obtain two equations relating the velocity of the 1.4-kg pendulum just before the collision with the velocities of the pendulums just after the collision. To obtain these equations use conservation of momentum and the condition that the collision is elastic.

• Use these equations to obtain expressions for each final velocity in terms of the masses of both pendulums and the initial velocity of the 1.4-kg pendulum.

• Use the expression referred to in the first hint to relate the final amplitudes with the final speeds. Solve for the final amplitudes.

10. Estimate the natural frequency of a man of mass 75 kg swinging from the end of a 5-m-long rope.

How to Solve It

• Assume that the man and the rope can be approximated as a simple pendulum.

• With this assumption, the man's mass is simply window dressing.

• The period of a simple pendulum is determined only by its length.

11. (a) How long is a simple pendulum whose period is 2.00 s? (Pendulums with a period of 2 s are common in tall clocks.) (b) If this pendulum has a mass of 0.2 kg and a total energy of 0.02 J, what is the amplitude of its oscillation?

How to Solve It

• The period of a simple pendulum is determined by its length.

• The total energy is equal to the maximum value of the kinetic energy. From this, you can get the maximum speed of the pendulum.

• Knowing the maximum speed and the frequency, which you can determine from the period, you can find the amplitude of the pendulum's swing.

12. A damped harmonic oscillator of natural frequency 180 Hz loses 5 percent of its energy in each cycle. (a) What is the Q factor for this oscillator? (b) When this oscillator is driven, what is the width Δf of its resonance curve?

How to Solve It

• The fractional energy loss in a cycle is equal to $2\pi/Q$.

• The fractional width $\Delta f/f_0$ of the resonance of the driven oscillator is equal to $1/Q$.

13. A child on a swing swings back and forth once every 3.75 s. The total mass of the child and the swing seat is 38 kg. At the bottom of the swing's arc, the child is moving at 1.82 m/s. (a) What is the total energy of the swing and the child? (b) If the Q factor is 25, what average power must be supplied to the swing to keep it moving with a constant amplitude? (Neglect the mass of the ropes.)

How to Solve It

• The total energy is equal to the maximum kinetic energy of the child and swing.

• Relate the Q factor with the fraction of energy that is dissipated in each cycle. Also, relate the average power, the energy dissipated per cycle, and the period.

• Obtain an expression for the average power and use it to calculate the average power that must be supplied to replace the dissipated energy.

14. A certain damped harmonic oscillator loses 5 percent of its energy in each full cycle of oscillation. By what factor must the damping constant be increased in order to damp it *critically*?

How to Solve It

• Relate the critical damping constant b_c with the mass and period.

• Obtain an equation relating the energy loss per cycle of a lightly damped oscillator with the energy, the damping constant, the mass, and the period.

• The requested factor equals the ratio b_c/b.

15. The rim of a 0.3-kg bicycle wheel 95 cm in diameter is hanging from a horizontal nail on the wall. If it is knocked slightly to one side and starts swinging, what is the period at which it oscillates?

How to Solve It

• Obtain an expression for the period of a physical pendulum.

• Obtain an expression for the moment of inertia of the rim about its axis of symmetry. This expression can be found in Table 9-1 on page 264 of the text. Use the parallel-axis theorem to relate the moment of inertia about an axis along the top of the nail with the moment of inertia about the symmetry axis.

• Solve for the period.

16. A nonuniform 400-g stick 120 cm long is found to have its center of mass 70 cm from one end. The stick is hung from a wall by a horizontal nail passing through a small hole 10 cm from the lighter end of the stick. When the stick is knocked gently to one side it oscillates with a period of 1.8 s. (*a*) What is the moment of inertia of the stick about the nail? (*b*) What is the moment of inertia about an axis passing through the center of mass of the stick and parallel to the nail?

How to Solve It
• Use the formula for the period of a physical pendulum to obtain an expression for the moment of inertia about the axis along the top of the nail.

• Use the parallel-axis theorem to relate the moment of inertia about an axis along the top of the nail with the moment of inertia about the parallel axis through the center of mass.

17. Oscillator number 1 has a peak resonance angular frequency of 19,500 s^{-1} and a Q factor of 20. Oscillator number 2 has a peak resonance angular frequency of 20,000 s^{-1} and a Q factor of 100. On the same graph sketch an estimate of the resonance curve for each oscillator showing the power delivered from a driving source versus the angular frequency of the source. On these sketches let the peak average power for each oscillator be the same.

How to Solve It
• Before sketching anything determine the widths of the resonance curve for each oscillator. The scale and offset for the angular frequency axis should be determined to accommodate these widths.

• Draw resonance curves with widths and peaks in line with the given information.

18. For the resonance curve shown in Figure 14-4 estimate the peak resonance frequency, the width of the resonance curve, and the Q factor.

Figure 14-4

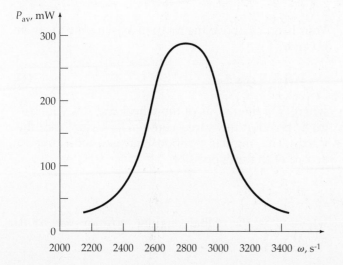

How to Solve It
• From the graph estimate the peak resonance frequency ω_0.

• Also from the graph, estimate the average peak power. Divide the average peak power by two and, at half the peak power, draw a line on the graph parallel to the angular frequency axis. This line will intersect the resonance curve at two points.

• The length of the line segment between the two points is called the width $\Delta\omega$ of the resonance curve. The Q factor is the ratio of the peak angular frequency ω_0 to the width $\Delta\omega$.

Solutions

1. (*a*) The formula for the angular frequency is

$$\omega = \sqrt{\frac{k}{m}}$$

$$= \sqrt{\frac{40\ \text{N/m}}{0.1\ \text{kg}}} = \underline{20\ \text{rad/s}}$$

(*b*) The formulas for the position and velocity as a function of time are

$$y(t) = A \cos(\omega t + \delta)$$
$$v(t) = -\omega A \sin(\omega t + \delta) \qquad (1)$$

The initial ($t = 0$) conditions are $y(0) = 0.06$ m and $v(0) = 0$. Substituting these values into Equation 1 gives

$$0.06\ \text{m} = A \cos \delta$$
$$0 = -\omega A \sin \delta$$

where A and ω are 0.06 m and 20 rad/s. These equations reduce to

$$\cos \delta = 1$$
$$\sin \delta = 0$$

The initial condition for velocity ($\sin \delta = 0$) implies $\delta = 0, \pm\pi, \pm 2\pi, \ldots$. The equation $\cos \delta = 1$ implies $\delta = 0, \pm 2\pi, \pm 4\pi, \ldots$. The values of δ that satisfy both of these conditons are $\delta = 0, \pm 2\pi, \pm 4\pi, \ldots$, and for convenience, we select $\delta = 0$.

$$y(t) = A \cos(\omega t + \delta)$$
$$= (0.06\ \text{m}) \cos[(20\ \text{rad/s})t + 0]$$
$$= \underline{(0.06\ \text{m}) \cos[(20\ \text{rad/s})t]}$$

(c) The initial ($t = 0$) conditions are $y(0) = 0.03$ m and $v(0) < 0$. Substituting these values into Equation 1 gives

$$0.03 \text{ m} = A \cos \delta$$

$$0 > -\omega A \sin \delta$$

where A and ω are 0.06 m and 20 rad/s. These conditions reduce to

$$\cos \delta = 0.5$$

$$\sin \delta > 0$$

The condition $\cos \delta = 0.5$ is satisfied if $\delta = \pi/3$ or if $\delta = 5\pi/3$ (if we impose the restriction $0 \le \delta < 2\pi$), and the condition $\sin \delta > 0$ is satsfied if $0 < \delta < \pi$. These conditions are simultaneously satisfied only if $\delta = \pi/3$. Thus

$$y(t) = \underline{(0.06 \text{ m}) \cos [(20 \text{ rad/s})t + \pi/3]}$$

2. (a) About 25 cm   (b) 0.72 s   (c) 1.4 Hz
   (d) 46 N/m

3. (a) In equilibrium, the net force on the hanging mass is

$$F_{net} = ky_0 - mg = 0$$

where y_0 is the extension of the spring when the block is at its equilibrium position. (See page 413 of the text.) Thus,

$$\frac{k}{m} = \frac{g}{y_0}$$

The angular frequency of the motion is

$$\omega = \sqrt{\frac{k}{m}} = \sqrt{\frac{g}{y_0}}$$

and the frequency is

$$f = \frac{\omega}{2\pi} = \frac{1}{2\pi} \sqrt{\frac{g}{y_0}}$$

$$= \frac{1}{2\pi} \sqrt{\frac{9.81 \ m/s^2}{0.02 \text{ m}}} = \underline{3.52 \text{ Hz}}$$

(b) The equation for the displacement from equilibrium may be written

$$y = A \cos \omega t$$

The displacement is zero when $\omega t = \pi/2, 3\pi/2, 5\pi/2,$ . . . . The equation for the velocity is

$$v = \frac{dy}{dt} = -\omega A \sin \omega t$$

When $\omega t = \pi/2, 3\pi/2, 5\pi/2, \ldots$, $\sin \omega t = \pm 1$, and the speed $|v|$ is

$$|v| = \omega A = 2\pi f A = 2\pi(3.52 \text{ Hz})(0.05 \text{ m})$$

$$= \underline{1.11 \text{ m/s}}$$

4. (a) 4930 N/m   (b) 0.200 s   (c) 1.57 m/s

5. (a) The formula for the angular frequency is

$$\omega = \frac{\Delta\theta}{\Delta t} = \frac{1.00 \text{ rev}}{10.0 \text{ s}} \left(\frac{2\pi \text{ rad}}{\text{rev}} \right) = \frac{\pi \text{ rad}}{5.00 \text{ s}}$$

$$= \underline{0.628 \text{ rad/s}}$$

(b) To find your height as a function of the time we first make a sketch (Figure 14-5) which includes both your initial position and your position at time t.

Figure 14-5

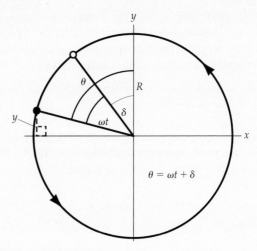

$$\theta = \omega t + \delta$$

Your height y above the rotation axis at time t is then given by

$$y(t) = R \cos \theta \qquad (1)$$

where R is the radius of the wheel and θ is your angular position measured counterclockwise from the y axis. The angular position increases with time in accord with the expression

$$\theta = \omega t + \delta$$

where δ is the initial angular position (at $t = 0$, $\theta = \delta$).

The initial height equals 3.00 m. Substituting this into Equation (1) we obtain

$$y_0 = R \cos \delta$$

so

$$\delta = \cos^{-1} \frac{y_0}{R} = \cos^{-1} \left(\frac{3.00 \text{ m}}{5.00 \text{ m}} \right) = 0.927 \text{ rad}$$

and

$$y(t) = R \cos (\omega t + \delta)$$
$$= (5.00 \text{ m}) \cos [(0.628 \text{ rad/s})t + 0.927 \text{ rad}]$$

(c) At $t = 0$ you are at the top of the Ferris wheel so $\delta = 0$ and

$$y(t) = R \cos (\omega t + \delta)$$
$$= (5.00 \text{ m}) \cos [(0.628 \text{ rad/s})t]$$

(d) At $t = 0$ you are 2.00 m below the rotation axis and moving upward so

$$\delta = \cos^{-1} \frac{y_0}{R} = \cos^{-1} \left(\frac{-2.00 \text{ m}}{5.00 \text{ m}} \right) = 114° \text{ or } 246°$$

Since you are moving upward we know the angle must be greater than 180°. This means that $\delta = 246° = 4.30$ rad and

$$y(t) = R \cos (\omega t + \delta)$$
$$= (5.00 \text{ m}) \cos [(0.628 \text{ rad/s})t + 4.30 \text{ rad}]$$

6. (a) 0.524 m/s   (b) 3.49 rad/s
   (c) $y(t) = (0.15 \text{ m}) \cos (3.49 \text{ s}^{-1})t$

7. The equation relating the energy E to the amplitude A of the motion

$$E = \tfrac{1}{2} kA^2$$

where k is the spring constant. To determine the spring constant from the information given we apply Newton's second law

$$F = ma$$
$$-kx = ma$$

where x is the displacement of the object from its equilibrium position and a is its acceleration. Solving for the acceleration we have

$$a = -\frac{k}{m} x$$

The acceleration is maximum when x is a minimum, and the minimum value of x is $-A$. Thus,

$$a_{max} = \frac{k}{m} A$$

so

$$k = \frac{m a_{max}}{A}$$

Substituting this expression into the energy equation we obtain

$$E = \tfrac{1}{2} kA^2 = \tfrac{1}{2} \frac{m a_{max}}{A} A^2$$
$$= \tfrac{1}{2} m a_{max} A = \tfrac{1}{2} (1 \text{ kg})(5 \text{ m/s}^2)(0.12 \text{ m})$$
$$= 0.300 \text{ J}$$

8. 20.1 Hz

9. The period and velocity of a pendulum are

$$T = 2\pi \sqrt{\frac{L}{g}}$$

and

$$v = -\omega A \sin (\omega t + \delta)$$

where $\omega = 2\pi/T$. The speed at the lowest point in its swing is the maximum magnitude of v, namely

$$v_{max} = \omega A = \frac{2\pi A}{T} = A \sqrt{\frac{g}{L}} \tag{1}$$

For an elastic collision, the relative speed of recession after the collision equals the relative speed of approach before the collision. Thus,

$$v_{2f} - v_{1f} = -(v_{2i} - v_{1i})$$

The initial speed of the 2.0-kg pendulum is zero, so $v_{2i} = 0$ and

$$v_{2f} - v_{1f} = v_{1i} \tag{2}$$

Conserving momentum we have

$$m_1 v_{1f} + m_2 v_{2f} = m_1 v_{1i} + m_2 v_{2i}$$
$$m_1 v_{1f} + m_2 v_{2f} = m_1 v_{1i} + m_2(0)$$
$$v_{1f} + \frac{m_2}{m_1} v_{2f} = v_{1i} \tag{3}$$

We now solve Equations (2) and (3) to obtain expressions for the final velocities. Adding Equations (2) and (3) and solving for v_{2f} we have

$$v_{2f} = \frac{2m_1}{m_1 + m_2} v_{1i}$$

Substituting this expression for v_{2f} in Equation (3) and solving for v_{1f} we have

$$v_{1f} = \frac{m_1 - m_2}{m_1 + m_2} v_{1i}$$

Substituting the magnitudes of these expressions for the final velocities into Equation (1) we obtain

$$A_{1f} = v_{1f}\sqrt{\frac{L}{g}}$$

$$= \left|\frac{m_1 - m_2}{m_1 + m_2}\right| v_{1i}\sqrt{\frac{L}{g}}$$

$$= \left|\frac{m_1 - m_2}{m_1 + m_2}\right| A_{1i} = \left|\frac{1.4 \text{ kg} - 2.0 \text{ kg}}{1.4 \text{ kg} + 2.0 \text{ kg}}\right|(0.06 \text{ m})$$

$$= \underline{0.0106 \text{ m} \text{ (or 1.06 cm)}}$$

$$A_{2f} = v_{2f}\sqrt{\frac{L}{g}}$$

$$= \frac{2m_1}{m_1 + m_2} v_{1i}\sqrt{\frac{L}{g}}$$

$$= \frac{2m_1}{m_1 + m_2} A_{1i} = \frac{2(1.4 \text{ kg})}{1.4 \text{ kg} + 2.0 \text{ kg}}(0.06 \text{ m})$$

$$= \underline{0.0494 \text{ m} \text{ (or 4.94 cm)}}$$

10. $\underline{0.223 \text{ Hz}}$

11. (a) The formula for the period of a simple pendulum is

$$T = 2\pi\sqrt{\frac{L}{g}}$$

so the length of the pendulum is

$$L = g\left(\frac{T}{2\pi}\right)^2 = (9.81 \text{ m/s}^2)\left(\frac{2 \text{ s}}{2\pi}\right)^2 = \underline{0.994 \text{ m}}$$

(b) The kinetic energy K is

$$K = \tfrac{1}{2}mv^2 = \tfrac{1}{2}m(-\omega A \sin \omega t)^2$$

The total energy equals the maximum kinetic energy, so

$$E = K_{max} = \tfrac{1}{2}mv_{max}^2 = \tfrac{1}{2}m(\omega A)^2 = \tfrac{1}{2}m\left(\frac{2\pi A}{T}\right)^2$$

where we have used the result $\omega = 2\pi/T$. Solving for the amplitude we have

$$A = \sqrt{\frac{2E}{m}}\frac{T}{2\pi} = \sqrt{\frac{2(0.02 \text{ J})}{0.2 \text{ kg}}}\frac{2 \text{ s}}{2\pi} = \underline{0.142 \text{ m}}$$

12. (a) $\underline{126}$ (b) $\underline{1.40 \text{ Hz}}$

13. (a) The total energy of the swing and child is

$$E = K_{max} = \tfrac{1}{2}mv_{max}^2$$
$$= \tfrac{1}{2}(38 \text{ kg})(1.82 \text{ m/s})^2 = \underline{62.9 \text{ J}}$$

(b) The Q factor is defined by

$$Q = 2\pi\frac{E}{|\Delta E|}$$

where $|\Delta E|$ is the decrease in mechanical energy per cycle through frictional dissipation. To keep the swing moving, this energy must be supplied from an external source, which provides average power $P_{av} = |\Delta E|/T$, where T is the period. Thus,

$$Q = 2\pi\frac{E/T}{|\Delta E|/T} = 2\pi\frac{E/T}{P_{av}}$$

or

$$P_{av} = 2\pi\frac{E}{QT} = 2\pi\frac{62.9 \text{ J}}{(25)(3.75 \text{ s})} = \underline{4.22 \text{ W}}$$

14. $\dfrac{b}{b_c} = \dfrac{4\pi}{0.05} = \underline{251}$

15. The formula for the period of a physical pendulum is

$$T = 2\pi\sqrt{\frac{I}{MgD}}$$

where I is the moment of inertia about the nail and D is the distance of the nail to the center of mass. For this pendulum D equals the radius R of the rim. The parallel axis theorem tells us that

$$I = I_{cm} + MR^2$$

where the moment of inertia about the parallel axis through the center of mass (see Table 9-1 on page 264 of the text) is

$$I_{cm} = MR^2$$

Therefore

$$I = 2MR^2$$

and the period is

$$T = 2\pi\sqrt{\frac{I}{MgD}} = 2\pi\sqrt{\frac{2MR^2}{MgR}}$$

$$= 2\pi\sqrt{\frac{2R}{g}} = 2\pi\sqrt{\frac{0.95 \text{ m}}{9.81 \text{ m/s}^2}} = \underline{1.96 \text{ s}}$$

The mass of the wheel turns out to be irrelevant.

16. (a) $\underline{0.193 \text{ kg} \cdot \text{m}^2}$ (b) $\underline{0.049 \text{ kg} \cdot \text{m}}$

17. The central frequencies of the resonance peaks are $\omega_{01} = 19{,}500 \text{ s}^{-1}$ and $\omega_{02} = 20{,}000 \text{ s}^{-1}$. Since $Q_1 = \omega_{01}/\Delta\omega_1$ and $Q_2 = \omega_{02}/\Delta\omega_2$, we have

$$\Delta\omega_1 = \frac{\omega_{01}}{Q_1} = \frac{19{,}500 \text{ s}^{-1}}{20} = 975 \text{ s}^{-1}$$

and

$$\Delta\omega_2 = \frac{\omega_{02}}{Q_2} = \frac{20{,}000 \text{ s}^{-1}}{100} = 200 \text{ s}^{-1}$$

It seems reasonable to draw the angular frequency graph with the angular frequency ranging from $18{,}500 \text{ s}^{-1}$ to $20{,}500 \text{ s}^{-1}$. This is done in Figure 14-6.

Figure 14-6

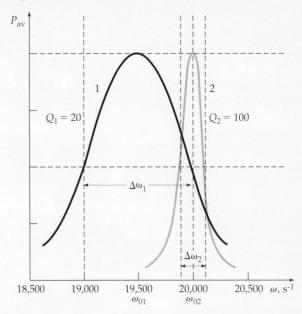

18. $\omega_0 = \underline{2720 \text{ s}^{-1}}$, $\Delta\omega = \underline{400 \text{ s}^{-1}}$, and $Q = \underline{6.8}$

Chapter 15

Wave Motion

I. Key Ideas

Wave motion is the transport of energy and momentum from one point to another without the transport of matter. Mechanical waves require a material medium through which to propagate (travel), whereas electromagnetic waves can propagate in a vacuum. For mechanical waves, a disturbance in a medium propagates because of the elastic properties of the medium.

15-1 Simple Wave Motion A wave in which the disturbance of the medium is perpendicular to the direction of propagation is called a **transverse wave**. Waves on a plucked string are examples of transverse waves. A wave in which the disturbance of the medium is parallel to the direction of propagation is called a **longitudinal wave**. Sound waves are longitudinal waves. Surface waves on water are a combination of the two types of waves.

Wave Pulses and Wave Speed If one end of a string under tension is given a flip, a wave pulse travels down the string. Energy and momentum are carried along the string, but the material of the string itself only twitches from side to side. The speed at which the pulse travels depends on the tension in the string and its mass density (mass per unit length). The size, shape, and location of a pulse is described by a **wave function** $y = f(x \pm vt)$, where y is the displacement of the string and v is the speed of the pulse. The plus sign in used to describe a pulse moving in the negative x direction whereas the minus sign is for one moving in the positive x direction. The **wave speed** v is given by

Speed of transverse waves on a taut string

$$v = \sqrt{\frac{F}{\mu}}$$

where F is the tension in the string and μ is its mass per unit length. For a pressure wave in a fluid, like sound traveling through air, v is given by

Speed of sound in a fluid

$$v = \sqrt{\frac{B}{\rho}}$$

where B is the bulk modulus of the fluid and ρ is its mass per unit volume. (These formulas are derived by applying Newton's laws to a segment of the string or fluid.) For sound waves in a gas such as air this becomes

Speed of sound in a gas

$$v = \sqrt{\frac{\gamma RT}{M}}$$

where T is the absolute temperature in kelvins (K), R is the universal gas constant ($R = 8.314$ J/mol·K), M is the molar mass of the gas (the number of kilograms per mole), and γ is a number that depends upon the molecular structure. For a monatomic gas $\gamma = 1.67$ and for a diatomic gas $\gamma = 1.40$.

The **wave equation** expresses Newton's second law applied to an infinitesimal segment of a taut string. This equation is

Wave equation

$$\frac{\partial^2 y}{\partial x^2} = \frac{\mu}{F} \frac{\partial^2 y}{\partial t^2}$$

15-2 Harmonic Waves If one end of a taut string is driven transversely in simple harmonic motion (displacement a sinusoidal function of time), a sinusoidal wave propagates along the string. Such a wave is called a harmonic wave. When a harmonic wave travels along the string, the shape of the string at any particular instant in time is that of a sine function.

157

The distance between two successive wave crests is called the wavelength λ. The wavelength is the distance after which the shape repeats itself. When a harmonic wave propagates along a string, each point on the string moves in simple harmonic motion with the same frequency f and period $T = 1/f$. During one period, the wave crest moves a distance equal to the wavelength. Then the speed of the wave is

Wave speed

$$v = \frac{\lambda}{T} = f\lambda = \frac{\omega}{k}$$

where we introduce the angular frequency ω, the frequency in radians per second:

Angular frequency

$$\omega = \frac{2\pi}{T}$$

and the **wave number** k, which has units of radians per meter:

Wave number

$$k = \frac{2\pi}{\lambda}$$

The wave function describing a harmonic wave traveling in the positive x direction with speed v is

Displacement of harmonic transverse wave

$$y(x, t) = A \sin(kx - \omega t)$$

where A is the **amplitude** of the wave. The argument of the sine function (θ is the argument of $\sin \theta$) is called the **phase**.

Next we consider the energy transported by waves on a string. The rate of energy transfer (power) at a point on a taut string is given by $P = \vec{F} \cdot \vec{u} = F_x u_x + F_y u_y$, where $\vec{u}$ is the velocity of the point. A taut string supports transverse waves where the point moves only in the transverse direction and $u_x = 0$. It follows that

Power for waves on a taut string

$$P = F_y u_y = F \frac{\partial y}{\partial x} \frac{\partial y}{\partial t}$$

where $u_y = \partial y / \partial t$ and $F_y = F \, \partial y / \partial x$. For a harmonic wave [$y = A \sin(kx - \omega t)$] this gives

Power for harmonic waves on a taut string

$$P = \mu \omega^2 A^2 v \cos^2(kx - \omega t)$$

where μv^2 has been substituted for the tension F. Since the average value of $\cos^2(kx - \omega t)$ is $\frac{1}{2}$, the average power at any point is

Average power for harmonic waves on a taut string

$$P_{av} = \tfrac{1}{2} \mu \omega^2 A^2 v$$

Note that the average power is proportional to the square of the amplitude.

When considering the energy of a small segment of string, we know part of the energy is kinetic, due to its transverse motion, and the other part is potential, due to its stretching. This stretching is proportional to the magnitude of the slope $\partial y / \partial x$, which, for a harmonic wave, is greatest when the segment passes through the equilibrium position where its speed is also greatest.

In **harmonic sound waves** the molecules of the medium undergo simple harmonic motion. The wave function for such a wave is

Displacement for harmonic sound waves

$$s(x, t) = s_0 \sin(kx - \omega t)$$

where s is the displacement from equilibrium of the molecules and s_0 is the amplitude of their simple harmonic motion. These displacements cause variations (compressions and rarefactions) in the density of the medium, and therefore in the pressure of the medium. The associated pressure wave function is

Pressure for harmonic sound waves

$$\begin{aligned} p(x, t) &= p_0 \sin(kx - \omega t - \pi/2) \\ &= -p_0 \cos(kx - \omega t) \end{aligned}$$

where p is the difference between the actual pressure and the equilibrium pressure and the pressure amplitude p_0 is the maximum value of p. The pressure is $90°$ ($\pi/2$ rad) out of phase with the displacement. The relation between the pressure amplitude and the displacement amplitude is

Pressure-amplitude/displacement-amplitude relation

$$p_0 = \rho \, \omega \, v \, s_0$$

15-3   *Waves in Three Dimensions: Intensity* When a pebble is dropped onto the surface of a pond, a few surface waves ripple outwardly in a circular pattern. When a small, steadily vibrating object is placed on the surface of the pond, a steady train of surface waves will ripple outwardly in a circular pattern. When the vibrator is immersed deeply in the water, the steady train of compression maxima and minima produced will ripple outward in a spherical pattern. A spherical surface moving outward with one of these compression maxima is called a **wavefront** (a wavefront is a surface of constant phase). The outward motion of the wavefronts can be illustrated by **rays**, which are lines directed perpendicularly to the wavefronts. For the spherical wavefronts described here, the rays are straight lines directed away from the wave source.

The waves transport energy in the direction of the rays. If the source (the vibrator) generates waves steadily for a long time, the rate at which energy flows through any fixed spherical surface centered at the source equals the rate at which energy is generated by the source. The energy transferred during one cycle, divided by the duration of a cycle, is called the wave power P_{av}, and the wave *intensity I* is the wave power per unit area. I is measured in units of watts per square meter. The intensity a distance r from the point source is

Intensity of a point source

$$I = \frac{P_{av}}{A} = \frac{P_{av}}{4\pi r^2}$$

where A is the surface area of a sphere of radius r. (In other words, for waves in three dimensions, the intensity varies inversely with the square of the distance from the source.)

If the incident wave is normal to the boundary, the power transferred through it is given by $P = \vec{F} \cdot \vec{u} = Fu$ where F is the force (pressure times area) exerted by the fluid and u is its velocity ($\partial s/\partial t$) at the boundary. It follows that the average power per unit area (intensity) is the pressure p times the velocity, averaged over one cycle. That is

Intensity of a harmonic sound wave

$$I = (pu)_{av} = \left(p\, \frac{\partial s}{\partial t} \right)_{av} = \tfrac{1}{2}\omega s_0 p_0 = \tfrac{1}{2}\rho\omega^2 s_0^2 v$$

The intensity must also equal the product of the average **energy density** η_{av} (energy per unit volume) and the wave speed v. If follows that

Energy density of a harmonic sound wave

$$\eta_{av} = \frac{I}{v} = \tfrac{1}{2}\rho\omega^2 s_0^2$$

One might expect that the sensation of loudness would vary directly with the intensity of sound. However, this is not the case. In fact, the sensation of loudness depends upon both intensity and frequency. However, the dependence of loudness on frequency will not be considered here. The sensation of loudness is found to vary more or less logarithmically with intensity, so the intensity level β is expressed in decibels (dB):

Intensity level

$$\beta = (10\ \mathrm{dB})\log_{10}\frac{I}{I_0}$$

where I is the intensity of a sound and I_0 is a reference level which we take to be the threshold of hearing:

Threshold of hearing

$$I_0 = 10^{-12}\ \mathrm{W/m^2}$$

The threshold of pain occurs at an intensity of about $1\ \mathrm{W/m^2}$. As the intensity varies from $10^{-12}\ \mathrm{W/m^2}$ to $1\ \mathrm{W/m^2}$, the intensity level varies from 0 to 120 dB. A doubling of intensity results in a 3-dB increase in intensity level.

15-4   *Waves Encountering Barriers* When a wave is incident on a boundary that separates two regions of different wave speed, part of the wave is reflected and part is transmitted. A pulse on a string that is attached to a more dense string is reflected at the boundary as an inverted wave (an inversion corresponds to a phase shift of 180°). However, if the second string is less dense than the first, the reflected pulse is not inverted. In either case, the transmitted pulse is not inverted. If the string is tied to a fixed point the reflected pulse is inverted.

When a sound wave traveling in air strikes a solid or liquid surface, the reflected and incident rays make equal angles with the normal to the boundary, and the angle between the normal and the transmitted ray differs from that between the normal and the incident ray. This change of direction (bending) of the transmitted ray is called **refraction**. The angle that the refracted ray makes with the normal is greater or less than that of the incident ray, depending on whether the wave speed in the second region is less or greater than the wave speed in the incident medium.

Diffraction In a homogeneous medium, such as air at constant density, sound waves travel in straight lines. In three dimensions a wavefront is a surface of constant phase, like a pressure maximum or minimum. At great distances from a compact source, a small part of a wavefront can be approximated by a plane (flat surface), and the wave propagates in a direction normal (perpendicular) to the wavefront. Such a wave is called a **plane wave**.

When a portion of a wavefront is blocked by an obstruction, the portions that are not blocked will spread out in all directions—including the region *behind* the obstruction. This spreading out, which in principle can be observed whenever a part of a wavefront is blocked, is called **diffraction**. Consider an obstruction consisting of a large screen that blocks waves incident upon it. If there is a hole (aperture) through the screen, and if the hole is just a fraction of a wavelength across, the portion of the wavefronts that pass through the hole will spread out—just as if the opening were a point source of waves. However, if the hole is many wavelengths across, the spreading (diffraction) of the wave into the region behind the screen will be hardly noticeable. The approximation that waves propagate along straight lines, that is without diffracting, is known as the **ray approximation**. Wave motion for which the wavelengths are small compared to the size of holes can be adequately described via this approximation.

To experience the diffraction of sound, listen to your stereo from another room. Place yourself so the sound must pass through a doorway and spread out in order to reach your ears. The doorway will be narrow compared to the wavelength of the low (bass) frequencies and wide compared to the high (treble) frequencies. Low-frequency notes (like those of the bassoons) will therefore diffract as they pass through the doorway and you will be able to hear them. However, you will not be able to hear high-frequency notes (like those of the piccolos) because they will not diffract.

15-5 *The Doppler Effect* When a wave source and receiver are moving relative to each other, the rate at which the compressions leave the source is not the same as the rate at which they arrive at the receiver. Thus, the frequency observed by the receiver differs from the frequency emitted by the source. This phenomenon is called the **Doppler effect**. The relationship between the frequency of the source and the wavelength can be expressed as

Emitted frequency

$$f_0 = \frac{v''}{\lambda}$$

where v'' is the speed of the wave relative to the source and λ is the wavelength. Similarly, the observed frequency f' and the wavelength are related by

Observed frequency

$$f' = \frac{v'}{\lambda}$$

where v' is the speed of the wave relative to the observer. The formula relating the observed and emitted frequencies, obtained by eliminating λ from these two equations, is

Doppler effect

$$f' = \frac{v \pm u_r}{v \mp u_s} f_0$$

where v is the speed of sound, u_r is the speed of the receiver, and u_s is the speed of the source. All speeds are relative to the propagating medium, which means Doppler shift problems are best done in a reference frame where there is no wind.

The correct choices for the signs are most easily obtained by remembering that the source moving toward the receiver produces an increase in frequency—as does the receiver moving toward the source. For example, if the receiver is moving toward the source and the source is moving toward the receiver, a plus sign is used in the numerator and a minus sign in the denominator. Notice that the observed frequency is increased when we have a smaller value for $v \mp u_s$ in the denominator. It is the upper sign in either the numerator or the denominator ("plus" in the numerator, "minus" in the denominator) that gives an increase in frequency. When the relative speed of approach u of the source and receiver is small compared to the speed of sound v, the **Doppler shift** in frequency can be written

Doppler shift

$$\frac{\Delta f}{f_0} \approx \pm \frac{u}{v} \qquad (u \ll v)$$

where the plus sign is used if the distance between the source and receiver is decreasing.

When a source moves at a speed greater than the speed of sound, a conical **shock wave** with the source at the apex forms behind the source at an angle θ with the path of the source. The angle θ between one side of the cone and the path of the source is related to the speed u of the source by the formula

Mach angle

$$\sin \theta = \frac{v}{u}$$

The ratio of the source speed u to the wave speed v is called the **Mach number**:

Mach number

$$\text{Mach number} = \frac{u}{v}$$

II. Numbers and Key Equations

Numbers

Universal gas constant

$$R = 8.314 \text{ J/mol·K}$$

Gas constants

$\gamma = 1.67$ (monatomic gas)

$\gamma = 1.40$ (diatomic gas)

Key Equations

Speed of transverse waves on a taut string

$$v = \sqrt{\frac{F}{\mu}}$$

Speed of sound in a fluid

$$v = \sqrt{\frac{B}{\rho}}$$

Speed of sound in a gas

$$v = \sqrt{\frac{\gamma R T}{M}}$$

Wave equation

$$\frac{\partial^2 y}{\partial x^2} = \frac{\mu}{F} \frac{\partial^2 y}{\partial t^2}$$

Wave speed

$$v = \frac{\lambda}{T} = f\lambda = \frac{\omega}{k}$$

Angular frequency

$$\omega = \frac{2\pi}{T}$$

Wave number

$$k = \frac{2\pi}{\lambda}$$

Displacement of harmonic transverse wave

$$y(x, t) = A \sin (kx - \omega t)$$

Power for waves on a taut string

$$P = F_y u_y = F \frac{\partial y}{\partial x} \frac{\partial y}{\partial t}$$

Power for harmonic waves on a taut string

$$P = \mu \omega^2 A^2 v \cos^2 (kx - \omega t)$$

Average power for harmonic waves on a taut string

$$P_{av} = \tfrac{1}{2}\mu \omega^2 A^2 v$$

Displacement for harmonic sound waves

$$s(x, t) = s_0 \sin (kx - \omega t)$$

Pressure for harmonic sound waves

$$\begin{aligned} p(x, t) &= p_0 \sin (kx - \omega t - \pi/2) \\ &= -p_0 \cos (kx - \omega t) \end{aligned}$$

Pressure-amplitude/displacement-amplitude relation

$$p_0 = \rho \, \omega \, v \, s_0$$

Intensity of a point source

$$I = \frac{P_{av}}{A} = \frac{P_{av}}{4\pi r^2}$$

Intensity of a harmonic sound wave

$$I = (pu)_{av} = \left(p \frac{\partial s}{\partial t} \right)_{av} = \tfrac{1}{2}\omega s_0 p_0 = \tfrac{1}{2}\rho \omega^2 s_0^2 v$$

Energy density of a harmonic sound wave

$$\eta_{av} = \frac{I}{v} = \tfrac{1}{2}\rho \omega^2 s_0^2$$

Intensity level

$$\beta = (10 \text{ dB})\log_{10} \frac{I}{I_0}$$

Threshold of hearing

$$I_0 = 10^{-12}\,\text{W}/\text{m}^2$$

Emitted frequency

$$f_0 = \frac{v''}{\lambda}$$

Observed frequency

$$f' = \frac{v'}{\lambda}$$

Doppler effect

$$f' = \frac{v \pm u_r}{v \mp u_s}f_0, f' = \frac{1 \pm u_r/v}{1 \mp u_s/v}$$

Doppler shift

$$\frac{\Delta f}{f_0} \approx \pm\frac{u}{v} \quad (u \ll v)$$

Mach angle

$$\sin\theta = \frac{v}{u}$$

Mach number

$$\text{Mach number} = \frac{u}{v}$$

III. Potential Pitfalls

The term "amplitude" has the same meaning for a sinusoidal mechanical wave as it does for simple harmonic motion. It is the maximum displacement of a portion of the medium from its undisturbed, equilibrium position. It is not the distance from crest to valley.

Transverse waves propagating along a string disturb the string; that is, they displace it. Be careful not to confuse the transverse velocity of a segment of the string due to the wave with the velocity of the wave disturbance itself. The velocity of the wave is the rate at which the pattern of disturbance is moving along the string. The two velocities are not directly related and, in fact, are directed at right angles with each other. (In a longitudinal wave the two velocities are in opposite directions fifty % of the time.)

The equation $v = \lambda/T = f\lambda$ is really just another way of stating that the speed equals the distance divided by the time. The wavelength is the distance and the frequency is the reciprocal of the period—the time it takes the disturbance to travel one wavelength. The wave speed v is determined by the properties of the medium that transports the wave and may or may not depend on the wave's frequency. The speed of sound in air, for example, is nearly independent of the frequency, whereas the speed of ripples on a pond depends strongly on the frequency.

It is really easy to make sign errors when doing Doppler-effect problems. Here is a case where simply memorizing formulas will not suffice. Remember, in the frequency formula given, u_r, u_s, and v are never negative because they are speeds, not velocities. A source moving in the direction of a receiver tends to produce an increase in the observed frequency, so the minus sign preceding u_s is used. A receiver moving in the direction of the source tends to produce an increase in frequency, so the plus sign preceding u_r is used. If both the source and receiver are moving, u_s and u_r are the speeds of the source and receiver relative to the medium, not to each other.

Although the terms intensity and intensity level are similar, they represent quite different quantities. The intensity I is the power per unit area, whereas the intensity level β equals $(10\,\text{db})\log_{10}(I/I_0)$, where I_0 is a reference intensity. We suggest that you memorize that 3 dB refers to an intensity ratio of 2.

To simplify physics problems we usually assume that a sound source is isotropic—that is, the intensity is broadcast uniformly in all directions. As you know, actual sound sources, such as a loudspeaker or a human voice, generally have intensities that are greater in the forward direction.

IV. True and False and Responses

True and False

_____ 1. A function of the form $y(x + vt)$ describes the shape or pattern represented by $y(x)$ moving with speed v in the $+x$ direction.

_____ 2. A harmonic wave is one in which the wave function has a sinusoidal shape.

_____ 3. The wavelength of a harmonic wave on a string is the distance along the string of one complete cycle of the wave.

_____ 4. The speed of a harmonic wave is the wavelength divided by the frequency.

_____ 5. Every element of a string on which a harmonic wave is propagating is undergoing simple harmonic motion.

_____ **6.** If two elements of a string on which a harmonic wave is propagating move in phase, the two elements must be an integral number of wavelengths apart.

_____ **7.** A wave motion in which the displacement of the propagating medium is parallel with the direction of propagation of the disturbance is called a longitudinal wave.

_____ **8.** For a given pressure, the speed of sound in air is independent of temperature.

_____ **9.** For a given temperature, the speed of sound in air is independent of pressure.

_____ **10.** The speed at which sound waves propagate in a material medium depends upon the speed of the sound source relative to the medium.

_____ **11.** In a medium transporting a harmonic sound wave, the pressure at a given point is 90° out of phase with the displacement at that point.

_____ **12.** The ratio of the pressure amplitude to the displacement amplitude of a sound wave depends upon the wave frequency.

_____ **13.** A wave pulse reflected at the end of a string is always inverted.

_____ **14.** A wave passing through a very small hole in a barrier propagates outward from the hole in all directions on the far side of the barrier, as though from a point source.

_____ **15.** The received frequency of a sound wave is always equal to the frequency of the source, even when there is relative motion between the source and the receiver.

_____ **16.** The Doppler shift of a wave frequency depends only upon the relative speed between the receiver and the propagating medium.

_____ **17.** The intensity of a harmonic wave is the power per unit area.

_____ **18.** The intensity level of a harmonic wave is the logarithm of the intensity.

_____ **19.** The decibel is a unit of intensity.

_____ **20.** When the source and the receiver are moving such that they are getting closer together, the received frequency is higher than the frequency of the source.

Responses

1. False. It describes the shape or pattern represented by $y(x)$ moving with speed v in the $-x$ direction.

2. True

3. True

4. False. The speed is the wavelength divided by the period, that is, the wavelength multiplied by the frequency.

5. True

6. True. The motions at distances separated by an integral number of wavelengths are identical.

7. True

8. False. It is proportional to the square root of the absolute temperature.

9. True

10. False. The speed at which sound waves propagate depends only on properties of the medium.

11. True

12. True

13. False. Reflections at a free end are not inverted.

14. True

15. False. That the frequency received and the frequency of the source differ when there is relative motion between the source and the receiver is known as the Doppler effect.

16. False. It depends on the speed of sound, the speed of the source, the speed of the receiver, on whether the source is moving toward the receiver or not, and on whether the receiver is moving toward the source or not.

17. True

18. False. The intensity level is ten times the logarithm, to the base ten, of the ratio of the intensity to the reference intensity.

19. False. The decibel is a unit of intensity level.

20. True

V. Questions and Answers

Questions

1. For a function to be a wave function it must satisfy certain criteria. One necessary criterion is that it satisfy the condition

$$y(x, t) = y_1(x - vt) + y_2(x + vt)$$

A string must be continuous. Therefore only continuous functions can be wave functions for waves on a string.

Which of the following functions can be wave functions for waves on a string?

(a) $y(x, t) = Ae^{-(x - vt)^2/2a^2}$

(b) $y(x, t) = Ae^{-x^2}e^{+bt}$

(c) $y(x, t) = 0 \qquad x \le vt - a$

$\quad y(x, t) = D \qquad vt - a \le x \le vt + a$

$\quad y(x, t) = 0 \qquad x \ge vt + a$

Figure 15-1

Pulse velocity

The function in (*a*) looks like the pulse shown in Figure 15-1 and the function in (*c*) is shown in Figure 15-2.

Figure 15-2

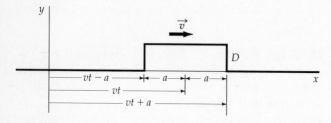

2. A Slinky is a children's toy that is just a long, loose-coiled spring. It is quite useful as a wave demonstrator. How would you generate longitudinal waves in a Slinky? What about transverse waves?

3. A tuning fork, which vibrates at a fixed frequency, is being waved back and forth along the direction from it to you. Describe what you hear. What will you hear if the fork is moved from side to side?

4. Consider a sound wave that propagates uniformly in all directions from a point source. How does the amplitude of the wave vary with distance from the source?

5. A nice, pleasant sound level at which to listen to music is 60 dB. Would 120 dB be twice as loud?

6. Imagine that you're listening to orchestral music being played on a stereo that is in another room down the hall. Will diffraction have any effect on the sound you hear?

7. Why is it that we can hear, but not see, around corners?

Answers

1. (*a*) The function

$$y(x, t) = Ae^{-(x-vt)^2/2a^2}$$

is a possible wave function, as y is a function of $x - vt$.

(*b*) The function

$$y(x, t) = Ae^{-x^2}e^{+bt} = Ae^{-x^2+bt}$$

cannot be expressed as the sum of a function of $x + vt$ and a function of $x - vt$. Therefore it cannot be a wave function.
(*c*) The function

$$y(x, t) = 0 \qquad x \le vt - a$$

$$y(x, t) = D \qquad vt - a \le x \le vt + a$$

$$y(x, t) = 0 \qquad x \ge vt + a$$

is a function of $x - vt$. This can be seen by subtracting vt from each term in each of the inequalities giving

$$y(x, t) = 0 \qquad x - vt \le -a$$

$$y(x, t) = D \qquad -a \le x - vt \le a$$

$$y(x, t) = 0 \qquad x - vt \ge a$$

It would represent a rectangularly shaped pulse of height D and width $2a$ traveling in the positive x direction with speed v. However, this function is discontinuous at the points $x = vt - a$ and $x = vt + a$. A string cannot be discontinuous, so this function cannot be a wave function.

Discontinuous functions change abruptly and thus are not possible wave functions. In textbooks, however, discontinuous functions are sometimes used as wave functions. You should think of these as continuous functions that are close approximations to the discontinuous functions. Figure 15-3 shows a continuous function that is a close approximation to the discontinuous function in (*c*).

Figure 15-3

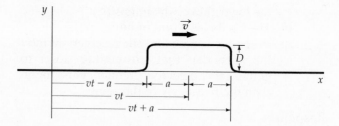

2. You can generate longitudinal waves in a Slinky very easily by tying one end of the spring to a wall and stretching it to a moderate degree. Moving the free end back and forth along the line of the spring will produce easily visible compression waves that travel at a few meters per second (see Figure 15-4*a*). Transverse waves are generated by moving the free end up and down (see Figure 15-4*b*.)

Figure 15-4

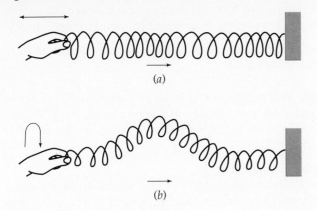

(a)

(b)

wavelengths of audible sound, but all are very large compared to the wavelengths of visible light. Thus, sound waves are significantly diffracted, but light waves are not.

VI. Problems and Solutions

Problems

3. You'll hear the pitch rise and fall as the tuning fork is moved toward or away from you, but you'll hear only a steady pitch as it is moved from side to side.

4. For a wave propagating uniformly in all directions in three dimensions the intensity decreases inversely with the square of the distance from the source. The intensity is also proportional to the square of the amplitude. Thus the amplitude must decrease inversely with the first power of the distance from the source.

5. No. In fact, just what "twice as loud" means isn't clear because loudness is a function of perception; but this isn't it. Typically, a tone sounds twice as loud when the intensity increases by a factor of ten, which corresponds to an intensity level increase of 10 dB. The 60 dB increase means the sound intensity is increased by a factor of 10^6 (or one million).

6. The diffraction of sound waves is the main reason you can hear "around corners" from the music room to the hall and then into your room. The smallest aperture involved is likely to be the width of a doorway, around 1 m. Sound waves with high frequencies and short wavelengths (much less than 1 m) diffract less than do waves with low frequencies and wavelengths much longer than 1 m. Thus, the high frequency sound waves reaching your ears are diminished more than are the low frequency sound waves. You will hear high notes, and particularly the high harmonics that give musical sounds their tone quality, less effectively, so the music will sound "flat" or "muffled" to you. *Remark:* Walls, drapes, and other surfaces selectively absorb some frequencies better than others. This will also affect what the listener hears.

7. The obstacles, apertures, and such that surround us in everyday circumstances—doorways, for instance—tend to be not very large compared to the

1. Far out at sea, your ship is traveling at a speed of 12 mi/h with respect to the water in the same direction as the waves. You notice that one swell passes you every 9 s. If the waves are 500 ft apart, what is the speed of the waves relative to the water?

How to Solve It
• The wavelength divided by the observed period gives you velocity of the waves relative to the ship.

• The velocity of the waves relative to the water equals the velocity of the waves relative to the ship plus the velocity of the ship relative to the water.

2. Far out at sea, your sailboat is traveling at a speed of 8 mi/h with respect to the water in the same direction as the waves. You notice that the boat moves so that it remains on top of a single wave crest. What is the speed of the waves relative to the water?

How to Solve It
• The velocity of the waves relative to the water equals the velocity of the waves relative to the boat plus the velocity of the boat relative to the water.

3. The wave function for a harmonic wave on a string is $y(x, t) = 0.002 \sin (31.4x - 628t + 1.57)$, where y and x are in meters and t is in seconds. (a) Find the amplitude, wavelength, frequency, period, and phase constant of this wave. (b) In what direction does this wave travel and what is its speed? (c) At $t = 0.02$ s find the phase at $x = 0.3$ m.

How to Solve It
• Use the formula for a harmonic wave function to identify the terms in the wave function.

• The phase of a harmonic wave is the argument of the sine function.

4. The wave function for a harmonic wave on a string is $y(x, t) = 0.001 \sin (62.8x + 314t - 1.57)$ where y and x are in meters and t is in seconds. (a) Find the amplitude, wavelength, frequency, period, and phase constant of this wave. (b) In what direction does this wave travel and what is its speed? (c) Find the phase difference between the points at $x = 0.50$ m and at $x = 0.55$ m at any time t.

How to Solve It
- Use the formula for a harmonic wave function to identify the terms in the wave function.

- The phase of a harmonic wave is the argument of the sine function.

5. Show by direct substitution that the function $y = (x - vt)^2$ is a solution of the wave equation.

How to Solve It
- Substitute the function into the wave equation and verify that its function satisfies the equation.

6. Show by direct substitution that the functions $y = \cos(x + vt)$ and $y = -7(x - vt)^2 + 4\cos(x + vt)$ each satisfy the wave equation.

How to Solve It
- Substitute the functions into the wave equation and verify that each function satisfies the equation.

7. The temperature inside a house is maintained at 20°C, so the speed of sound in the air within the house is 340 m/s. On a cold day the speed of sound in the air outside the house is 334 m/s. A tuning fork in the house is struck, and starts vibrating at 440 Hz. A boy outside the house hears the sound from the tuning fork. What frequency does he hear? Determine the wavelength of the wave both in the air inside the house and in the air outside the house.

How to Solve It
- Determine the frequency at which the compressions of the sound wave enter the cold outside air.

- Determine the frequency of the sound in the cold outside air.

- The speed of the wave equals the frequency divided by the wavelength.

8. The extreme range of the human singing voice is roughly from 53 Hz (A flat below bass low C) to 1267 Hz (E flat above soprano high C). Find the range of wavelengths that corresponds to this frequency range.

How to Solve It
- The speed of sound is the frequency multiplied by the wavelength.

9. Air ($\gamma = 1.40$) has a molar mass of 28.9 g/mol and helium ($\gamma = 1.67$) has a molar mass of 4.00 g/mol. (*a*) At room temperature (20°C = 293 K), calculate the speed of sound in air and in helium. (*b*) In each medium calculate the frequency of a sound with a wavelength of 2.2 m.

How to Solve It
- There is a formula for the speed of sound waves in a gas in terms of temperature, molar mass, and γ.

- The speed of sound is the frequency multiplied by the wavelength.

10. You stand on the edge of a canyon 420 m wide and clap your hands. An echo comes back to you 2.56 s later. What is the temperature of the air? For air, $\gamma = 1.40$.

How to Solve It
- Calculate the speed of the sound from the data given. Don't forget that the sound has to go across the canyon and back.

- There is a formula for the speed of sound waves in a gas in terms of temperature, molar mass, and γ. Solve for the temperature.

11. Suppose each of the two loudspeakers of my stereo system is delivering 1 watt of average power in the form of audible sound in my den. Also suppose that this power is being delivered uniformly in all directions. (*a*) If I'm sitting 2.5 m away from the speakers, what is the intensity of the sound I hear directly from the speakers (that is, neglecting the sound reflected from the walls of the room)? (*b*) To what intensity level does this correspond?

How to Solve It
- Calculate the direct sound intensity from the speakers from their power output and the receiver's distance from them.

- Note that, in reality, the sound level would be higher than this, due to reflections from the walls unless the walls were specially constructed to be nonreflecting.

- Convert the intensity you calculated in part (*a*) to an intensity level (in dB).

12. Assume that a barking dog can put out (on the average) 1 mW of sound power and that the sound propagates uniformly in all directions. What is the sound intensity level (in dB) 6 m away from a pen containing eight barking dogs?

How to Solve It
- Intensity is power per unit area.

- Don't forget that there is more than one dog. Are these sources coherent?

- Calculate the intensity and convert it to an intensity level.

13. A woman with perfect pitch, standing next to a railroad track on a still day, is amused to notice that the whistle of an approaching train is sounding a true concert A (440 Hz). After the train has passed her and is receding, the pitch has dropped to a true F natural (349 Hz). (*a*) How fast was the train going? (*b*) What was the actual frequency of the whistle?

How to Solve It
* The first frequency is heard when the train is coming straight toward the woman. Write a formula for the received frequency in terms of the unknown source frequency, the source speed, the receiver's speed, and the speed of sound.

* Repeat this for the second frequency, which is heard when the train is receding from the woman at the same source speed.

* Eliminate the actual whistle frequency f_0 from these two equations, and you will be left with a relation that can be solved for the velocity ratio u/v.

* When you have solved for the velocity u of the train, use that in either of the equations to find f_0.

14. My car's horn sounds a tone of frequency 250 Hz. If I am driving directly at you at 25 m/s, blowing my horn, on a hot, still day (93°F), (*a*) what is the wavelength of the sound that reaches you? (*b*) What frequency do you hear?

How to Solve It
* The only thing out of the ordinary is that you have to recalculate the speed of sound at the temperature given.

* The wavelength is decreased by the factor $(1 - u_s/v)$ because the sound source is coming at you.

15. On a still day, a car moves at a speed of 40 km/h directly toward a stationary wall. Its horn emits a sound at a frequency of 300 Hz. (*a*) If the velocity of sound is 340 m/s, find the frequency at which the sound waves hit the wall. (*b*) The waves reflect off the wall and are received by the driver of the car. What frequency does she hear?

How to Solve It
* It's simplest to do this problem in steps. What is the frequency of the sound "received" by the wall?

* Sound of this same frequency (f_1, say) is reflected back toward the oncoming car.

* The driver of the oncoming car is now a moving observer receiving waves that originate from a stationary source with frequency f_1. What frequency does she hear?

16. On a still day, a car traveling at 100 km/h sounds its horn, which has a frequency of 250 Hz. What frequency is heard by a receiver who is proceeding at 60 km/h in the same direction (*a*) directly ahead of the car and (*b*) directly behind it?

How to Solve It
* In this problem both the source and the receiver are moving. Use the formula relating the received frequency f' with the frequency of the source f_0. In part (*a*) the source is moving toward the receiver, an action that tends to cause an upward shift in the received frequency. Use the appropriate sign in the denominator to provide this upward frequency shift. Furthermore, the receiver is moving away from the source, an action that tends to cause a downward shift in the received frequency. Use the appropriate sign in the numerator to provide this downward frequency shift.

* In part (*b*) the source is moving away from the receiver and the receiver is moving toward the source. Use the signs that tend to provide frequency shifts in the appropriate direction(s).

Solutions

1. The period you observe on board the ship is 9 s, so the velocity v' of the waves relative to the ship is

$$v' = \frac{\lambda}{T} = \frac{500 \text{ ft}}{9 \text{ s}} = 55.6 \text{ ft/s}$$

The velocity of the waves relative to the water is

$$v = v' + u$$

where u is the velocity of the ship relative to the water. Thus,

$$v = 55.6 \text{ ft/s} + (12 \text{ mi/h})(5280 \text{ ft/mi})(1\text{h}/3600 \text{ s})$$

$$= \underline{73.2 \text{ ft/s}}$$

2. $\underline{8 \text{ mi/h}}$

3. The wave function is $y(x, t) = 0.002 \sin (31.4x - 628t + 1.57)$, where y and x are in meters, t is in seconds, and the quantity 1.57 is in radians.
(*a*) The formula for the wave function of a harmonic wave is $y(x, t) = A \sin (kx - \omega t + \delta)$ where A is the amplitude, k is the wave number, ω is the angular frequency, and δ is the phase constant. The ampli-

tude, wavelength, frequency, period, and phase constant are, respectively,

$$A = \underline{0.002 \text{ m}}$$

$$\lambda = \frac{2\pi}{k} = \frac{2\pi}{31.4 \text{ m}^{-1}} = \underline{0.200 \text{ m}}$$

$$f = \frac{\omega}{2\pi} = \frac{628 \text{ s}^{-1}}{2\pi} = \underline{100 \text{ Hz}}$$

$$T = \frac{1}{f} = \frac{1}{100 \text{ Hz}} = \underline{0.0100 \text{ s}}$$

$$\delta = \underline{1.57 \text{ rad}}$$

(b) The wave travels in the positive x direction with speed

$$v = \lambda f = (0.200 \text{ m})(100 \text{ Hz}) = \underline{20.0 \text{ m/s}}$$

(c) The phase θ is the argument of the sine function so

$$\theta = kx - \omega t + \delta$$
$$= (31.4 \text{ m}^{-1})(0.3 \text{ m}) - (628 \text{ s}^{-1})(0.02 \text{ s}) + 1.57$$
$$= \underline{-1.57 \text{ rad}}$$

4. (a) $A = \underline{0.001 \text{ m}}$, $\lambda = \underline{0.100 \text{ m}}$, $f = \underline{50.0 \text{ Hz}}$, $T = \underline{0.0200 \text{ s}}$, and $\delta = \underline{-1.57 \text{ rad}}$
(b) $\underline{5 \text{ m/s in the negative } x \text{ direction}}$
(c) $\underline{3.14 \text{ rad}}$

5. Substituting $y = (x - vt)^2$ in the wave equation gives

$$\frac{\partial^2 y}{\partial x^2} = \frac{1}{v^2} \frac{\partial^2 y}{\partial t^2}$$

$$\frac{\partial^2 (x - vt)^2}{\partial x^2} = \frac{1}{v^2} \frac{\partial^2 (x - vt)^2}{\partial t^2}$$

$$\frac{\partial [2(x - vt)]}{\partial x} = \frac{1}{v^2} \frac{\partial [-2v(x - vt)]}{\partial t}$$

$$2 = 2$$

which verifies that the function $y = (x - vt)^2$ satisfies the wave equation.

6. Each function satisfies the wave equation.

7. The frequency of the sound in the warm inside air is 440 Hz, the same as the frequency of the tuning fork. The compressions leave the warm air and enter the cold air at a frequency 440 Hz. Therefore the frequency in the cold air is also 440 Hz. The frequency that the boy hears is $\underline{440 \text{ Hz}}$.
    The wavelength in the warm air is

$$\lambda = \frac{v}{f} = \frac{340 \text{ m/s}}{440 \text{ Hz}} = \underline{0.773 \text{ m}}$$

and the wavelength in the cold air is

$$\lambda = \frac{v}{f} = \frac{334 \text{ m/s}}{440 \text{ Hz}} = \underline{0.759 \text{ m}}$$

8. $\underline{6.40 \text{ m to } 26.8 \text{ cm}}$

9. (a) The speed of sound in a gas is

$$v = \sqrt{\frac{\gamma RT}{M}}$$

At a temperature of 20°C = 293 K, the speed of sound in air is

$$v_{air} = \sqrt{\frac{1.40(8.31 \text{ J/mol·K})(293 \text{ K})}{0.0289 \text{ kg/mol}}} = \underline{343 \text{ m/s}}$$

and the speed of sound in helium is

$$v_{He} = \sqrt{\frac{1.67 (8.31 \text{ J/mol·K})(2.3 \text{ K})}{0.00400 \text{ kg/mol}}} = \underline{1010 \text{ m/s}}$$

(b) The frequency of a wave of wavelength λ is

$$f = \frac{v}{\lambda}$$

so for $\lambda = 2.20$ m

$$f_{air} = \frac{343 \text{ m/s}}{2.20 \text{ m}} = \underline{156 \text{ Hz}}$$

and

$$f_{He} = \frac{1010 \text{ m/s}}{2.20 \text{ m}} = \underline{459 \text{ Hz}}$$

10. A chilly $\underline{-5.7°C \text{ (267 K)}}$

11. (a) The intensity at a distance r from a loudspeaker delivering 1 W of average power is

$$I = \frac{P_{av}}{4\pi r^2} = \frac{1 \text{ W}}{4\pi(2.5 \text{ m})^2} = 1.27 \times 10^{-2} \text{ W/m}^2$$

Because the two loudspeakers are not coherent, the intensity for two speakers is twice the intensity for one speaker. Thus, the total intensity is $\underline{2.55 \times 10^{-2} \text{ W/m}^2}$.
    (b) The corresponding intensity level is

$$\beta = 10 \log_{10} \frac{I}{I_0}$$

$$= 10 \log_{10} \left(\frac{2.55 \times 10^{-2} \text{ W/m}^2}{10^{-12} \text{ W/m}^2} \right) = \underline{104 \text{ dB}}$$

This is an extremely loud noise. One or two watts can chase you out of the room! You don't buy amplifiers that will put out lots of watts because you need that much acoustical power; rather, it's because good-quality loudspeakers have low efficiency.

12. 72 dB

13. (a) The formula relating the received frequency f' with the frequency of the source f_0 is

$$f' = \frac{(1 \pm u_r/v)}{(1 \mp u_s/v)} f_0$$

The receiver (the woman) is standing still, so $u_r = 0$.

Initially the source (the whistle) is moving toward the receiver. This motion results in an increase in the received frequency f_1', so the minus sign in the denominator is used, and

$$f_1' = \frac{f_0}{(1 - u_s/v)} \tag{1}$$

After the train passes, the whistle is moving away from the receiver, which results in a decrease in the received frequency f_2' so this time the plus sign in the denominator is used:

$$f_2' = \frac{f_0}{(1 + u_s/v)}$$

We now eliminate the frequency of the source from these two equations and solve for the speed u_s of the train. The ratio of the received frequencies is

$$\frac{f_1'}{f_2'} = \frac{1 + u_s/v}{1 - u_s/v}$$

Solving for u_s we have

$$f_1' - f_1' u_s/v = f_2' + f_2' u_s/v$$

$$(f_1' + f_2')\frac{u_s}{v} = f_1' - f_2'$$

$$u_s = \frac{f_1' - f_2'}{f_1' + f_2'} v = \frac{440\ \text{Hz} - 349\ \text{Hz}}{440\ \text{Hz} + 349\ \text{Hz}} (340\ \text{m/s})$$

$$= 39.2\ \text{m/s}$$

(b) Substituting the value of u_s into Equation (1), we find the frequency of the whistle is

$$f_0 = f_1'\left(1 - \frac{u_s}{v}\right) = (440\ \text{Hz})\left(1 - \frac{39.2\ \text{m/s}}{340\ \text{m/s}}\right)$$

$$= 389\ \text{Hz}$$

14. (a) 1.31 m   (b) 269 Hz

15. (a) For convenience we convert the speed of the car to meters per second

$$v_c = (40\ \text{km/h})\frac{1000\ \text{m/km}}{3600\ \text{s/h}} = 11.1\ \text{m/s}$$

The formula relating the received frequency f' with the frequency of the source f_0 is

$$f' = \frac{(1 \pm u_r/v)}{(1 \mp u_s/v)} f_0$$

The wall receives the sound from the car horn at a frequency f_1 of

$$f_1 = \frac{f_0}{(1 - v_c/v)} = \frac{300\ \text{Hz}}{1 - \dfrac{11.1\ \text{m/s}}{340\ \text{m/s}}} = 310\ \text{Hz}$$

where we have set $u_r = 0$, $u_s = v_c$, and have used the minus sign in the denominator because when the source moves in the direction of the receiver the received frequency is shifted upward.

(b) The frequency of the sound reflecting off the wall is the same as if the wall were a sound source with a frequency equal to f_1, the frequency at which the wall receives the sound from the horn. In the formula relating the received frequency f_2 with the source frequency f_1 the speed of the source (the wall) u_s is zero, the speed of the receiver u_r equals the speed of the car v_c, and the plus sign in the numerator is used because when the receiver moves toward the source the received frequency is shifted upward. Thus,

$$f_2 = f_1(1 + v_c/v) = (310\ \text{Hz})\left(1 + \frac{11.1\ \text{m/s}}{340\ \text{m/s}}\right)$$

$$= 320\ \text{Hz}$$

16. (a) 259 Hz   (b) 242 Hz

Chapter 16

Superposition and Standing Waves

I. Key Ideas

16-1 Superposition of Waves As a wave travels along a string, the transverse displacement y of the string varies with the position x along the string and time t. The function $y = y(x, t)$ is called the **wave function**. The wave function $y_1(x - vt)$ gives the transverse displacement of the string for a wave moving in the positive x direction with speed v. The wave function $y_2(x + vt)$ gives the transverse displacement of the string for a wave moving in the negative x direction with speed v.

When two waves traveling along the same string superpose (overlap), the resultant displacement y of the string is the algebraic sum of the displacements y_1 and y_2 that would occur if only one or the other of the waves were present. The relation $y = y_1 + y_2$ is called the **principle of superposition**. Each wave propagates as if the other weren't there. Pulses traveling in opposite directions pass right through each other, emerging with their original size and shape.

The most general wave function $y(x, t)$ can always be expressed as the superposition of two wave functions describing waves traveling in opposite directions:

Wave function

$$y(x, t) = y_1(x - vt) + y_2(x + vt)$$

The superposition of harmonic waves is called **interference**. Consider two harmonic waves of equal amplitude and frequency that are simultaneously traveling along the same string. If the crests of one of these waves exactly superpose the crests of the other, the difference between the phases of the two waves is zero. Such waves are said to be in phase and the resultant wave will have an amplitude equal to the sum of the amplitudes of the original waves. This phenomenon is called **constructive interference**. On the other hand, when the crests of one of these waves exactly superpose the troughs of the other, the differ-

ence between the phases of the two waves is 180° (π rad) and the resultant wave has an amplitude equal to zero—the difference between the amplitudes of the original waves. This is called **destructive interference**. When two harmonic waves interfere destructively they are said to be 180° out of phase. For interference to occur the waves do not have to be either in phase or 180° out of phase. They do, however, have to have a phase difference that remains fixed (is constant). For two waves y_1 and y_2 with an arbitrary phase difference δ, the two waves interfere such that the wave function y of the resultant wave is

Interference of harmonic waves

$$y = y_1 + y_2$$
$$= A \sin(kx - \omega t) + A \sin(kx - \omega t + \delta)$$
$$= 2A \cos\frac{\delta}{2} \sin\left(kx - \omega t + \frac{\delta}{2}\right)$$

where $2A \cos(\delta/2)$ and $\delta/2$ are the amplitude and phase constant of the resultant harmonic wave. An analogous equation describes the resultant wave when two sound waves interfere. For a sound wave the y representing the transverse displacement of the medium is replaced either by s for the longitudinal displacement or by p for pressure.

Beats When two harmonic waves with the same amplitude, but with different angular frequencies ω_1 and ω_2, overlap, the resulting pressure at a given point may be expressed in terms of the difference in angular frequency $\Delta\omega$ and the average angular frequency ω_{av}:

Superposition of harmonic waves at a given point

$$p = p_1 + p_2 = p_0 \sin\omega_1 t + p_0 \sin\omega_2 t$$
$$= 2p_0 \cos\left(\tfrac{1}{2}\Delta\omega t\right) \sin\omega_{av}t$$

where $\Delta\omega = \omega_1 - \omega_2$ and $\omega_{av} = (\omega_1 + \omega_2)/2$. The resultant wave can also be expressed in terms of the frequencies by substituting $2\pi\,\Delta f$ and $2\pi f_{av}$ for $\Delta\omega$ and ω_{av}. When the difference in frequency is small compared with the average frequency, the term $2\pi\frac{1}{2}$ $\Delta f\, t$ varies slowly in comparison with the term $2\pi f_{av}t$. The resulting pressure at a fixed point varies with the frequency f_{av} and with an amplitude of $2p_0$ $\cos(2\pi\frac{1}{2}\Delta f\, t)$. This amplitude varies in time with a frequency of $\Delta f/2$.

The intensity of a sound wave is proportional to the square of the amplitude. Using the trigonometric identity $\cos^2\theta = \frac{1}{2} + \frac{1}{2}\cos 2\theta$ it can be shown that because the amplitude varies with frequency $\Delta f/2$, the square of the amplitude, and thus the intensity, varies with frequency Δf. Therefore, the ear hears the intensity periodically getting louder and then quieter at a frequency Δf. This frequency is called the **beat frequency** f_{beat}:

Beat frequency

$$f_{beat} = \Delta f = f_1 - f_2$$

The interference of two waves of different but nearly equal frequencies produces a tone whose intensity varies alternately between loud and soft—a phenomenon known as **beats**.

Phase Difference due to Path Difference A common cause of phase difference between two waves is the difference in path length between each source and the point of interest. The formula relating the phase difference δ with the associated path-length difference Δx is

Phase difference–path-length difference relation

$$\delta = (2\pi\text{ rad})\frac{\Delta x}{\lambda} = (360°)\frac{\Delta x}{\lambda}$$

Coherent sources are sources that vibrate so that the phase difference between the two sources remains fixed. Coherent sound sources can be produced by two loudspeakers that are driven by the same electrical signal. Sources that vibrate so that the phase difference is not constant but varies randomly over time are called **incoherent sources**. Two violins playing the same music in an orchestra are incoherent sources.

Even when two coherent sources vibrate in phase, the waves are typically not in phase when they arrive at a given point of interest. Suppose there are two loudspeakers vibrating coherently and in phase. If you are half a wavelength farther from one of the speakers then the other, the sound from the two speakers will arrive at your ears 180° out of phase. At such a location you hear nothing because the interference is destructive. (We are considering only the sound that travels directly from the speaker to you, neglecting any that reaches you after reflecting off a wall or some other object.) Destructive interference occurs in locations where the difference in path lengths is an odd number of half wavelengths. (Constructive interference occurs any time this difference is an even number of wavelengths.) A region where destructive interference occurs is called a **node**.

16-2 Standing Waves Waves traveling along a taut string with one or both ends fixed reflect at the ends of the string. For harmonic waves of certain specific frequencies, the waves reflecting back and forth superpose to form a stationary vibration pattern called a **standing wave**. When one end of the string is vibrated at one of these frequencies it responds "enthusiastically." This enthusiastic response is called **resonance**, and the frequencies at which the string resonates (vibrates in a standing wave) are called the **resonance frequencies** of the string. On a string that is vibrating in a standing wave there are locations called **nodes** that remain stationary. Between each pair of adjacent nodes is a location of maximum amplitude called an **antinode**.

The standing-wave patterns for a string that is fixed at both ends are shown in Figure 16-1.

Figure 16-1

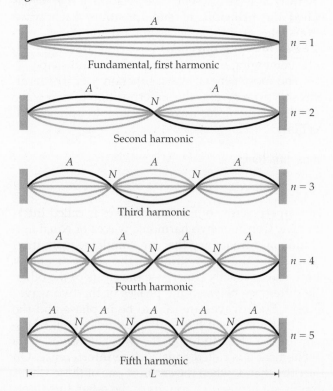

Fundamental, first harmonic

Second harmonic

Third harmonic

Fourth harmonic

Fifth harmonic

The lowest resonance frequency for such a string is called the **fundamental frequency** f_1. When the string vibrates at its fundamental frequency the wave pattern produced is referred to as the **fundamental** mode of vibration or the **first harmonic**. The wave pattern produced when the string vibrates at the second lowest resonance frequency f_2 is called the **second harmonic**, and the pattern when it vibrates at the nth lowest frequency f_n is the **nth harmonic**. For these standing waves, the resonance frequency f_n of the nth harmonic is related to the fundamental frequency f_1 by

Resonance frequencies, both ends fixed

$$f_n = nf_1 \qquad n = 1, 2, 3, \ldots$$

where

Fundamental frequency

$$f_1 = \frac{v}{2L} = \frac{1}{2L}\sqrt{\frac{F}{\mu}}$$

For a flexible string fixed at both ends, the wavelength λ_n of the nth harmonic is related to the length L of the string by the equation

Standing-wave condition, both ends fixed

$$L = n\frac{\lambda_n}{2} \qquad n = 1, 2, 3, \ldots$$

The wavelength of a standing wave is the wavelength of the harmonic waves that interfere to form the resultant standing wave.

In the terminology often used in music, the frequencies of the second and higher harmonics are called **overtones**. Thus the frequency of the second harmonic is called the *first overtone*, that of the third harmonic is called the *second overtone*, and so forth. The resonance frequencies of the string are called its **natural frequencies**, and each vibrational pattern is called a **mode of vibration**. A sequence of natural frequencies that are integral multiples of a fundamental frequency is called a **harmonic series**, and the individual natural frequencies are called **harmonics**.

Standing waves on a taut string can also be produced on a string with one end fixed and the other end free. The standing-wave patterns for such a string are shown in Figure 16-2.

Figure 16-2

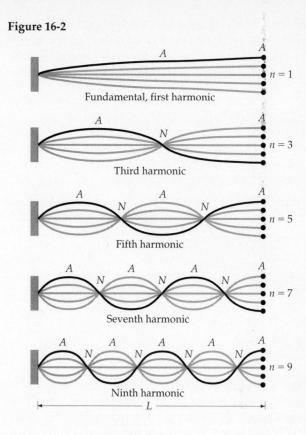

Fundamental, first harmonic — $n = 1$

Third harmonic — $n = 3$

Fifth harmonic — $n = 5$

Seventh harmonic — $n = 7$

Ninth harmonic — $n = 9$

The pertinent relations are

Resonance frequencies, one end free

$$f_n = nf_1 \qquad n = 1, 3, 5, \ldots$$

Fundamental frequency

$$f_1 = \frac{v}{4L} = \frac{1}{4L}\sqrt{\frac{F}{\mu}}$$

Standing-wave condition, one end free

$$L = n\frac{\lambda_n}{4} \qquad n = 1, 3, 5, \ldots$$

Note that a string with one end free resonates only at the odd multiples of the fundamental frequency. That is, only the odd harmonics occur.

To realize the condition just described we must provide a connection that maintains tension F in the string, yet allows the "free" end to slide up and down with negligible friction. This can only occur in an ideal world. The analogous arrangement (an open-ended organ pipe) for sound waves is more readily achieved.

Standing Sound Waves Standing sound waves have much in common with the standing waves of a vibrating string. If air is confined in a tube of length L with both ends closed, any standing waves in the air in the tube will have displacement nodes (points of zero displacement amplitude) at both ends of the tube. Thus,

Standing-wave condition, both ends closed

$$L = n\frac{\lambda_n}{2} \qquad n = 1, 2, 3, \ldots$$

and

Fundamental frequency, both ends closed

$$f_1 = \frac{v}{2L} = \frac{1}{2L}\sqrt{\frac{B}{\rho}}$$

Resonant frequencies, both ends closed

$$f_n = nf_1 \qquad n = 1, 2, 3, \ldots$$

where λ is the wavelength, f_n is the frequency, and v is the speed of sound in air. The standing-wave frequency spectrum of the air in a tube with both ends open is the same as that for the air in a tube with both ends closed. This happens because the air in a tube with both ends open has a displacement antinode (maximum) at each end, which means that for any standing wave in either of these tubes the length of the tube must equal an integral number of half wavelengths.

In a tube with only one end open to the atmosphere, the standing waves of the air in the tube have a displacement antinode at the open end. This displacement antinode is also a pressure node since the pressure at the open end is fixed at the pressure throughout the room, usually atmospheric pressure. Thus, for a standing wave in such a tube

Standing-wave condition, one end open

$$L = n\frac{\lambda_n}{4} \qquad n = 1, 3, 5, \ldots$$

and

Resonant frequencies, one end open

$$f_n = nf_1 \qquad n = 1, 3, 5, \ldots$$

Fundamental frequency, one end open

$$f_1 = \frac{v}{4L} = \frac{1}{4L}\sqrt{\frac{B}{\rho}}$$

The wave function for the nth harmonic standing wave is given by

Standing-wave function

$$y_n(x, t) = A_n(x) \cos(\omega_n t + \delta_n)$$

where $A_n(x) = A_n \sin k_n x$ is the amplitude which depends upon x,

$$\omega_n = 2\pi f_n = n2\pi f_1 = n\omega_1$$

is the angular frequency, and

$$k_n = \frac{2\pi}{\lambda_n} = n\frac{2\pi}{\lambda_1} = nk_1$$

gives the wave number for the nth mode.

16-3   The Superposition of Standing Waves   For a vibrating string of finite length, the motion of the string can always be treated as a superposition of standing waves. The wave function is a sum of the standing wave functions:

Superposition of standing wave functions

$$y(x, t) = \sum_n A_n \sin k_n x \cos(\omega_n t + \delta_n)$$

where A_n, ω_n, δ_n, and k_n are the amplitude, angular frequency, phase constant, and wave number of the nth harmonic.

16-4   Harmonic Analysis and Synthesis   When an air column, such as the air in a clarinet or trombone, vibrates, it usually vibrates not as a single standing wave but as the superposition of two or more standing waves. A plot of the pressure versus time for the sound produced is called a **waveform**. Waveforms can be analyzed to determine the amplitudes, frequencies, and phase constants of the harmonics present. This analysis is called **harmonic analysis** or **Fourier analysis**. The inverse of harmonic analysis—the construction of the original periodic waveform from its harmonic components—is called **harmonic synthesis**.

16-5   Wave Packets and Dispersion   A periodic (repetitive) waveform, like the sound of a sustained note on a clarinet, is a superposition of harmonic waves whose frequency distribution is discrete (not continuous). A waveform that is not periodic, such as the sound from a firecracker, can be considered a superposition of many harmonic waves whose frequency distribution is continuous. Such a superposition of harmonic waves, which produce a pulse, is called a **wave packet**. For a wave packet, the range of angular frequencies $\Delta\omega$ and the time interval Δt for the packet to pass any point are such that

Temporal wave packet relation

$$\Delta\omega\,\Delta t \approx 1$$

The analogous relation for the spatial parameters is

Spatial wave packet relation

$$\Delta k\,\Delta x \approx 1$$

where Δk is the range of wave numbers and Δx is the spatial length of the packet. If a wave packet is to sustain its size and shape, the speed of all the harmonic waves that make up the packet must be the same. This will occur only in a nondispersive medium; that is, a medium in which the speed of a harmonic wave is independent of its wavelength or frequency. In a **dispersive medium**, the speed of harmonic waves varies with frequency. The average speed of the individual harmonic waves that make up a wave packet is called the **phase velocity**, and the velocity of the packet itself is called the **group velocity**.

II. Numbers and Key Equations

Numbers

There are no new numbers for this chapter.

Key Equations

Wave function

$$y(x, t) = y_1(x - vt) + y_2(x + vt)$$

Interference of harmonic waves

$$
\begin{aligned}
y &= y_1 + y_2 \\
&= A \sin (kx - \omega t) + A \sin (kx - \omega t + \delta) \\
&= 2A \cos \frac{\delta}{2} \sin \left(kx - \omega t + \frac{\delta}{2} \right)
\end{aligned}
$$

Superposition of harmonic waves at a given point

$$
\begin{aligned}
p &= p_1 + p_2 = p_0 \sin \omega_1 t + p_0 \sin \omega_2 t \\
&= 2p_0 \cos \left(\tfrac{1}{2}\Delta\omega t \right) \sin \omega_{av} t
\end{aligned}
$$

Beat frequency

$$f_{beat} = \Delta f = f_1 - f_2$$

Phase difference–path-length difference relation

$$\delta = (2\pi\,\text{rad})\frac{\Delta x}{\lambda} = (360°)\frac{\Delta x}{\lambda}$$

Standing-wave patterns for string, both ends fixed

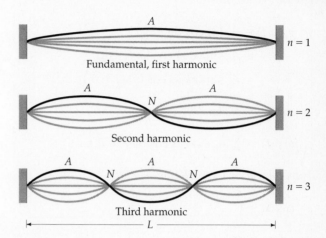

Standing-wave condition, both ends fixed (closed) or both ends free (open)

$$\lambda_n = \frac{2L}{n} \qquad n = 1, 2, 3, \ldots$$

Resonance frequencies, both ends fixed (closed) or both ends free (open)

$$f_n = \frac{v}{\lambda_n} = n\frac{v}{2L} = nf_1 \qquad n = 1, 2, 3, \ldots$$

Standing-wave patterns for string, one end free

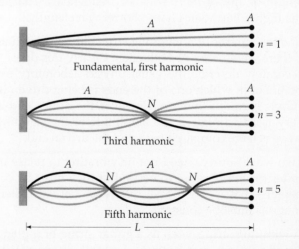

Standing-wave condition, one end fixed (closed) and one end free (open)

$$\lambda_n = n\frac{4L}{n} \qquad n = 1, 3, 5, \ldots$$

Resonant frequencies, one end fixed (closed) and one end free (open)

$$f_n = \frac{v}{\lambda_n} = n\frac{v}{4L} = nf_1 \qquad n = 1, 3, 5, \ldots$$

Standing-wave function

$$y_n(x, t) = A_n(x) \cos(\omega_n t + \delta_n)$$
$$= A_n \sin(k_n x) \cos(\omega_n t + \delta_n)$$
$$= A_n \sin(nk_1 x) \cos(n\omega_1 t + \delta_n)$$

Superposition of standing wave functions

$$y(x, t) = \sum_n A_n \sin k_n x \cos(\omega_n t + \delta_n)$$

Temporal wave packet relation

$$\Delta\omega\, \Delta t \approx 1$$

Spatial wave packet relation

$$\Delta k\, \Delta x \approx 1$$

III. Possible Pitfalls

Destructive interference means that the two interfering waves are 180° out of phase with one another and so will tend to cancel. The cancellation is total, however, only if the two waves have equal amplitudes.

The fundamental frequency of a taut string is the lowest frequency at which it can resonate. Correspondingly, the fundamental has the longest wavelength of any of the resonance frequencies. That is, the higher harmonics have shorter wavelengths.

The frequencies of a vibrating string are calculated as if there were nodes at both ends. Clearly this doesn't accurately describe the usual physics laboratory experiment in which one of the ends is being driven in simple harmonic motion. However, the driven end is very nearly a node if the amplitude of the standing wave is large compared to that of the driven end.

Two wave sources need not be vibrating in phase in order for interference to occur, but they must be coherent—that is, the phase difference between their motions must be constant.

A vibrating string, such as a guitar string or a piano string, can act as a source of sound waves of the same frequency in air as on the string. But notice that the wavelength of the sound waves and the wavelength of the standing waves on the string are not the same. This is because the wave speeds for transverse waves on the string and for sound waves in air are different.

A tube with both ends closed does not correspond to any practical musical instrument. In the case of a flute or an organ pipe with the end open, *both* ends are open. In these cases the wavelengths are the same as if both ends were *closed*.

IV. True and False and Responses

True and False

_____ 1. If two wave pulses traveling in the opposite directions along the same string meet, they reflect off each other.

_____ 2. When one end of a taut string is shaken transversely at one of the string's natural frequencies there is an antinode at the point where the string is attached to the shaker.

_____ 3. Standing waves on a taut string that is fixed at each end have wavelengths that are integral multiples of the length of the string.

_____ 4. When a string is vibrating in a standing wave, every element of the string, except for those at nodes, is undergoing simple harmonic motion.

_____ 5. When a string is vibrating in a standing wave, the motion at one antinode is in phase with the motion at all other antinodes.

_____ 6. When a standing wave exists in the air within a tube that has an open end, a pressure node can be found just beyond the open end of the tube.

_____ 7. The fundamental frequency of standing waves in a tube with one end open, or in a tube approximately twice as long with both ends open, is the same.

_____ 8. The fundamental frequency of standing waves in a tube with both ends open, or a tube of approximately equal length with both ends closed, is the same.

_____ 9. The complete destructive interference of two harmonic waves requires that they have the same amplitude.

_____10. For the destructive interference of two harmonic waves from two point sources that vibrate in phase to occur, the path-length difference must be an integral number of wavelengths.

_____11. At a maximum in the interference pattern of the waves from two identical point sources, the intensity is double the intensity due to either source alone.

_____ **12.** Two sources are coherent if the phase difference between them is fixed.

_____ **13.** If piano strings of frequencies 257 Hz and 261 Hz are sounded together, you will hear 2 beats per second.

_____ **14.** A periodic waveform can be synthesized by the superposition of its harmonic components.

_____ **15.** In a nondispersive medium the speed of a harmonic wave depends upon its frequency.

Responses

1. False. The waves "pass through" each other.

2. False. The amplitude of vibration of the shaker is much less than the amplitude of the string's motion at an antinode.

3. False. The wavelengths are given by $2L/n$, where L is the length of the string.

4. True

5. False. At adjacent antinodes the motions are $180°$ out of phase. That is, when the displacement of the string at one antinode is $+A$, the displacement at the next antinode is $-A$.

6. True

7. True

8. True

9. True

10. False. The path lengths must differ by $(n - \frac{1}{2})\lambda$, where n is a positive integer and λ is the wavelength.

11. False. It is the amplitude that is double the amplitude due to the wave from one source alone. However, because the intensity is proportional to the square of the amplitude, the intensity is four times greater than the intensity due to the wave from one source alone.

12. True

13. False. The beat frequency equals the difference in the frequencies, which is 4 beats per second.

14. True

15. False. In a nondispersive medium all waves travel at the same speed.

V. Questions and Answers

Questions

1. Two long strings, one heavy the other light, are joined end to end and then stretched under tension between two trees. A harmonic wave travels in the direction from the light string to the heavy string. Are the wave speed, the frequency, and the wavelength the same in the heavy string as they are in the light string?

2. The wave pulse shown in Figure 16-3 travels along a stretched string. What is the velocity of the element of string that is momentarily at the top of the pulse? What is the direction of the velocities of the elements of string in the leading half of the pulse? The trailing half? Sketch both the displacement y and the velocity v_y of the string elements as a function of position x along the string.

Figure 16-3

Pulse velocity

3. In Figure 16-4, two wave pulses (one erect and one inverted) of the same size and shape approach one another along a string under tension. A short time later, the pulses superpose (overlap). When the pulses completely superpose, destructive interference occurs and the string is momentarily flat. At the moment when the string is flat, where is the energy that the pulses were carrying? Sketch both the displacements y and velocities v_y of the string elements as a function of position x along the string at the moment when the string is flat.

Figure 16-4

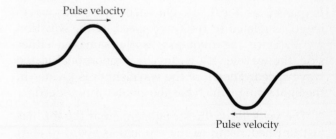

Pulse velocity

Pulse velocity

4. Most orchestral instruments are based on the resonance frequencies of either a stretched string or air in a pipe. What happens to these frequencies as the temperature increases?

5. The fundamental frequency of an air column in a pipe closed at both ends is 240 Hz; the fundamental frequency is 117.5 Hz if the pipe is open at one end. Why is the ratio of the frequencies approximately

2:1? Why is it not exactly 2:1? What would you expect to be the fundamental frequency of the pipe if it is open at both ends?

6. What happens to the frequency of a clarinet if the clarinetist breathes out pure helium?

7. Which of the properties that determine the fundamental frequency of a vibrating system (a string, air in a pipe, or the like) is normally varied in playing a musical instrument at different frequencies?

8. How can the phenomenon of beats be used to tune musical instruments?

9. When we play chords on the piano, several notes (frequencies) are sounded at once. Why don't we hear beats?

10. It's hard to set up a really convincing demonstration of two-source interference using sound waves in a lecture room. You can set up two loudspeakers vibrating in phase, and as you move from one to another part of the room, you can hear variations in the loudness of the sound, but they aren't terribly striking. Why not?

11. One end of a stretched string is vibrated to produce a standing wave on the string. What is it that determines the amplitude of the standing wave?

Answers

1. The wave speed equals $\sqrt{F/\mu}$, where F is the tension and μ is the mass per unit length. The tension is the same on either side of the knot joining the strings (or else the knot would accelerate toward the side with the highest tension). Therefore the wave speed is greatest in the light string.

The two string elements adjacent to the knot, one on either side, move up and down at the same frequency as the knot. On each string the wavelength is related to the wave speed by the equation $v = \lambda f$. Since the frequency is the same on either side, the wavelength is directly proportional to the wave speed. Therefore the wavelength is greater in the light string, which has the greater wave speed.

2. The velocity of the string element at the top has just finished rising to the top and is about to start descending to its equilibrium position. Its velocity is momentarily zero. The string elements in the leading half of the pulse are rising, so their velocities are directed upward. The string elements in the trailing half of the pulse are returning to their equilibrium positions, so their velocities are directed downward. Figure 16-5a shows the displacement y of the string as a function of position x along the string; Figure 16-5b shows the velocity v_y of the string as a function of position x.

Figure 16-5

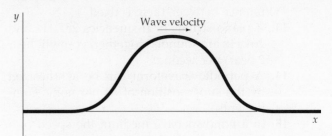

(a) Displacement of string as a function of position x along the string

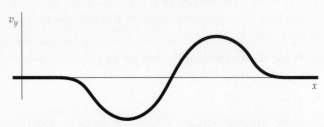

(b) Velocity of the string as a function of position x along the string

3. Figure 16-6a shows the displacement y of the string as a function of position x along the string; as the pulses approach each other, Figure 16-6b shows the velocity v_y of the string as a function of position x.

Figure 16-6

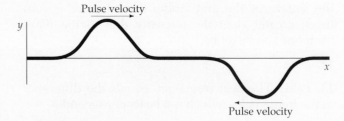

(a) Displacement of string as a function of position x along the string

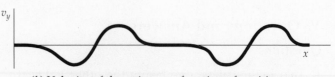

(b) Velocity of the string as a function of position x along the string

When the pulses superpose, the resultant displacement and velocity of the string are the algebraic sums of the displacements and velocities of the individual pulses. At the moment the pulses completely superpose, Figure 16-7 shows the displacement y and velocity v_y of the string as a function of the position x along the string. At the moment the pulses exactly superpose, the energy being carried by the two pulses is in the kinetic energy of the string.

Figure 16-7

(*a*) Displacement of string as a function of position x along the string

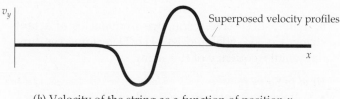

(*b*) Velocity of the string as a function of position x along the string

4. The frequencies of the wind instruments increase because the speed of sound in air increases with increasing temperature. Since strings expand a little as the temperature goes up, the tension in a string decreases somewhat; the wave speed and fundamental frequency therefore decrease. Thus, wind instruments go sharp and string instruments go flat as they warm up. This is why an orchestra warms up first and then tunes.

5. The ratio is about 2:1 because the length of the pipe is a half wavelength when both ends are closed and a quarter-wavelength when one end is open. It's not exactly 2:1 because the displacement antinode is not exactly at the open end but a little outside, so the effective air column is a little longer than the actual pipe. With both ends open, the fundamental is around 240 Hz again. (In fact it would be a little less than 240 Hz—perhaps 232 Hz— because the ends are open.) The fundamental frequency depends not on whether the ends are open or closed, but more on whether the two ends are the same or different.

6. All the frequencies that the clarinet produces are higher by the ratio of the speed of sound in helium to that in air, which is about 2.5 times. Thus all its tones sound more than an octave higher—provided that the clarinetist tightens his lips.

7. Most often, it is the length of the system that is varied. You finger the violin string to change its vibrating length, and you open or close holes or insert extra lengths of pipe on a wind instrument to "end" the air column at different places. Length modification is secondary, however, in playing some brass instruments. With them, you select among the harmonic frequencies of a pipe of fixed length by how you hold your mouth.

8. The phenomenon of beats is a clearly audible indicator of a small frequency difference. For example, when you simultaneously sound two strings that are supposed to have the same frequency, any difference in frequency produces beats that you can hear. Then you adjust the tension in one string, decreasing the beat frequency until no beats are heard.

9. In a way we do hear them. In this case the beat or difference frequencies are tens to hundreds of hertz, and we can't hear them as intensity variations. However, due to the way the ear responds to sound waves, "tones" at the difference frequencies are present in the combination of sounds we hear and, in some circumstances, are easily distinguished.

10. In almost any room, a large part of the sound you are hearing comes not directly from the sources but has been reflected (one or more times) from the walls. It is only the part coming directly from the sources in which you perceive strong interference maxima and minima, so these are usually not very noticeable.

11. The amplitude of the standing wave grows until the power being dissipated in the string plus the acoustic power being transferred from the string to the air equals the power being delivered to the string by the signal source.

VI.   Problems and Solutions

Problems

1. A string has a tension of 10 N and a mass per unit length of 5 g/m. Two harmonic waves traveling in the same direction on this string, each with a 20-cm wavelength and a 1-cm amplitude, differ in phase by 90°. Determine the amplitude and the power of the wave that results when these waves interfere.

How to Solve It

• When two waves on the same string have the same wavelength, they also have the same frequency.

• When two harmonic waves of the same amplitude, frequency, wavelength, and velocity interfere, the amplitude of the resultant wave can be expressed in terms of the phase difference between the two waves.

• The power depends on the angular frequency, the amplitude, the wave speed, and the linear mass density.

2. A string has a tension of 40 N and a mass per unit length of 5 g/m. Two harmonic waves of equal amplitude, each with a wavelength of 15 cm, that are traveling in the same direction differ in phase by 120°. When these waves interfere the amplitude of the resultant wave is 1 cm. Determine the amplitude of each of the original waves and the power of the resultant wave.

How to Solve It
• When two waves on the same string have the same wavelength, they also have the same frequency.

• When two harmonic waves of the same amplitude, frequency, wavelength, and velocity interfere, the amplitude of the resultant wave can be expressed in terms of the phase difference between the two waves.

• The power depends on the angular frequency, the amplitude, the wave speed, and the linear mass density.

3. A string is stretched between fixed supports 0.70 m apart, and the tension is adjusted until the fundamental frequency of the string is 264 Hz. If the string has a mass of 4.2 g, what is the tension?

How to Solve It
• The speed of a harmonic wave equals the product of the frequency and the wavelength.

• The wavelength of a string that is fixed at both ends and vibrating at its fundamental frequency is twice the length of the string.

• The speed of a wave on a string is determined by the string's linear density (mass per unit length) and its tension.

4. The lowest pitch on the standard piano is A (27.5 Hz). On my piano (a small console) the 85-cm-long "string" that sounds this pitch has a linear mass density of 0.235 g/cm. Assuming the string is vibrating in its fundamental mode, what is the tension in the string?

How to Solve It
• Piano strings are fastened at both ends, so the wavelength is twice the length of the string.

• The speed of a wave on a string is determined by the string's linear density (mass per unit length) and its tension.

5. Three successive resonance frequencies of a taut string are 273, 364, and 455 Hz. Determine the fundamental frequency of the string.

How to Solve It
• For a string with both ends fixed the difference between successive resonance frequencies equals the fundamental frequency. For a string with one end free the difference between successive resonance frequencies equals twice the fundamental frequency.

• To determine whether the string has both ends fixed or has one end free use the formulas for the resonance frequencies and compare the ratio of successive resonance frequencies of a string with both ends fixed with the ratio of successive resonance frequencies of a string with one end free. These ratios can be used to determine which case we have here.

• The fundamental frequency can be determined from the difference between successive resonance frequencies.

6. Three successive resonance frequencies of a taut string are 363, 605, and 847 Hz. Determine the fundamental frequency of the string.

How to Solve It
• Follow the approach for Problem 5.

7. (*a*) A man is sitting in a room directly between two loudspeakers 5 m apart that are vibrating in phase. He is 1.8 m from the nearer speaker, at point A in Figure 16-8. If the lowest frequency at which he observes maximum destructive interference is 122 Hz, what is the speed of sound in air? (*b*) If, instead, he listens from point B, what is the lowest frequency at which he will observe destructive interference?

Figure 16-8

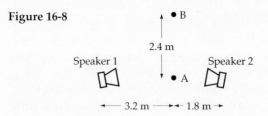

How to Solve It
• When the man observes maximum destructive interference, the waves reaching him are one-half cycle out of phase. (The phase difference could be 3/2 or 5/2 cycles or so on, but we are concerned with the lowest frequency).

• The phase difference in each case arises from a difference in path length from each source to the listener, which must be half a wavelength since the interference is destructive.

- The velocity of a harmonic wave is its frequency multiplied by its wavelength.

8. Two loudspeakers radiate sound waves in phase at a frequency of 100 Hz. A listener is 8.5 m from one speaker and 13.6 m from the other. Either speaker alone would produce a sound intensity level of 75 dB at the position of the listener. (*a*) What is the sound intensity level at the listener due to both speakers? (*b*) What is the intensity level at the listener if the same two speakers are 180° out of phase?

How to Solve It
- In part (*a*), the phase difference is due to the difference in path length.

- Changing the speakers so that they are out of phase by 180° just changes the phase difference of the two waves at the position of the listener by half a cycle.

- Remember that when two interfering harmonic waves are in phase, their amplitudes simply add.

9. A loudspeaker that is producing sound with a wavelength much longer than the diameter of the speaker acts as a point source; that is, the sound radiates more or less uniformly in all directions. If the wavelength is very small compared to the diameter of the speaker, on the other hand, the sound travels straight ahead from the speaker in approximately a straight line. If each speaker in a portable stereo is 18 cm in diameter, for what frequency is the sound wavelength (*a*) ten times and (*b*) one-tenth of the loudspeaker diameter?

How to Solve It
- The factor of 10 in each case is just to put an arbitrary numerical value on "much larger than" or "very small compared to."

- The speed of a harmonic wave is its frequency multiplied by its wavelength.

10. Two loudspeakers 3 m apart and a few meters away from you are producing sound of the same frequency. At about what frequency can you tell by the sound that there are two sources rather than one?

How to Solve It
- A wave cannot be used to observe or locate details much smaller than one wavelength.

- At what frequency do sound waves have a wavelength of 3 m?

11. Two identical strings on a piano are tuned to concert A (440 Hz). Each is under a tension of 1300 N. Over the course of time, one string loosens to the point that, when you strike the two strings simulta-neously; you hear beats every 1.1 s. By how much has the tension in the loose string decreased?

How to Solve It
- From the beats, determine how much the fundamental frequency of the loosened string differs from 440 Hz.

- The frequency has decreased because the transverse wave speed on the string has decreased.

- Knowing how much the wave speed on the string has decreased, you can find how much the tension in it has decreased.

12. When a violin string is played simultaneously with a tuning fork of frequency 264.0 Hz, beats occurring once every 0.65 s are heard in the sound. Tightening the string slightly causes the beats to disappear. What was the initial frequency of the violin string?

How to Solve It
- When you tighten a string, you increase its fundamental frequency.

- Thus, the original frequency of the string was less than 264.0 Hz, and tightening the string has increased its frequency to 264.0 Hz.

- The initial beat frequency is the difference between the frequency of the tuning fork and the initial frequency of the violin string.

13. The fundamental frequency of a violin string 30 cm long is sounded next to the open end of a closed organ pipe 41 cm long. The strongest standing sound wave in the pipe occurs when the string is vibrating at the pipe's fundamental frequency. This happens when the tension in the string is 220 N. What is the mass of the string?

How to Solve It
- What is the fundamental frequency for a closed organ pipe 0.41 m long? Remember, a closed organ pipe is open at one end.

- If a string 30 cm long has this frequency for a fundamental, find the wave speed on the string.

- From the wave speed and the tension the string is under, find its mass.

14. The maximum range of human hearing is about 20 Hz to 20 kHz. What is the longest organ pipe that could have a fundamental frequency within this range?

How to Solve It
- The longest wavelength corresponds to the lowest frequency.

- Thus, we want an organ pipe whose fundamental frequency is 20 Hz.

- Would the pipe have to be longer if it were open at both ends or closed at one end?

Solutions

1. The wave function of the resultant wave is

$$y = A \sin(kx - \omega t) + A \sin(kx - \omega t + \delta)$$

$$= 2A \cos \frac{\delta}{2} \sin\left(kx - \omega t + \frac{\delta}{2}\right)$$

where $2A \cos(\delta/2)$ is the amplitude A_r of the resultant wave. Therefore

$$A_r = 2A \cos \frac{\delta}{2} = 2(1 \text{ cm}) \cos \frac{90°}{2} = \sqrt{2} \text{ cm}$$

$$= \underline{1.41 \text{ cm}}$$

The resultant wave is a harmonic wave, so the power transmitted by it is

$$P = \tfrac{1}{2}\mu\omega^2 A_r^2 v$$

where $v = \sqrt{F/\mu}$ and $\omega = kv = 2\pi v/\lambda$. Therefore

$$P = \tfrac{1}{2}\mu\left(\frac{2\pi v}{\lambda}\right)^2 A^2 v = \tfrac{1}{2}\mu\left(\frac{2\pi A}{\lambda}\right)^2 \left(\frac{F}{\mu}\right)^{3/2}$$

$$= \tfrac{1}{2}(0.005 \text{ kg/m})\left[\frac{2\pi(\sqrt{2} \text{ cm})}{20 \text{ cm}}\right]^2 \left(\frac{10 \text{ N}}{0.005 \text{ kg/m}}\right)^{3/2}$$

$$= \underline{44.1 \text{ W}}$$

2. The amplitude is 1.0 cm and the power is 314 W.

3. The wavelength λ_1 of the fundamental of a taut string with both ends fixed is twice the length L of the string, and the wave speed v equals the square root of the ratio of the tension F to the linear mass density μ. The wave speed also equals the product of the frequency and the wavelength:

$$v = \lambda_1 f_1$$

$$\sqrt{\frac{F}{\mu}} = \lambda_1 f_1 = 2L f_1$$

so

$$F = \frac{m}{L}(2L f_1)^2 = 4mL f_1^2$$

$$= 4(0.0042 \text{ kg})(0.70 \text{ m})(264 \text{ Hz})^2 = \underline{820 \text{ N}}$$

4. 51.4 N

5. For a string fixed at both ends the resonance frequencies are

$$f_n = n f_1 \qquad n = 1, 2, 3, \ldots$$

where f_1 is the fundamental frequency. Thus the ratio of successive resonance frequencies is

$$\frac{f_{n+1}}{f_n} = \frac{(n+1)f_1}{n f_1} = \frac{n+1}{n} \qquad n = 1, 2, 3, \ldots$$

For a string fixed at one end and free at the other the resonance frequencies are

$$f_n = n f_1 \qquad n = 1, 3, 5, \ldots$$

where f_1 is the fundamental frequency. Thus the ratio of successive resonance frequencies is

$$\frac{f_{n+2}}{f_n} = \frac{(n+2)f_1}{n f_1} = \frac{n+2}{n} \qquad n = 1, 3, 5, \ldots$$

We are given three successive resonance frequencies $f_a = 273$, $f_b = 364$, and $f_c = 455$ Hz, which have ratios

$$\frac{f_b}{f_a} = \frac{364}{273} = \frac{4}{3}$$

and

$$\frac{f_c}{f_b} = \frac{455}{364} = \frac{5}{4}$$

Because both of these ratios are of the form $(n+1)/n$ we conclude the string has both ends fixed. The fundamental frequency of a string with both ends fixed is the difference between any two successive resonance frequencies. Therefore

$$f_1 = f_{n+1} - f_n = 364 \text{ Hz} - 273 \text{ Hz} = \underline{91 \text{ Hz}}$$

6. 121 Hz

7. (a) The path-length difference of the two sound waves is

$$\Delta s = s_2 - s_1 = 3.2 \text{ m} - 1.8 \text{ m} = 1.4 \text{ m}$$

Since the man observes complete destructive interference, the path-length difference must be equal to one-half wavelength. Thus,

$$\lambda = 2 \Delta s = 2(1.4 \text{ m}) = 2.8 \text{ m}$$

The speed of sound is therefore

$$v = f\lambda = (122 \text{ Hz})(2.8 \text{ m}) = \underline{342 \text{ m/s}}$$

(b) In the second case, the path-length difference must (again) be one-half wavelength. The distance of point B from speaker 1 (see Figure 16-9) is

$$d = \sqrt{(3.2 \text{ m})^2 + (2.4 \text{ m})^2} = 4.0 \text{ m}$$

and the distance from speaker 2, found in the same way, is 3.0 m. Thus the wavelength of the sound is 2.0 m. The speed of the sound hasn't changed, so the lowest frequency for destructive interference is given by

$$f = \frac{v}{\lambda} = \frac{342 \text{ m/s}}{2.0 \text{ m}} = \underline{171 \text{ Hz}}$$

Figure 16-9

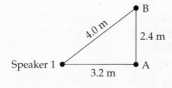

8. (a) The intensity is <u>zero</u>; this can't be expressed on the decibel scale.

   (b) <u>81 dB</u> (which means the intensity is four times that of one speaker alone).

9. For a harmonic wave

$$v = f\lambda$$

so

$$f = \frac{v}{\lambda}$$

(a) For a wavelength ten times the diameter of the loudspeaker,

$$f = \frac{v}{10D} = \frac{340 \text{ m/s}}{10(0.18 \text{ m})} = \underline{189 \text{ Hz}}$$

(b) For a wavelength one-tenth the diameter of the loudspeaker,

$$f = \frac{v}{0.1D} = \frac{340 \text{ m/s}}{0.1(0.18 \text{ m})} = \underline{18.9 \text{ kHz}}$$

10. Somewhere around <u>110 Hz</u>.

11. The frequency of the loosened string is

$$f' = f_0 - f_{\text{beat}} = 440 \text{ Hz} - \frac{1}{1.1 \text{ s}} = 439.09 \text{ Hz}$$

For a string fixed at both ends, the fundamental frequency $f = v/2L$. Since the length L is fixed, the wave velocity on the string and the fundamental frequency are directly proportional:

$$\frac{f'}{f_0} = \frac{v'}{v_0} = \sqrt{\frac{F'/\mu}{F/\mu}} = \sqrt{\frac{F'}{F}}$$

Solving for the tension in the loosened string, we obtain

$$F' = \left(\frac{f'}{f_0}\right)^2 F = \left(\frac{439.09 \text{ Hz}}{440 \text{ Hz}}\right)^2 (1300 \text{ N})$$

$$= 0.9959(1300 \text{ N}) = 1294.6 \text{ N}$$

Thus, the tension has decreased by 1300 N − 1294.6 N = <u>5.4 N</u>.

The extra significant figures in f' and N' are needed to get a precise result in the *change* of tension.

12. <u>262.5 Hz</u>

13. For a closed organ pipe 0.41 m long, the fundamental wavelength is

$$\lambda_1 = 4L = 4(0.41 \text{ m}) = 1.64 \text{ m}$$

If the speed of sound is 340 m/s, the fundamental frequency is

$$f_1 = \frac{v}{\lambda_1} = \frac{340 \text{ m/s}}{1.64 \text{ m}} = 207 \text{ Hz}$$

This is the fundamental frequency of the pipe and, therefore, of the stretched string that excites it. The wave speed on the string is thus

$$v = f_1\lambda_1 = f_1(2L) = (207 \text{ Hz})(2)(0.30 \text{ m})$$

$$= 124 \text{ m/s}$$

The wave speed on a stretched string is

$$v = \sqrt{F/\mu}$$

where F is the tension in the string and μ its linear density (mass per unit length). Thus,

$$v^2 = \frac{F}{\mu} = \frac{F}{m/L} = \frac{FL}{m}$$

and

$$m = \frac{FL}{v^2} = \frac{(220 \text{ N})(0.30 \text{ m})}{(124 \text{ m/s})^2} = \underline{4.29 \times 10^{-3} \text{ kg}}$$

14. It is <u>8.5 m long, open at both ends</u>.

Chapter 17

Wave-Particle Duality and Quantum Physics

I. Key Ideas

The first part of the twentieth century marked one of the most creative times of humankind. Einstein developed the theory of special relativity in 1905 and the theory of general relativity in 1916. The ideas of quantum physics started in 1900 with Planck's explanation of blackbody radiation. Einstein extended Planck's quantum ideas to explain the photoelectric effect in 1905, the same year he put forth the theory of special relativity. Ideas built upon ideas about how nature behaves. The fields of quantum physics, relativity, cosmology, and other areas of modern physics were given birth and nurtured by Einstein, Planck, Bohr, Fermi, Dirac, and many others, most of whom were awarded Nobel prizes for their work. Much of the work in the beginning of modern physics centered around the nature of light.

17-1   Light   Light—is it a wave or a particle? The answer depends upon the results of experiments conducted on light. Waves are spread out in space and bend around corners (diffraction) and interfere with each other, producing constructive and destructive interference. The energy of waves is spread out in space. In contrast, particles are localized in space and travel on well-defined lines until they collide with something, after which they again travel on well-defined lines.

Newton (1642–1727) believed that light was composed of particles, a belief that was accepted for over a century. However, many experiments conducted in the 1800s suggested that Newton's particle picture of light was incorrect. Thomas Young in 1801 passed light from a single source through two closely spaced parallel slits and produced an interference pattern on a screen. Augustin Fresnel (1788–1827) developed the mathematical theory of light on the basis of a wave picture, and showed that all experimental results agreed with his wave analysis.

James Clerk Maxwell in 1860 put forth his theory of electromagnetism, which predicted the existence of electromagnetic waves that propagate at the speed of light, $c \approx 3 \times 10^8$ m/s. The implication is that light is an electromagnetic wave, differing from other electromagnetic waves like television and radio only in wavelength. The wavelength of electromagnetic light waves is in the visible region of about 400 nm to about 700 nm (1 nm = 10^{-9} m). These and many other experiments seemed to show conclusively that light is a wave and not a particle.

17-2   The Particle Nature of Light: Photons   At the beginning of the 1900s, various experiments were performed that could not be explained with a wave picture of light. Instead, the experiments suggested that light energy comes in discrete amounts carried by localized packets called **photons**. Einstein, in explaining the photoelectric effect (next section), first put forth the idea that each photon has an energy E given by

Einstein equation for photon energy

$$E = hf = hc/\lambda$$

where $f = c/\lambda$ is the frequency of each photon, and h is **Planck's constant**

Planck's constant

$$h = 6.626 \times 10^{-34} \text{ J·s} = 4.136 \times 10^{-15} \text{ eV·s}$$

where the standard energy unit Joules (J) is related to the nonstandard energy unit electron volts (eV) by 1 eV = 1.60×10^{-19} J.

The Photoelectric Effect   The first experiment showing the photon nature of light was the **photoelectric effect**. The apparatus for a photoelectric experiment is shown in Figure 17-1. Light of a variable but known frequency f strikes a metal surface C in an evacuated tube, causing electrons to be emitted from C with various kinetic energies. After traversing a short distance in the tube, the electrons are

Figure 17-1

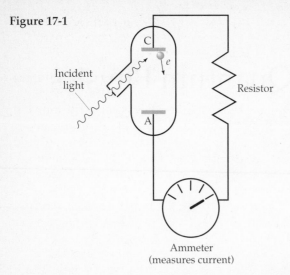

Ammeter
(measures current)

collected by a second metal plate A. The electrons collected at the plate A constitute a current that is measured by the ammeter.

In the photoelectric experiment, the current is measured as a function of the frequency and the intensity of the incident light. In addition, the maximum kinetic energy K_{max} is also measured. The photoelectric experiment has five main results:

1. Electrons emitted from the surface C have kinetic energies ranging from zero to a maximum value $K_{max} = (\frac{1}{2}mv^2)_{max}$.

2. K_{max} does not depend on the intensity of the incident light.

3. K_{max} depends linearly on the frequency of the incident light according to a relationship known as **Einstein's photoelectric equation**:

Einstein's photoelectric equation

$$K_{max} = (\tfrac{1}{2}mv^2)_{max} = hf - \phi$$

The constant ϕ, called the **work function**, is equal to the energy needed to remove the least tightly bound electrons from the surface of C and is, therefore, a characteristic of the particular metal composing C.

4. Photons with frequencies below a smallest **threshold frequency** f_t (or with wavelengths longer than a longest **threshold wavelength** λ_t) do not have enough energy to eject electrons from the surface C, no matter how intense the incident light. This means that for f_t and below (and λ_t and above), $K_{max} = 0$, which, from Einstein's photoelectric equation gives

Threshold frequency and wavelength

$$\phi = hf_t = \frac{hc}{\lambda_t}$$

5. The current is observed a very short time, of the order of 10^{-9} s, after the incident light strikes C, no matter how low the intensity of the light.

The classical picture of light as being composed of electromagnetic waves does not explain the results of the photoelectric experiment. Classically, the energy in a beam of light depends only on the light's intensity—the frequency of the incident light plays no role at all in energy balance. The electrons in surface C absorb more energy from more intense light and eventually, after a significant time lapse, leave C with kinetic energies ranging from zero to infinity.

Einstein explained *all* the experimental results of the photoelectric experiment by assuming that the light incident on the surface C is composed of photons, all with the same energy $E = hf$, where h is the same constant determined by Planck in his blackbody analysis about five years previously. Each photon interacts with an electron in the emitter C in an all-or-nothing fashion, either giving all its energy to an electron or not interacting at all.

Electrons leave C with varying kinetic energies because some lose energy as they travel through the material of C. An electron at the surface of C requires the smallest amount of energy—equal to the work function ϕ of the material—in order to be released. Such an electron is emitted with the maximum kinetic energy K_{max}, given in Einstein's photoelectric equation, which is equal to the energy hf of the absorbed photon minus the energy ϕ required to release the least tightly bound electron from the surface of C.

Thus, Einstein's quantum picture of light explains how the maximum kinetic energy of an emitted electron depends on the light's frequency, why there is a threshold frequency f_t for which $K_{max} = 0$, and why the experimentally observed threshold frequency depends on the work function of the material through $hf_t = \phi$.

In the picture of light as photons, higher intensity corresponds to more photons in a light beam; and more photons eject more electrons from C. If the frequency of the incident photons stays constant, the maximum kinetic energy of the ejected electrons does not change as the intensity of the photons varies, which agrees with experimental results. The all-or-nothing transfer of energy from a photon to an electron also accounts for the exceedingly short time it takes for a current to be observed after light strikes C.

Compton Scattering In 1923, Compton provided additional evidence that the photon picture of light is correct by his relativistic analysis of a "billiard ball" type of experiment between a photon and an effectively free electron.

In classical theory, the energy and momentum of an electromagnetic wave are related by $E = pc$, or $p = E/c$. Applying the quantum relationship $E = hf$ together with the wave relationship $c = f\lambda$ gives the momentum of a photon as $p = hf/c = h/\lambda$,

Momentum of a photon in terms of its wavelength

$$p = h/\lambda$$

along with the energy of a photon as $E = hc/\lambda$.

In a Compton experiment, a photon with initial energy $E_1 = hc/\lambda_1$ and momentum $p_1 = h/\lambda_1$ is sent on a collision course with an electron that is at rest. After an elastic collision, the photon moves off at an angle θ with its original direction with a different energy $E_2 = hc/\lambda_2$ and different momentum $p_2 = h/\lambda_2$, and the electron moves at an angle ϕ with the original direction of the photon, as shown in Figure 17-2.

Figure 17-2

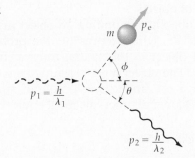

Compton applied relativistic conservation of energy and momentum to the collision, which is beyond the scope of our present studies. The end result is that the wavelength λ_2 of the photon scattered at angle θ after the collision is related to the wavelength λ_1 of the incident photon before the collision by the **Compton equation**

Compton equation

$$\lambda_2 - \lambda_1 = \frac{h}{m_e c}(1 - \cos\theta)$$

The quantity multiplying the term $(1 - \cos\theta)$ is called the **Compton wavelength**

Compton wavelength

$$\lambda_C = \frac{h}{m_e c} = \frac{hc}{m_e c^2} = \frac{1240\ \text{eV·nm}}{5.11 \times 10^5\ \text{eV}}$$
$$= 2.43 \times 10^{-3}\ \text{nm} = 2.43 \times 10^{-12}\ \text{m} = 2.43\ \text{pm}$$

For Compton scattering from a particle other than an electron, you must use the mass of that particle in the above expressions.

17-4   Electrons and Matter Waves   We have seen that a photon has the particle attributes of energy $E = hf$ and momentum $p = h/\lambda$. In addition, light shows wave behavior in its interference and diffraction.

The de Broglie Hypothesis   In 1924, after consulting Einstein, de Broglie boldly suggested in his doctoral dissertation that if photons behave both like waves and particles, then perhaps electrons, which were regarded as "particles," could also have this dual character.

De Broglie used the momentum expression $p = h/\lambda$ for a photon to define the wavelength λ of an electron:

De Broglie wavelength of an electron

$$\lambda = h/p$$

where $p = mv$ is the (nonrelativistic) momentum of the electron. The expression for the frequency of the wave associated with an electron is the same as that for a photon:

Electron wave frequency

$$f = E/h$$

where E is the energy of the electron.

Subsequently, the wave nature of electrons, protons, and neutrons was substantiated experimentally.

Electron Interference and Diffraction   If electrons are waves, it should be possible to perform experiments exhibiting this wave nature. This was first done in 1927 by C. J. Davisson and L. H. Germer at the Bell Telephone Laboratories. Davisson and Germer studied electron scattering from a nickel target. They found that the electron intensity in the scattered beam as a function of scattering angle showed maxima and minima corresponding to the exact de Broglie wavelength associated with the energy of the incident electron beam.

In the same year, G. P. Thomson (son of J. J. Thomson, who discovered the existence of the electron) demonstrated electron diffraction by sending electron beams through thin metal foils. The radii of the resulting diffraction pattern of concentric circles agreed with the de Broglie wavelength associated with the energy of the incident electron beam.

An important application of de Broglie waves associated with electrons is the electron microscope, which employs electrons to "see" objects at scales far smaller than microscopes using visible light. Also, diffraction is now routinely observed with "particles" other than electrons, such as neutrons.

Standing Waves and Energy Quantization
When an object such as an electron is confined in a certain spatial region, standing waves occur for only certain wavelengths and frequencies consistent with the boundary conditions. This is similar to the effect of boundary conditions on standing waves in a string. As will be discussed in more detail in Chapter 36, Schrödinger and others showed, around 1928, that the application of boundary conditions to de Broglie waves led to a new fundamental description of nature called **quantum theory**, **quantum mechanics**, or **wave mechanics**, built around the concept of a wave function that satisfies an equation known as Schrödinger's equation. A consequence of quantum theory is that the energy of a bound system is quantized, that is, the energy of the system can not be continuous, but can have only certain discrete values.

17-5 *The Interpretation of the Wave Function*
Classical physics is built on the notion that it is possible to determine the location of a particle such as an electron with unlimited precision; thus, its trajectory—its location as a function of time—can be known exactly.

Quantum mechanics paints an entirely different picture of the entity that was regarded classically as a "particle." According to quantum mechanics, we can determine only the probability of finding a particle, such as an electron, in a small space around a point. In one dimension, quantum mechanics specifies the probability of finding an electron somewhere in a given spatial interval dx, rather than exactly at some point x.

The probability of finding a particle in an interval is proportional to the size of the interval. If you double the size of dx, the probability of finding the particle in the interval doubles. There is never any certainty that you will find an electron in the interval—there is only a probability. The one thing that can be said with certainty is that you will never find an electron in an interval where the probability is zero.

In a probability description, the **probability density** $P(x)$ gives the likelihood of finding a particle in the vicinity of one value of x rather than another value. The probability of finding a particle in a spatial interval dx that straddles a point located at x is

Probability of finding an object in a spatial interval dx that straddles a point located at x

Probability = $P(x) dx$

In the quantum mechanical picture, the quantity that determines the probability density $P(x)$ is called the **wave function** $\psi(x)$. $P(x)$ and $\psi(x)$ are related by

Relation between the probability density and the wave function

$$P(x) = \psi^2(x)$$

Thus, $\psi^2 dx$ is the probability of finding a particle in a spatial interval dx that straddles a point located at x:

Probability of finding an object in a spatial interval dx that straddles a point located at x

Probability = $\psi^2(x) dx$

The functional form of the wave function $\psi(x)$ is determined from the Schrödinger wave equation, which will be described in Chapter 36. You can think of $\psi(x)$ as the entity that exhibits the wave-like properties of de Broglie waves, which result in the wave properties of objects.

The probability of finding an object in the interval dx is $\psi^2 dx$, which represents the probability of finding an object in the spatial interval between x and $x + dx$. The object must certainly be somewhere between $x = -\infty$ and $x = +\infty$, so the sum of the probabilities over all intervals must equal 1:

Normalization condition

$$\int_{-\infty}^{\infty} \psi^2 dx = 1$$

This defines a **normalization condition** that the wave function $\psi(x)$ must satisfy. In addition, the wave function $\psi(x)$ must approach zero as x approaches plus or minus infinity.

17-6 *Wave–Particle Duality*
At times objects exhibit wave properties and at other times particle properties. This is known as **wave–particle duality**. Both the wave and particle pictures are necessary for a complete understanding of an object, but you cannot observe both aspects simultaneously in a single experiment, which is the **principle of complementarity**, first put forth by Bohr in 1928.

The Two-slit Experiment Revisited
In Young's two-slit experiment, visible light is incident on two parallel closely-spaced slits, and an interference pattern of alternating bright and dark bands is observed on a screen placed beyond the slits. For a given slit separation, the separation between the bright and dark bands depends upon the wavelength of the light.

If the experiment is repeated with a beam of electrons instead of visible light, a similar interference pattern of bright and dark bands of electrons is observed on a screen or film. The separation between the bright and dark bands depends upon the de Broglie wavelength associated with the energy of the electron beam. One can watch the buildup of the bands as each individual electron strikes the screen. The buildup is determined by the probability of detecting an electron on the screen, which is proportional to $\psi^2(x)$.

The Uncertainty Principle   In a thought experiment, when we say we have located a particle at a given position x, we really mean we have determined that its position is somewhere in the interval between x and $x + \Delta x$, where Δx is the uncertainty in the position x. Similarly, if we measure the momentum p of a moving object, we really mean that we have determined that the object's momentum is somewhere in the interval between p and $p + \Delta p$, where Δp is the uncertainty in the momentum p.

In classical physics, Δx and Δp can in principle be individually made to be arbitrarily small. In fact, a basic premise in describing the motion of a classical particle is that rigorously $\Delta x = 0$ and $\Delta p = 0$, so that one can talk unambiguously about the *exact* instantaneous position of a particle along with its *exact* instantaneous velocity or momentum.

When uncertainties are analyzed taking quantum considerations into account, one finds that the position and momentum of an object can not be specified with infinite precision. Suppose, for example, you want to determine the position of an electron by "looking" at it. To do this measurement, you must bounce a photon off the electron. The interaction of the photon with the electron as the bouncing takes place causes the electron's velocity and momentum to change by an unknown amount.

A rigorous treatment of the quantum-mechanical measurement process, first put forth by Werner Heisenberg in 1927, shows that in a simultaneous determination of position and momentum of a particle such as an electron, the uncertainty in the momentum, Δp, and position, Δx, are related by the **uncertainty principle**:

Heisenberg's uncertainty principle for position and momentum

$$\Delta p \, \Delta x \geq \tfrac{1}{2}\hbar$$

where $\hbar$ (read h bar) $= h/2\pi$.

The equality gives intrinsic lower limits, and represents the best uncertainty that is possible. If you try to locate the position of a particle more and more precisely by making Δx smaller and smaller, the uncertainty Δp in the particle's momentum becomes larger and larger, and vice versa. In any actual experiment, additional experimental uncertainties produce values of Δp and Δx that are larger than the intrinsic lower

limits resulting from wave–particle duality, giving rise to the inequality in the uncertainty relationship.

A Particle in a Box   A "particle in a box" refers to a particle of mass m that is confined to a one-dimensional region of length L, and can never be found outside the boundaries of the box. Very importantly, the energies of a particle in a box are quantized—only certain discrete values are possible. Although somewhat artificial, a particle in a box exhibits many properties such as discrete energies similar to those of an electron bound within an atom, or a proton inside a nucleus.

Let the box be between $x = 0$ and $x = L$. Since the particle is confined to the box, it can never be found outside the boundaries of the box. This means that the wave function ψ is zero outside the boundaries of the box, that is,

The wave function must be zero outside the confines of the box

$$\psi = 0 \quad \text{for } x \leq 0 \text{ and for } x \geq L$$

A consequence of the **boundary conditions** that $\psi = 0$ at $x = 0$ and $x = L$ is that the energy of the particle can not be continuous, as in classical theory. Rather, the particle's energy can only take on one of the quantized (discrete) values given by

Allowed energies for a particle in a box

$$E_n = n^2 \frac{h^2}{8mL^2} = n^2 E_1 \qquad n = 1, 2, 3, \ldots$$

where the lowest or **ground-state energy** is given by

Ground-state energy for a particle in a box

$$E_1 = \frac{h^2}{8mL^2}$$

The integers $n = 1, 2, 3, \ldots$ are called **quantum numbers**. The allowed energies E_n are shown in the **energy level diagram** of Figure 17-3.

Figure 17-3

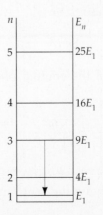

Thus, the smallest possible value of the energy is not zero, as in classical theory. Rather, the smallest (ground state) energy, called the **zero-point energy**, is given by the expression for E_1. This means that a particle in a box can never be at rest, but must always be moving with one of the allowed energies given by E_n. Since E_1 varies as $1/L^2$, the smaller the confines of the box, the larger is the zero-point energy.

The relation of quantized energies to the size of a box is an expression of the uncertainty principle. We know the particle is inside the box, but we do not know exactly where the particle is inside the box, so the uncertainty in the particle's position is $\Delta x = L$. The particle's energy can be any of the quantized values $E_n = n^2 E_1$. A measure of the uncertainty in the particle's energy can be taken as $\Delta E = E_1$. The relationship between a particle's momentum and energy is $p = \sqrt{2mE}$, so the uncertainty Δp in momentum corresponding to the uncertainty in energy is $\Delta p = \sqrt{2m\Delta E} = \sqrt{2mE_1} = \sqrt{2m(h^2/8m\,\Delta x^2)} = h/(2\,\Delta x)$, from which $\Delta x\,\Delta p = h/2$, which is a statement of the Heisenberg uncertainty principle for a particle in a box.

It is possible for a system that has quantized energies, such as a particle in a box, to make a transition from an initial energy state E_i to a different final energy state E_f. If E_i is higher than E_f ($E_i > E_f$) a photon will be emitted, while if E_i is lower than E_f ($E_i < E_f$) a photon will be absorbed. From conservation of energy, the energy of the emitted or absorbed photon is equal to the energy difference between the initial and final states of the system:

Energy of a photon emitted or absorbed when a system makes a transition from an initial to a final energy state

$$hf = hc/\lambda = E_i - E_f$$

As an example with a particle in a box, a transition from state 3 to the ground state ($n = 1$) is shown by the vertical arrow in Figure 17-3. The frequency of the emitted photon is given by

$$hf = E_i - E_f = E_3 - E_1$$
$$= 3^2\frac{h^2}{8mL^2} - 1^2\frac{h^2}{8mL^2} = \frac{h^2}{mL^2}$$

The wavelength of the emitted photon can then be found:

$$\lambda = \frac{c}{f} = \frac{hc}{E_i - E_f} = \frac{hc}{E_3 - E_1} = \frac{mcL^2}{h}$$

Standing Wave Functions   The wave functions ψ_n obtained by solving the Schrödinger equation for a particle in a box satisfying the normalization condition, which will be done in Chapter 36, are the same form as for a vibrating string:

Wave function for a particle in a box lying between $x = 0$ and $x = L$

$$\psi_n = \sqrt{\frac{2}{L}}\sin\frac{n\pi x}{L}\qquad n = 1, 2, 3, \ldots$$

The corresponding probability distribution $P_n = \psi_n^2$ is

Probability distribution for a particle in a box lying between $x = 0$ and $x = L$

$$P_n = \psi_n^2 = \frac{2}{L}\sin^2\frac{n\pi x}{L}\qquad n = 1, 2, 3, \ldots$$

Plots of ψ_n and ψ_n^2 are shown in Figures 17-4 and 17-5 respectively for the ground state ($n = 1$) and the first two excited states ($n = 2, 3$). For each standing wave function, the particle has an energy $E_n = n^2\dfrac{h^2}{8mL^2} = n^2 E_1$ as given above, and shown in Figure 17-3.

Figure 17-4

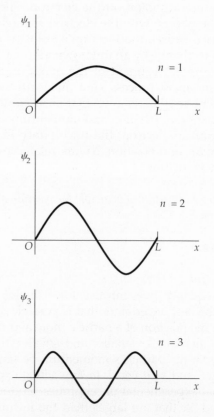

Figure 17-5

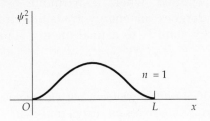

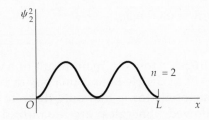

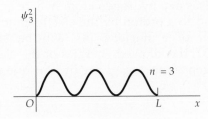

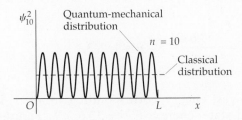

For very large values of the quantum number n, the maxima of the probability distribution $P_n = \psi_n^2$ are so closely spaced that P_n can hardly be distinguished from its neighboring value P_{n+1}. The fact that the probability distribution is nearly constant across the whole box means that the particle is nearly equally likely to be found anywhere in the box, which is the same as the classical result. An example of this is shown in Figure 17-5 for $n = 10$. This is an example of **Bohr's correspondence principle**:

In the limit of very large quantum numbers, the classical calculation and the quantum calculation must yield the same results.

17-8 Expectation Values
A typical problem in classical mechanics is the specification of the exact position of a particle as a function of time. In con-trast, according to quantum mechanics it is intrinsically impossible to specify exactly the position of a particle at any given time. The most we can know about the position of a particle is the relative probability of finding it in one interval or another.

However, if you measure the position x of a particle in a large number of identical experiments, you will obtain some average value of x, designated by $<x>$. This average value is called the **expectation value** and is related to the wave function ψ by

Expectation value of x

$$<x> = \int x\psi^2(x)\,dx$$

If, instead of the position x, you look at the expectation value of any function $f(x)$, you observe

Expectation value of an arbitrary function

$$<f(x)> = \int f(x)\psi^2(x)\,dx$$

17-9 Energy Quantization in Other Systems
When you learn how to solve the Schrödinger equation for quantum mechanical systems in Chapter 36, a key component will be the **potential energy function** $U(x)$ of the system you are looking at. For a particle in a box located between $x = 0$ and $x = L$, which we have looked at above, the potential energy function $U(x)$ is given by

Potential energy function for a particle in a box

$$U(x) = 0 \qquad 0 < x < L$$
$$U(x) = \infty \qquad x < 0 \quad \text{or} \quad x > L$$

The interpretation of the potential energy function $U(x)$ is that the particle can be found only between the coordinate values $x = 0$ and $x = L$. The particle will never be found outside the box in the region $x < 0$ or $x > L$.

The Harmonic Oscillator The **harmonic oscillator** is a physically more realistic system than a particle in a box. With a harmonic oscillator, a particle of mass m is attached to a spring of force constant k and undergoes small oscillations of amplitude A about a fixed equilibrium point located at $x = 0$. The classical potential-energy function $U(x)$ for a harmonic oscillator is

Potential energy function for a classical harmonic oscillator

$$U(x) = \tfrac{1}{2}kx^2 = \tfrac{1}{2}m\omega_0^2 x^2$$

where $\omega_0 = \sqrt{k/m}$ is the natural angular frequency in rad/s of the harmonic oscillator. The natural oscillational frequency f_0 in vibrations per second is related to ω_0 by $\omega_0 = 2\pi f_0$, so that

Natural oscillational frequency of a classical harmonic oscillator

$$f_0 = \frac{1}{2\pi}\sqrt{k/m}$$

Classically, the object oscillates between $x = +A$ and $x = -A$.

The allowed energies for a particle in a quantum-mechanical box are quantized. Similarly, a quantum-mechanical analysis of a harmonic-oscillator particle (see Chapter 36) shows that the allowed energies are quantized, with the discrete energies E_n given by

Allowed discrete energies for a particle in a harmonic oscillator potential

$$E_n = (n + \tfrac{1}{2})hf_0 \qquad n = 0, 1, 2, 3, \ldots$$

where

Ground state energy for a harmonic oscillator particle

$$E_0 = \tfrac{1}{2}hf_0$$

is the lowest (ground state) energy of the harmonic oscillator. The allowed energy levels are shown in Figure 17-6, along with the potential energy function $U(x)$. Note that the energy levels for a harmonic oscillator shown in Figure 17-6 are evenly spaced, as compared to the unevenly-spaced energy levels for a particle in a box shown in Figure 17-3.

As with a particle in a box, the frequency f of a photon emitted when a harmonic-oscillator particle undergoes an energy transition from an initial energy state E_i to a final energy state E_f is given by

Frequency f of an emitted photon when a harmonic oscillator particle undergoes an energy transition

$$hf = E_i - E_f = (n_i - n_f)hf_0$$

The Hydrogen Atom   A main result of a quantum analysis of a particle bound in a box or bound in a harmonic oscillator is that only discrete energies are allowed. Energies are quantized and are described by quantum numbers n. A similar situation arises in a hydrogen atom, where an electron is bound to a proton by the electrostatic force of attraction of a nucleus. As will be shown in Chapter 37, the allowed energies of the electron in a hydrogen atom are given by the quantized values

Allowed energies of an electron in a hydrogen atom

$$E_n = -\frac{13.6 \text{ eV}}{n^2} \qquad n = 1, 2, 3, \ldots$$

The lowest energy, corresponding to $n = 1$, is the ground-state energy $E_1 = -13.6$ eV. Figure 17-7 shows the energy-level diagram for a hydrogen atom.

Figure 17-6

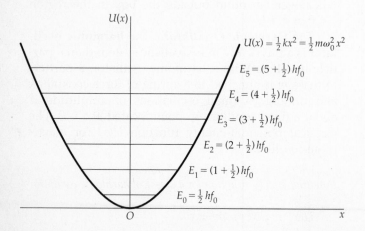

Figure 17-7

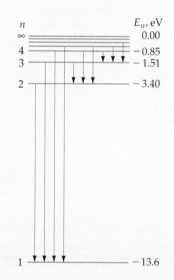

As with a particle in a box and a harmonic oscillator, the frequency f of a photon emitted when an electron in a hydrogen undergoes an energy transition from an initial energy state E_i to a final energy state E_f is quantized, and is given by

Frequency f of an emitted photon when an electron in a hydrogen atom undergoes an energy transition

$$hf = E_i - E_f = \left[-\frac{13.6 \text{ eV}}{n_i^2} \right] - \left[-\frac{13.6 \text{ eV}}{n_f^2} \right]$$
$$= (13.6 \text{ eV}) \left[\frac{1}{n_f^2} - \frac{1}{n_i^2} \right]$$

Transitions from higher to lower energy states are indicated by the vertical arrows on Figure 17-7.

Other atoms are more complicated than hydrogen, but they exhibit discrete energy levels similar to those of hydrogen, with ground-state energies of the order of -1 to -10 eV.

II. Numbers and Key Equations

Numbers

Planck's constant

$$h = 6.626 \times 10^{-34} \text{ J·s} = 4.136 \times 10^{-15} \text{ eV·s}$$

$$\hbar = h/2\pi = 1.05 \times 10^{-34} \text{ J·s}$$
$$= 0.658 \times 10^{-15} \text{ eV·s}$$

$$hc = 19.865 \times 10^{-26} \text{ J·m} = 1240 \text{ eV·nm}$$

Compton wavelength for an electron

$$\lambda_C = 2.43 \times 10^{-12} \text{ m} = 2.43 \text{ pm}$$

Ground-state energy for a hydrogen atom

$$E_1 = 13.6 \text{ eV}$$

Key Equations

Photon energy

$$E = hf = hc/\lambda$$

Einstein's photoelectric equation

$$K_{max} = (\tfrac{1}{2}mv^2)_{max} = hf - \phi$$

Threshold frequency and wavelength

$$\phi = hf_t = \frac{hc}{\lambda_t}$$

Momentum of a photon in terms of its wavelength

$$p = h/\lambda$$

Compton equation

$$\lambda_2 - \lambda_1 = \frac{h}{m_e c} (1 - \cos \theta)$$

De Broglie wavelength of an electron

$$\lambda = h/p$$

Relation between the probability density and the wave function

$$P(x) = \psi^2(x)$$

Probability of finding an object in a spatial interval dx that straddles a point located at x

$$\text{Probability} = P(x)\, dx = \psi^2(x)\, dx$$

Normalization condition

$$\int_{-\infty}^{\infty} \psi^2 \, dx = 1$$

Heisenberg's uncertainty principle for position and momentum

$$\Delta p \, \Delta x \geq \tfrac{1}{2}\hbar$$

Energy of a photon emitted or absorbed when a system makes a transition from an initial to a final energy state

$$hf = hc/\lambda = E_i - E_f$$

Allowed energies for a particle in a box

$$E_n = n^2 \frac{h^2}{8mL^2} = n^2 E_1 \qquad n = 1, 2, 3, \ldots$$

Ground-state energy for a particle in a box

$$E_1 = \frac{h^2}{8mL^2}$$

Wave function for a particle in a box lying between $x = 0$ and $x = L$

$$\psi_n = \sqrt{\frac{2}{L}} \sin \frac{n\pi x}{L} \qquad n = 1, 2, 3, \ldots$$

Probability distribution for a particle in a box lying between $x = 0$ and $x = L$

$$P_n = \psi_n^2 = \frac{2}{L}\sin^2\frac{n\pi x}{L} \qquad n = 1, 2, 3, \ldots$$

Expectation value of x

$$<x> = \int x\psi^2(x)\,dx$$

Expectation value of an arbitrary function

$$<f(x)> = \int f(x)\psi^2(x)\,dx$$

Natural oscillational frequency of a classical harmonic oscillator

$$f_0 = \frac{1}{2\pi}\sqrt{k/m}$$

Allowed energies for a quantum-mechanical harmonic oscillator

$$E_n = (n + \tfrac{1}{2})hf_0 \qquad n = 1, 2, 3, \ldots$$

Ground state energy for a quantum-mechanical harmonic oscillator

$$E_0 = \tfrac{1}{2}hf_0$$

Allowed energies of an electron in a hydrogen atom

$$E_n = -\frac{13.6\text{ eV}}{n^2} \qquad n = 1, 2, 3, \ldots$$

III. Potential Pitfalls

Don't simply plug numbers into the photoelectric equation in working photoelectric effect problems. Understand the energy transfers that are going on.

Do not confuse the probability density with probability. The probability density $P(x)$ is a function defined at each point x. Probability refers to the probability of finding an object in a small spatial interval dx that straddles a point x; for example, the probability of finding an object in an interval between x and $x + dx$. The probability density is related to the probability by $P(x)\,dx = $ probability.

Understand that the probability density $P(x)$ is related to the square of the wave function $\psi(x)$ by $P(x) = \psi^2(x)$.

You might think that there is something unreal about a wave function ψ. However, according to quantum mechanics, a wave function is as real a description of an object (such as an electron) as we can get. After all, how real is a point particle in classical physics?

A free electron by itself is not quantized—it can have any arbitrary kinetic energy. The quantization of the energy of an electron occurs only when the electron is confined in some way, say as a particle in a box. When confined, quantization of energy arises from boundary conditions, which can occur only if there are boundaries. The three types of confinement discussed in this chapter are a particle in a box, a harmonic oscillator, and a hydrogen atom.

Be sure you understand the differences in the energy level structures of a particle in a box, a harmonic oscillator, and a hydrogen atom. In particular, make sure you understand how the energy levels depend on the quantum number n for each system.

A main result of quantized systems such as a particle in a box, a harmonic oscillator, and the hydrogen atom is that the energies of the particle in the system is quantized. Make sure you understand that the energy $E = hf$ of an emitted or absorbed photon equals the difference of the energies of the energy levels between which the particle makes a transition.

IV. True and False and Responses

True and False

_____ 1. In a photoelectric effect experiment, as you shine light of a fixed wavelength and increasing intensity on the emitting surface C, the kinetic energies of the electrons emitted from C also increase.

_____ 2. In a photoelectric effect experiment, as you shine light of a fixed wavelength less than the threshold wavelength with larger and larger intensity on the emitting surface C, the current measured at the collecting surface A increases.

_____ 3. For a photoelectric effect experiment with light of a given wavelength less than the threshold wavelength, the larger the work function of the emitting surface C the smaller the maximum kinetic energy of the electrons ejected from C.

_____ 4. The larger the wavelength of a photon, the more energy it has.

_____ 5. The Compton wavelength of a proton is shorter than the Compton wavelength of an electron.

_____ 6. The larger the scattering angle in a Compton scattering experiment, the larger the change in the wavelength of the scattered photon.

_____ 7. The de Broglie wavelength of a neutron is smaller than the de Broglie wavelength of an electron that has the same momentum.

_____ 8. As you increase the kinetic energy of an electron, its de Broglie wavelength increases.

_____ 9. The probability of finding an object in an interval between x and $x + dx$ is directly proportional to the size dx of the interval.

_____10. The larger the probability density, the more likely you will find an object in a specified interval dx.

_____11. The probability of finding an object in a given interval dx is proportional to the value of the wave function at the location of dx.

_____12. For a given system, such as a particle in a box or a hydrogen atom, the larger the amplitude of the wave function, the larger will be the quantum number n for an energy level.

_____13. You cannot detect both the wave and the particle aspects of an object simultaneously.

_____14. According to the uncertainty principle, the more uncertain that a particle's momentum is, the more uncertain will be its position.

_____15. Quantized energy levels arise because of confinement of a particle.

_____16. In a quantum mechanical harmonic oscillator, the frequency of a photon emitted in a transition from $n = 3$ to $n = 2$ is equal to the frequency of a photon emitted in a transition from $n = 2$ to $n = 1$.

_____17. In a hydrogen atom, the frequency of a photon emitted in a transition from $n = 3$ to $n = 2$ is equal to the frequency of a photon emitted in a transition from $n = 2$ to $n = 1$.

_____18. On an energy level diagram for a harmonic oscillator, the larger the value of n the higher the energy of the energy level.

_____19. On an energy level diagram for a hydrogen atom, the larger the value of n the lower the energy of the energy level.

_____20. When a particle is confined to a certain region of space in a system, the energy of an emitted photon equals the energy difference between the quantized initial and final energy levels of the system through which the particle makes a transition.

Responses

1. False. The maximum kinetic energy of the emitted electrons depends only on the frequency of the incident light, as given by Einstein's photoelectric equation. More intense light releases more electrons but does not change their maximum kinetic energy.

2. True

3. True

4. False. The energy varies as $E = hf = hc/\lambda$, so the larger the wavelength, the smaller the energy.

5. True

6. True

7. False. Because the de Broglie wavelength λ equals h/p, particles with the same momentum have the same wavelength.

8. False. The larger its kinetic energy, the larger an electron's velocity v. Because $\lambda = h/mv$, the de Broglie wavelength decreases as an electron's velocity increases with increasing kinetic energy.

9. True

10. True

11. False. The probability is proportional to the *square* of the wave function.

12. False. For a given system, the quantum numbers of the system's energy levels are related to the size of the system, and have nothing to do with the amplitude of the wave function.

13. True

14. False. The more uncertain that a particle's momentum is, the *more precisely* we can locate its position.

15. True

16. True

17. False. In a hydrogen atom, the frequency of an emitted photon varies as $f \propto [1/n_f^2 - 1/n_i^2]$, so the frequency of a photon emitted in a transition from $n = 3$ to $n = 2$ will be different from the frequency of a photon emitted from a transition from $n = 2$ to $n = 1$. (Which frequency will be larger?)

18. True

19. False. The energy levels in a hydrogen atom are given by $E_n = -(13.6 \text{ eV})/n^2$ (note the minus sign), so the larger the value of n, the higher is the value of the energy level (see Figure 17-7).

20. True

V. Questions and Answers

Questions

1. In a photoelectric experiment, why is there a linear relationship between the frequency of incident light and the maximum kinetic energy of emitted electrons?

2. Explain why there is a threshold frequency in a photoelectric experiment.

3. In a photoelectric experiment, what effect does changing the material of the emitting surface C to a new material with a higher work function have on the emitted electrons?

4. In a Compton scattering experiment, how does the change in the wavelength of the scattered photon vary with the scattering angle?

5. How does the de Broglie wavelength of a particle vary with its velocity (neglecting relativistic effects)?

6. An electron is accelerated from rest, acquiring a kinetic energy K. How does the de Broglie wavelength of the electron depend on K?

7. What two quantities determine the probability of locating an object?

8. How is the wave function related to probability?

9. Describe how energy levels depend upon the quantum number n for (a) a particle in a box, (b) a harmonic oscillator, and (c) a hydrogen atom.

10. Describe an experiment that measures the wave and particle aspects of an object simultaneously.

Answers

1. In the photoelectric effect, a photon transfers its energy $E = hf$ to an electron in the emitting material C. The electrons at the surface of the emitting material C are least tightly bound and come off with a maximum kinetic energy equal to the energy hf absorbed from the photon minus the work function ϕ, which is the energy necessary to release an electron from the emitter's surface. This energy balance is described by Einstein's photoelectric equation, $hf = K_{\max} - \phi$, which shows there is a linear relationship between the frequency f of the incident light and the maximum kinetic energy $K_{\max}$ of the ejected electrons.

2. A threshold photon energy $E_t = hf_t = hc/\lambda_t$ equal to the work function ϕ is required to eject the least tightly bound electrons from the surface of the emitting material C: $hc/\lambda_t = \phi$. A photon with less energy than this does not eject any electrons.

3. The larger the work function ϕ of the emitting surface C, the more energy is required to release electrons from the emitter's surface. Hence, for incident photons of a given frequency, and therefore a given energy, the larger the work function ϕ, the lower the maximum kinetic energy of the emitted electrons. This is seen from the energy balance in Einstein's photoelectric equation $hf = K_{\max} - \phi$.

4. As the scattering angle θ increases, the change in the wavelength of the scattered photon increases because $\lambda_2 - \lambda_1$ is proportional to $(1 - \cos\theta)$.

5. The de Broglie wavelength λ equals h/p. For nonrelativistic situations, $p = mv$, so $\lambda = h/mv$.

6. The velocity of the electron is related to its kinetic energy by $\frac{1}{2}mv^2 = K$, or $v = (2K/m)^{1/2}$. The de Broglie wavelength is $\lambda = h/mv = h/[m(2K/m)^{1/2}] = h/(2mK)^{1/2}$.

7. The size dx of the spatial interval, and the probability density $P(x)$. The probability of finding an object in an interval dx is given by $P(x)\, dx$.

8. The probability density $P(x)$ is the square of the wave function $\psi(x)$: $P(x) = \psi^2(x)$. The probability of finding a particle in an interval dx is $P(x)\, dx = \psi^2(x)\, dx$.

9. (a) $E_n \propto n^2$, (b) $E_n \propto n$, (c) $E_n \propto -1/n^2$.

10. Such an experiment does not exist. According to Bohr's principle of complementarity, which follows from the uncertainty principle, you can not measure both the wave and the particle aspects of an object in any single experiment.

VI. Problems and Solutions

Problems

1. In a photoelectric effect experiment, it is found that the maximum kinetic energy of emitted electrons for 400-nm light is 2.02 eV. Find the threshold wavelength for the emitter surface.

How to Solve It
• Write the energy balance for the transfer of energy from an incident photon to an electron in the emitting surface, as given by Einstein's photoelectric equation.

• Solve for the unknown quantity, in this case the work function ϕ for the surface.

• Use the relationship between the work function and the threshold wavelength, $\phi = hc/\lambda_t$.

2. The work function for potassium is 2.21 eV. In a photoelectric effect experiment using a potassium emitter, the maximum kinetic energy of the emitted electrons is observed to be 2.82 eV. What is the wavelength of the incident light?

How to Solve It
• Write the energy balance for the transfer of energy from an incident photon to an electron in the emitter's surface, as given by Einstein's photoelectric equation.

• Solve for the unknown quantity, in this case the wavelength.

3. Find the wavelength of a 0.6-MeV photon after scattering at an angle of 70° in a Compton scattering experiment.

How to Solve It
• Find the wavelength λ_1 from the photon energy expression $E = hc/\lambda$.

• Find the wavelength λ_2 from the Compton relationship.

4. What is the energy of the electron in Problem 3 after the scattering?

How to Solve It
• Subtract the final photon energy from the initial photon energy to get the energy of the scattered electron.

5. A typical energy of neutrons in a nuclear reactor is around 0.04 eV. What is the de Broglie wavelength of such a neutron?

How to Solve It
• Determine the momentum from $K = p^2/2m$.

• Calculate the de Broglie wavelength from $\lambda = h/p$.

6. What must be the energy of an electron for it to have the same de Broglie wavelength as the neutron in Problem 5?

How to Solve It
• Reverse the steps of the procedure in Problem 5.

7. Find the probability of finding a particle in the central region $L/4 < x < 3L/4$ of a one-dimensional box of length L when the particle is in the second excited state.

How to Solve It
• Determine the mathematical expression for the wave function $\psi(x)$.

• From the wave function determine the probability distribution $P(x)$ from $P(x) = \psi^2(x)$, and form the probability of finding a particle in the interval dx as the product $P(x)\,dx$.

• If $P(x)\,dx$ can be taken as constant or nearly constant over the given interval Δx, the answer is $P(x)$ Δx. If, as in this problem, $P(x)$ is not constant over the given interval, you must integrate the product $P(x)\,dx$ over the interval to obtain the probability of finding the object in the interval:

$$\text{probability} = \int_{x}^{x+\Delta x} P(x)\,dx$$

8. A particle in a one-dimensional box of length L is in its first excited state. If you start from $x = 0$, at what value of x will there be a 25% probability of finding the particle in this interval?

How to Solve It
• Follow the same procedure as in Problem 7 using $n = 2$.

• Keeping in mind that the probability equals 1 for finding the particle in the interval from $x = 0$ to $x = L$, draw a picture of the wave function and the square of the wave function to help you guess a simple solution to a complicated transcendental equation.

9. The size of a nucleus is about 10^{-14} m. Treating the nucleus as a one-dimensional particle in a box, find the ground-state energy of a proton and an electron if each particle were bound in the nucleus.

How to Solve It
• Substitute the given values of m and L into the ground-state energy expression $E_1 = h^2/8mL^2$.

10. For the proton and electron in Problem 9, find the wavelength of the photon emitted in a transition from the first excited state to the ground state.

How to Solve It
• Calculate the energy of the second energy level from $E_n = n^2 E_1$.

• Use the values of E_1 from Problem 9 and calculate the wavelengths from $hc/\lambda = E_i - E_f$.

11. For a harmonic oscillator, find the ratio of the frequency of a photon emitted from an $n = 3$ to $n = 1$ transition to the frequency of a photon emitted from an $n = 2$ to $n = 1$ transition.

How to Solve It
• Calculate each frequency from $hf = E_i - E_f$, with $E_n = (n + \frac{1}{2})hf_0$.

12. A 2-kg object is attached to the end of a spring with a force constant 400 N/m and oscillates without friction with an amplitude of 8 cm. If the energy of this classical harmonic oscillator were quantized according to quantum mechanical rules, find the value of the corresponding quantum number.

How to Solve It
• Equate the classical expression for the energy of a harmonic oscillator

$$E = \tfrac{1}{2}m\omega_0^2 A^2 = \tfrac{1}{2}kA^2$$

($\omega_0 = (k/m)^{1/2}$ and A is the amplitude) to the quantum-mechanical energy expression for a harmonic oscillator

$$E_n = (n + \tfrac{1}{2})hf_0$$

and solve for n.

13. The ground-state wave function for a quantum mechanical harmonic oscillator is $\psi(x) = Ce^{-x^2/2A^2}$, where C and A are constants for a given harmonic oscillator. Find the expectation value of x^2.

How to Solve It
• Substitute the relation $f(x) = x^2$ into the expectation value relation

$$<f(x)> = \int f(x)\psi^2(x)\, dx$$

• Evaluate the integral over the entire range of the variable x. Use integration tables if necessary.

14. Find how C is related to A so that the ground-state quantum mechanical oscillator wave function $\psi(x) = Ce^{-x^2/2\,A^2}$ is normalized.

How to Solve It
• Substitute the wave function into the normalization condition

$$\int_{-\infty}^{\infty} \psi^2(x)\, dx = 1$$

If necessary use standard integration tables to evaluate the integral.

Solutions

1. Using standard units, substitute the given data into Einstein's photoelectric equation

$$K_{max} = hc/\lambda - \phi$$

to obtain

$$(2.02\text{ eV})(1.6 \times 10^{-19}\text{ J/eV})$$
$$= \frac{(6.626 \times 10^{-34}\text{ J·s})(3 \times 10^8\text{ m/s})}{400 \times 10^{-9}\text{ m}} - \phi$$

$$\phi = 1.738 \times 10^{-19}\text{ J} = 1.08\text{ eV}$$

The same answer can be obtained more easily by using more convenient units:

$$2.02\text{ eV} = \frac{1240\text{ eV·nm}}{400\text{ nm}} - \phi$$

$$\phi = \underline{1.08\text{ eV}}$$

From the expression for the threshold wavelength,

$$\phi = hc/\lambda_t,$$

$$1.08\text{ eV} = \frac{1240\text{ eV·nm}}{\lambda_t}$$

$$\lambda_t = \underline{1148\text{ nm}}$$

2. $\underline{247\text{ nm}}$

3. The wavelength λ_1 of the incident photon is found from $E = hc/\lambda$.

$$0.6 \times 10^6\text{ eV} = \frac{1240\text{ eV·nm}}{\lambda_1}$$

$$\lambda_1 = 0.00207\text{ nm} = 2.07\text{ pm}$$

The wavelength λ_2 of the scattered photon is found from the Compton relation

$$\lambda_2 - \lambda_1 = \frac{hc}{\lambda}(1 - \cos\theta)$$

$$\lambda_2 - 2.07\text{ pm} = 2.43\text{ pm}(1 - \cos 70°)$$

$$\lambda_2 = \underline{3.67\text{ pm}}$$

4. $\underline{0.262\text{ MeV}}$

5. Find the momentum of the neutron from

$$K = \frac{p^2}{2m} \qquad p = \sqrt{2mK}$$

Find the de Broglie wavelength from

$$\lambda = \frac{h}{p} = \frac{h}{\sqrt{2mK}} = \frac{hc}{\sqrt{2mc^2\,K}}$$

$$\lambda = \frac{1240\text{ eV·nm}}{\sqrt{2(940 \times 10^6\text{ eV})(0.04\text{ eV})}}$$

$$\lambda = 0.143\text{ nm} = \underline{0.14\text{ nm}}$$

6. $\underline{77\text{ eV}}$ (using the rounded-off value 0.14 nm)

7. The wave functions for a particle in a one-dimensional box of length L are

$$\psi_n = \sqrt{\frac{2}{L}}\sin\frac{n\pi x}{L} \qquad n = 1, 2, 3, \ldots$$

For the second excited state, $n = 3$, so the wave function for our problem is $\psi_3 = \sqrt{2/L}\sin 3\pi x/L$. The given interval is $L/4 < x < 3L/4$, and the probability of finding the particle in this interval is

$$\text{probability} = \int \psi^2(x)\, dx = 2/L\int_{L/4}^{3L/4}\sin^2\frac{3\pi x}{L}\, dx$$

The integral can be evaluated using standard integration tables, which give

$$\int \sin^2 ax \, dx = \frac{x}{2} - \frac{1}{4a} \sin 2ax$$

to get

$$\text{probability} = \frac{2}{L}\left[\frac{x}{2} - \frac{L}{12\pi}\sin\frac{6\pi x}{L}\right]_{L/4}^{3L/4}$$

$$\text{probability} = \underline{0.394}$$

Thus there is about a 39 percent chance of finding the particle in the central region. It spends more time in the turnaround zones.

8. $\underline{L/4}$

9. For a proton, $m = 938$ MeV/c^2, and for an electron, $m = 0.511$ MeV/c^2. Substituting these values into the expression for the ground-state energy of a particle in a box,

$$E_1 = \frac{h^2}{8mL^2} = \frac{(hc)^2}{8mc^2L^2}$$

gives for the proton

$$E_1 = \frac{(1.24 \times 10^{-12} \text{ MeV·m})^2}{8 \times 938 \text{ MeV} \times (10^{-14} \text{ m})^2}$$

$$E_1 = \underline{2.05 \text{ MeV}}$$

and for the electron

$$E_1 = \frac{(1.24 \times 10^{-12} \text{ MeV·m})^2}{8 \times 0.511 \text{ MeV} \times (10^{-14} \text{ m})^2}$$

$$E_1 = \underline{3,760 \text{ MeV}}$$

10. Proton: $\underline{2.02 \times 10^{-4} \text{ nm}}$;
electron: $\underline{1.10 \times 10^{-7} \text{ nm}}$

11. For each of the transitions, we have

$$n = 3 \text{ to } n = 1, hf_3 = E_3 - E_1$$
$$= (3 + \tfrac{1}{2})hf_0 - (1 + \tfrac{1}{2})hf_0 = 2hf_0, f_3 = 2f_0$$

$$n = 2 \text{ to } n = 1, hf_2 = E_2 - E_1$$
$$= (2 + \tfrac{1}{2})hf_0 - (1 + \tfrac{1}{2})hf_0 = hf_0, f_2 = f_0$$

$$\underline{f_3/f_2 = 2}$$

The "harmony" in a harmonic oscillator arises from the fact that all $f_n = f_0$ for $\Delta n = \pm 1$.

12. $\underline{n = 9 \times 10^{32}}$ (This enormous number is an illustration of Bohr's correspondence principle, where in the limit of very large quantum numbers classical and quantum calculations must yield the same result.)

13. Substituting $f(x) = x^2$ into the expectation value expression,

$$<f(x)> = \int f(x)\psi^2 \, dx$$

$$<x^2> = C^2 \int_{-\infty}^{\infty} x^2 e^{-x^2/A^2} \, dx$$

From standard integration tables you can find

$$\int_0^{\infty} x^2 e^{-ax^2} \, dx = \frac{1}{4}\sqrt{\frac{\pi}{a^3}}$$

so

$$<x^2> = \tfrac{1}{2}C^2 A^3\sqrt{\pi}$$

14. $\underline{C = A^{-1/2}(\pi)^{-1/4}}$

Chapter 18

Temperature and the Kinetic Theory of Gases

I. Key Ideas

18-1 Thermal Equilibrium and Temperature **Thermodynamics** is the study of temperature and energy exchange. When two objects are in **thermal contact**, energy in the form of heat flows from the warmer object to the cooler object. This normally results in the warmer object cooling down and the cooler object warming up. As this process continues, both the rate of cooling of the warmer object and the rate of warming of the cooler object become less and less. When each of the objects is neither warming up nor cooling down the objects have reached a steady-state situation called **thermal equilibrium**. If two objects are each in thermal equilibrium with a third object, they are in thermal equilibrium with each other. This is called the **zeroth law of thermodynamics**:

> If two objects are each in thermal equilibrium with a third, then they are in thermal equilibrium with each other.

Two objects are defined to have the same *temperature* if they are in thermal equilibrium with each other. The zeroth law, as we will see, enables us to define a temperature scale.

Temperature is the measure of hotness or coldness of a substance. Warmer objects are at a higher temperature than cooler objects, and objects that are in thermal equilibrium are at the same temperature. When objects at different temperatures are in thermal contact, the direction of heat flow is always from the warmer object to the cooler object.

18-2 The Celsius and Fahrenheit Temperature Scales A physical property that changes with temperature is called a **thermometric property**. Any thermometric property can be used as the basis of a thermometer. A common example is the thermal expansion of a liquid sealed in a glass envelope. For such a thermometer, a temperature scale is established by assigning values at which the thermometer is in thermal equi-

librium with reproducible states of some system, such as water at its ice (freezing) and steam (boiling) points. The interval between these points is then divided up into equal increments called degrees.

For the **Celsius scale**, a value of zero degrees Celsius (0°C) is assigned to the ice point of water and a value of 100 degrees Celsius (100°C) is assigned to its steam point at atmospheric pressure. A Celsius thermometer is constructed by selecting a convenient thermometric property, measuring that property at the ice and steam points, and dividing the difference between these measured values into one hundred equally spaced increments. The length L_t of a metal bar is a thermometric property that can be used to construct a Celsius scale. The Celsius temperature t_C is then given by

Celsius scale

$$t_C = \frac{L_t - L_0}{L_{100} - L_0} \times 100°$$

where L_0 and L_{100} are the lengths of the bar when it is in thermal equilibrium with an ice bath and a steam bath, respectively.

In the **Fahrenheit scale**, still commonly used in the United States, a value of 32 degrees Fahrenheit (32°F) is assigned to the ice point of water and 212 degrees Fahrenheit (212°F) is assigned to the steam point at atmospheric pressure. The relation between a Fahrenheit temperature t_F and a Celsius temperature t_C is

Fahrenheit–Celsius conversion

$$t_C = \tfrac{5}{9}(t_F - 32°)$$

18-3 Gas Thermometers and the Absolute Temperature Scale A drawback to most thermometers is that the temperature readings of different kinds of thermometers do not agree, except at the defining temperatures, because different thermometric properties do not change with temperature in the same way. One class of thermometers that do agree over a wide range of temperatures is **constant-volume gas thermometers**, for which pressure is the thermometric property. If any gas at low pressure is confined to a constant volume, its pressure will increase with increasing temperature over a wide range of temperatures. The Celsius temperature t_C of such a thermometer is given by

Celsius scale, gas thermometer

$$t_C = \frac{P_t - P_0}{P_{100} - P_0} \times 100°$$

where P_0 and P_{100} are the pressures of the gas when it is in thermal equilibrium with an ice bath and a steam bath, respectively. For gases at very low densities, all constant-volume gas thermometers give the same value of the temperature, independent of the particular gas used. When the pressure in a constant-volume gas thermometer is extrapolated to zero, it approaches zero as the temperature approaches $-273.15°C$. This temperature is known as absolute zero.

The ice and steam points of water are less easy to reproduce precisely than the **triple point of water**, which is the single state at which ice, water, and steam can coexist in equilibrium. It occurs at a pressure of 4.58 mmHg and a temperature of 0.01°C. The **ideal-gas temperature scale** is defined by making the triple-point temperature 273.16 kelvins (K). Using a constant-volume gas thermometer this gives

Ideal-gas temperature scale

$$T = \frac{273.15 \text{ K}}{P_3} P$$

where P_3 is the pressure when the gas is at the temperature of the triple point of water and the **kelvin** (K) is a degree unit that is the same size as the Celsius degree. The ideal-gas temperature scale is identical with another scale, the absolute temperature scale (also called the **Kelvin scale**), which is defined independent of the properties of any substance. The symbol T is used when referring to the absolute temperature. To convert from degrees Celsius to kelvins, we simply add 273.15:

Celsius–kelvin conversion

$$T = t_C + 273.15$$

18-4   *The Ideal-Gas Law*   Experimentally, when a confined gas kept at low pressure P and at a constant temperature T is either compressed or allowed to expand, the product of the pressure of the gas and its volume remains constant. This observation is known as **Boyle's law**. Also, at low pressures the absolute temperature of a gas kept at a constant volume V is proportional to its pressure. These relations are contained in the **ideal-gas law**, the equation of state for gases at low pressures:

Ideal-gas law

$$PV = NkT$$

where N is the number of molecules and k is **Boltzmann's constant**.

Boltzmann's constant

$$k = 1.381 \times 10^{-23} \text{ J/K} = 8.617 \times 10^{-5} \text{ eV/K}$$

An amount of gas is often expressed in moles. A **mole** (mol) of any substance is the amount of that substance that contains **Avogadro's number** N_A of atoms or molecules, defined as the number of carbon atoms in 12 grams of ^{12}C:

Avogadro's number

$$N_A = 6.022 \times 10^{23} \text{ molecules/mol}$$

If we have n moles of a substance, the number of molecules is then $N = nN_A$. The ideal-gas law is then

$$PV = nN_A kT = nRT$$

where $R = N_A k$ is called the **universal gas constant**. Its value, which is the same for all gases, is

Universal gas constant

$$R = N_A k = 8.314 \text{ J/mol·K} = 0.08206 \text{ L·atm/mol·K}$$

The mass of one mole of a chemical element or compound is called its **molar mass** M, for which the customary unit is g/mol. The molar mass of ^{12}C is, by definition, 12 g/mol or 12×10^{-3} kg/mol. Approximate molar masses of some elements and compounds are

Species	H	H_2	He	C	N	O	N_2	O_2	CO_2
Molar mass (g/mol)	1	2	4	12	14	16	28	32	44

A given amount of a substance, such as 1 kg of water, is placed in a closed, otherwise empty container where the substance is maintained at a constant pressure and volume, and the container is maintained at a constant temperature. Eventually the substance and the container reach thermal equilibrium, at which time the temperature of the substance and the container are equal. After thermal equilibrium is established it is found that no further changes in the sample—such as changes from liquid to solid or from liquid to gas—can be observed. However, scientists understand that even after thermal equilibrium is reached, and the temperature, pressure, and volume of the sample remain constant, changes continue to occur within the sample. These changes occur at the microscopic level, a level which is *not* directly observable. At the microscopic level, matter consists of numerous molecules that are in constant, random motion. Consequently, the positions of the individual molecules vary with time, and so, at least at the microscopic level, the sample is constantly changing.

The state of the sample is determined only by its observable (macroscopic) properties. Because microscopic variations of the sample do not always result in observable changes, they do not necessarily result in a change in its state. Using instruments such as a pressure gauge, a graduated cylinder, a thermometer, and your senses, the pressure, volume, temperature, and phase (solid, liquid, or gaseous) of the sample can be observed. Thus, any changes in pressure, volume, temperature, or phase represent a change in the state of the sample.

The equation $PV = nRT$, which relates macroscopic variables P, V, and T for a given amount of a gas, is called an **equation of state**. It is the equation of state for a sample of gas at low density. The state of the sample is specified if any two of the three variables P, V, and T are specified.

18-5 *The Kinetic Theory of Gases* An ideal gas is one in which the molecules are separated, on the average, by distances that are large compared with their diameters, and they exert no forces on each other except when they collide. The properties of an ideal gas can be understood in terms of a simple kinetic model. We can picture the gas as consisting of noninteracting molecules in motion and the pressure as the result of the collisions of the gas molecules with the walls of their container. Applying the impulse–momentum theorem to calculate the pressure resulting from these collisions gives

Pressure of an ideal gas

$$PV = 2N(\tfrac{1}{2}mv_x^2)_{av}$$

where P is the pressure, V is the volume containing N molecules, m is mass of a gas molecule, and v_x is the x component of its velocity. The ideal-gas law ($PV = NkT$) together with this formula gives $kT = m(v_x^2)_{av}$. Furthermore, combining the relations $v^2 = v_x^2 + v_y^2 + v_z^2$ and $(v_x^2)_{av} = (v_y^2)_{av} = (v_z^2)_{av}$ (the first obtained via the Pythagorean theorem and the second via symmetry) we get $(v^2)_{av} = 3(v_x^2)_{av}$. It follows that $kT = m(v_x^2)_{av} = \tfrac{1}{3}m(v^2)_{av}$ or

Average translational kinetic energy

$$K_{av} = (\tfrac{1}{2}mv^2)_{av} = \tfrac{3}{2}kT$$

where K_{av} is the average translational kinetic energy of a gas molecule of mass m.

Thus the absolute temperature of the gas is a measure of the average translational kinetic energy of its molecules. The total translational kinetic energy K of n moles of a gas containing N molecules is

Total translational kinetic energy

$$K = N(\tfrac{1}{2}mv^2)_{av} = \tfrac{3}{2}NkT = \tfrac{3}{2}nN_AkT = \tfrac{3}{2}nRT$$

We can use these results to estimate the speeds of the molecules in a gas by solving for the **root mean square** (rms) speed, which is the square root of $(v^2)_{av}$:

Root mean square speed

$$v_{rms} = \sqrt{(v^2)_{av}} = \sqrt{\frac{3kT}{m}} = \sqrt{\frac{3RT}{N_Am}} = \sqrt{\frac{3RT}{M}}$$

where $M = N_Am$ is the molar mass.

The translational kinetic energy of a molecule of a gas can be expressed as the sum of three terms. That is, $\tfrac{1}{2}mv^2 = \tfrac{1}{2}mv_x^2 + \tfrac{1}{2}mv_y^2 + \tfrac{1}{2}mv_z^2$. If the gas is in equilibrium, these three terms are, on average, equal. This sharing of the energy equally between the three terms in the translational kinetic energy is a special case of the **equipartition theorem**, a result that follows from classical statistical mechanics. Each component of position and momentum (including angular position and angular momentum) that appears as a squared term in the expression for the energy of the system is called a **degree of freedom**. More generally, the equipartition theorem states that

When a substance is in equilibrium, there is an average energy of $\tfrac{1}{2}kT$ per molecule or $\tfrac{1}{2}RT$ per mole associated with each degree of freedom.

The average distance λ a gas molecule travels between collisions is called its **mean free path**. For an ideal gas the expression for the mean free path is

Mean free path

$$\lambda = \frac{1}{\sqrt{2}\, n_v \pi d^2}$$

where n_v is the number density (the number of molecules per unit volume) and d is the molecular diameter. If v_{av} is the average speed, the average distance traveled between collisions is $\lambda = v_{av}\tau$, where τ is the collision time (the mean time between collisions).

It is desirable to statistically connect the details of molecular motions to the macroscopic thermodynamic parameters. To do this involves making use of the Maxwell-Boltzmann distribution functions. The **Maxwell–Boltzmann speed distribution function** contains information concerning the distribution of molecular speeds. In a gas of N molecules, the number that have speeds in the range between v and $v + dv$ is dN, given by

$$dN = N\, f(v)\, dv$$

where $f(v)\, dv$ is the fraction dN/N that have speeds between v and $v + dv$. The Maxwell–Boltzmann speed distribution function $f(v)$ can be derived using statistical mechanics. The result is

Maxwell–Boltzmann speed distribution function

$$f(v) = \frac{4}{\sqrt{\pi}}\left(\frac{m}{2kT}\right)^{3/2} v^2 e^{-mv^2/2kT}$$

The distribution of the translational kinetic energies of the molecules in a gas is described by the **Maxwell–Boltzmann energy distribution function**. This function is

Maxwell–Boltzmann energy distribution function

$$F(E) = \frac{2}{\sqrt{\pi}}\left(\frac{1}{kT}\right)^{3/2} E^{1/2} e^{-E/kT}$$

where $E = \frac{1}{2}mv^2$. In the language of statistical mechanics, the energy distribution is considered to be the product of two factors: one, called the **density of states**, is proportional to $E^{1/2}$; the other is the probability of a state being occupied, which is $e^{-E/kT}$ and is called the **Boltzmann factor**.

The Maxwell–Boltzmann speed and energy distribution functions are continuous fractional distribution functions. They are valid in the classical approximation, that is, they are valid if the quantum nature of the energy distribution can be ignored. One property of such functions, called the **normalization condition**, is that the sum of the fractions equals 1. That is $\int f(x)\, dx = 1$, where f is any continuous fractional distribution function. Furthermore, the average value of any function $g(x)$ is given by

$$[g(x)]_{av} = \int g(x)\, f(x)\, dx$$

II. Numbers and Key Equations

Numbers

Boltzmann's constant

$$k = 1.381 \times 10^{-23} \text{ J/K} = 8.617 \times 10^{-5} \text{ eV/K}$$

Avogadro's number

$$N_A = 6.022 \times 10^{23} \text{ molecules/mol}$$

Universal gas constant

$$R = N_A k = 8.314 \text{ J/mol·K} = 0.08206 \text{ L·atm/mol·K}$$

Key Equations

Celsius scale

$$t_C = \frac{L_t - L_0}{L_{100} - L_0} \times 100°$$

Fahrenheit–Celsius conversion

$$t_C = \tfrac{5}{9}(t_F - 32°)$$

Celsius scale, gas thermometer

$$t_C = \frac{P_t - P_0}{P_{100} - P_0} \times 100°$$

Ideal-gas temperature scale

$$T = \frac{273.15 \text{ K}}{P_3}\, P$$

Celsius–kelvin conversion

$$T = t_C + 273.15$$

Ideal-gas law

$$PV = NkT = nRT$$

Average translational kinetic energy

$$K_{av} = (\tfrac{1}{2}mv^2)_{av} = \tfrac{3}{2}kT$$

Total translational kinetic energy

$$K = N(\tfrac{1}{2}mv^2)_{av} = \tfrac{3}{2}NkT = \tfrac{3}{2}nN_AkT = \tfrac{3}{2}nRT$$

Root mean square speed

$$v_{rms} = \sqrt{(v^2)_{av}} = \sqrt{\frac{3kT}{m}} = \sqrt{\frac{3RT}{N_Am}} = \sqrt{\frac{3RT}{M}}$$

Mean free path

$$\lambda = \frac{1}{\sqrt{2}n_v\pi d^2}$$

Maxwell–Boltzmann speed distribution function

$$f(v) = \frac{4}{\sqrt{\pi}}\left(\frac{m}{2kT}\right)^{3/2} v^2 e^{-mv^2/2kT}$$

Maxwell–Boltzmann energy distribution function

$$F(E) = \frac{2}{\sqrt{\pi}}\left(\frac{1}{kT}\right)^{3/2} E^{1/2}e^{-E/kT}$$

III. Potential Pitfalls

Be careful not to confuse a specific value of temperature with a change in temperature or a temperature range. When dealing with Celsius or Fahrenheit degrees, °C or °F should be used to denote a specific temperature and C° or F° should be used to denote either a change in temperature or a temperature range; no such distinction is possible with the kelvin.

Many of the equations in physics that include a temperature factor are valid only when temperature is expressed in kelvins. Be sure you know for which equations this is true. Of course, if you are dealing with a temperature difference, a kelvin and a Celsius degree are equal.

At a given temperature, molecules of all gases have the same average translational kinetic energy. Their average and root mean square speeds will not be the same, however, because the molecular mass of different gases are not the same.

In applying the ideal-gas law, be sure that all quantities, including the gas constant R, are in consistent units. Trying to work with pressure in torr, volume in cubic meters, and R in L·atm/mol·K will only give you a headache.

If a gas law is used for proportional calculations, such as Boyle's law ($P_1V_1 = P_2V_2$), you can use any units you like, but be sure to use absolute pressure, not gauge pressure. Of course, the units have to be the same on both sides of the equation.

IV. True and False and Responses

True and False

_____1. If both object A and object B are in thermal equilibrium with object C, then object A must be in thermal equilibrium with object B.

_____2. All Celsius thermometers must agree that the steam point of water is 100°C.

_____3. All Celsius thermometers must agree that normal body temperature is about 37°C.

_____4. All low-pressure, constant-volume gas thermometers agree that normal body temperature is about 37°C.

_____5. The molar mass of a sample of a gas is the mass of one mole of the gas.

Responses

1. True

2. True

3. False. Celsius thermometers must agree that the ice point and boiling point of water are 0 and 100°C, respectively, but they do not have to be in agreement for any other temperatures because not all thermometric properties vary with temperature in the same way.

4. True

5. True

V. Questions and Answers

Questions

1. In what way is the Celsius scale more convenient than the Kelvin scale for everyday use? In what way is the Kelvin scale more suitable then the Celsius scale for scientific use?

2. What change must be made in the temperature of an ideal gas to halve the rms speed of its molecules?

3. Two different ideal gases are at the same temperature. How do the rms speeds of their molecules compare? What about the average translational kinetic energy of their molecules? The average molecular speed varies inversely with the square root of the molar mass.

4. What properties should a physical system possess in order to serve as a good thermometer?

5. What is the mass of a mole of carbon monoxide gas (molecular formula CO)? How many molecules are there in one mole?

Answers

1. On the Celsius scale, the temperatures that we ordinarily deal with tend to be numbers between -15 and $+40$. These are more convenient to use and remember than temperature values of several hundred kelvins would be. The Kelvin scale is more suitable for scientific use in that many formulas are less complex when the Kelvin scale is used. Also, the Kelvin scale is independent of the thermometric properties of any particular substance.

2. The average molecular translational kinetic energy, which is proportional to the rms speed squared, must be reduced to $\frac{1}{4}$ its original value in order to halve the rms speed. Because this energy is proportional to the absolute temperature for an ideal gas, the absolute temperature must be reduced to $\frac{1}{4}$ its original value.

3. If two ideal gases are at the same temperature, they have the same average translational kinetic energy per molecule. If they do not happen to have the same molecular mass, however, then the molecules of the lighter gas will, on average, be moving at higher speeds.

4. It depends a lot on what the thermometer is supposed to be good for, but there are some general requirements that apply to most cases. The thermometric property used—length, electrical resistance, pressure, or whatever—should be easily measurable itself and should vary linearly, or nearly so, with absolute temperature. Otherwise, the thermometer will be usable only for a limited number of applications. In most cases, the thermometer will need to be physically small so that it will quickly come to thermal equilibrium with the system to be measured, and also so that it will affect that system as little as possible. Other considerations will arise depending on the particular situation.

5. The molar mass of CO is $12 + 16 = 28$, so a mole of CO has a mass of 28 g. There are 6.02×10^{23} molecules (Avogadro's number) in a mole of anything.

VI. Problems and Solutions

Problems

In some of these problems you will need the molar mass of one or more gases. These can be determined from the atomic mass values given in Appendix C of the text.

1. A certain constant-volume gas thermometer reads a pressure of 88 torr at the temperature of the triple point of water. (*a*) What pressure will the thermometer read at a temperature of 310 K? (*b*) What is the temperature when the thermometer reads a pressure of 70 torr?

How to Solve It
• For a constant-volume gas thermometer, the absolute temperature is proportional to the pressure reading.

• Part (*b*) is just the same calculation in reverse.

2. Mercury freezes at $-39°C$. Express this temperature on both the Fahrenheit and Kelvin scales.

How to Solve It
• A kelvin and a Celsius degree are the same size, so to convert from degrees Celsius to kelvins you need only to add 273 to the Celsius temperature.

• A formula relating the Celsius and Fahrenheit scales is given in the Key Ideas.

3. On a morning when the thermometer reads 52°F, you check the pressure in your bicycle tires using your pressure gauge, and the gauge reading is 75 lb/in². Later in the day, after riding several miles on hot pavement, the temperature of the air in the tires reaches 125°F. Assuming that the volume of the tires hasn't changed, what is the pressure in them now?

How to Solve It
• Convert the temperatures to kelvins.

• What the gauge reads is (reasonably enough) gauge pressure. To get absolute pressure, add the value of atmospheric pressure, 14.7 lb/in².

• Use the ideal gas law to relate the initial and final temperatures and pressures.

• There is no need to convert pressures to SI units because you are dealing with pressure ratios.

4. (*a*) If 3 mol of an ideal gas occupy a volume of 40 L at a pressure of 1 atm, what is the temperature of the gas? (*b*) If the gas is heated at constant pressure to 273 K, what volume does it occupy now?

How to Solve It
• Use the ideal-gas law directly to find the temperature in part (*a*). Watch the units! Express R in units that make your life easier.

• In part (*b*), use the same equation to find the volume, given the new temperature.

5. The gas in intergalactic space is mostly atomic hydrogen at a temperature of 3 K and a pressure of the order of 10^{-21} atm. How many hydrogen atoms are there per cubic centimeter? (*Note:* It is interesting to compare this result with that of the next example, which deals with a very good laboratory vacuum.)

How to Solve It
• The number of moles in a sample of gas can be calculated from the ideal-gas law.

• Each mole of atomic hydrogen corresponds to Avogadro's number of atoms.

6. A high-vacuum pump reduces the pressure in a container to 10^{-8} torr. If the temperature is 20°C, how many gas molecules per cubic centimeter are there in the container?

How to Solve It
• Convert the values for P, V, and the gas constant R to compatible units. Unit conversions play an important role in this sort of problem.

• Solve the ideal-gas law for n/V, the number of moles per unit volume. Each mole corresponds to Avogadro's number of molecules.

• Convert the result to the number of molecules per cubic centimeter.

7. The mass of a certain gas sample that occupies 3 L at 20°C and 10 atm pressure is found to be 55 g. What is this gas?

How to Solve It
• If we can calculate its molar mass then perhaps we can identify the gas.

• Use the ideal-gas law and obtain an expression for the number of moles of gas in the sample.

• The molar mass is the ratio of the mass to the number of moles. Calculate the molar mass. There are then only a few possibilities for what the gas might be.

8. A rigid, high-pressure gas cylinder has a mass of 21.22 kg when empty. Its interior volume is 1.33 L. Its mass is 21.61 kg after it is filled with nitrogen gas (N_2) at room temperature (20°C). What is the pressure in the filled cylinder?

How to Solve It
• From the given data, find the mass of nitrogen in the cylinder.

• Relate the number of moles to the mass of the nitrogen in the cylinder and the molar mass of nitrogen.

• Solve the ideal-gas law for the pressure.

9. Two gases present in the atmosphere are water vapor (H_2O) and argon. What is the ratio of their rms speeds?

How to Solve It
• To determine the molecular masses look up the appropriate atomic masses in Appendix C of the text. Argon is a monatomic gas; that is, there is one atom per molecule.

• The gases have the same temperatures, and so the average translational kinetic energy per molecule is the same for both species. Equate the kinetic energies and solve for the ratio of the rms speeds.

10. Naturally occurring uranium contains a rare isotope with an atomic mass of 235 and a common isotope with an atomic mass of 238. Uranium reacts with fluorine (atomic weight 19) to form the gas uranium hexafluoride (UF_6). What is the ratio of the rms speed of the UF_6 gas molecules containing atoms of uranium 238 to the rms speed of the gas molecules containing atoms of uranium 235?

How to Solve It
• Determine the molar masses of UF_6 molecules containing each species of uranium atoms.

• The average translational kinetic energy per molecule is the same for both species. Equate the kinetic energies and solve for the ratio of the rms speeds.

11. The surface of the sun consists primarily of monatomic hydrogen gas at a temperature of 6000 K. Compare the rms speed of a hydrogen atom at the surface of the sun with the escape speed from the sun's surface. The sun's radius is 6.96×10^8 m and its mass is 1.99×10^{30} kg.

How to Solve It
• Find an expression for the escape velocity at the surface of the sun from Chapter 11.

• Find an expression for the rms speed of hydrogen atoms.

• The ratio of these speeds is significant in astrophysics. The sun consists mostly of hydrogen. Unless the rms speed is much less than the escape speed, the sun's gravity is not strong enough to hold it together.

12. Like the molecules of all monatomic gases, the molecules of neon have no rotational or vibrational kinetic energy, only translational. (*a*) What is the total kinetic energy of the molecules in 1 L of neon gas (atomic mass of 22) at 1 atm pressure? (*b*) If the gas is expanded at constant temperature to a volume of 2 L, by how much does the total kinetic energy change?

How to Solve It

• The average kinetic energy of a gas molecule is $\frac{3}{2}kT$.

• Use the ideal-gas law to calculate the number of neon molecules present.

• The total translational kinetic energy equals the number of molecules times the average translational kinetic energy per molecule.

• Do all the algebra first. See how things cancel.

• How will this change during an expansion to 2 L at constant temperature?

13. One way to compare the root mean square speed v_{rms} with the mean speed v_{av} of the molecules of a gas is to find the ratio of the two speeds. Find this ratio using the result

$$\int_0^\infty x^3 e^{-a^2 x^2}\, dx = \frac{1}{2a^4}$$

How to Solve It

• The rms speed of a gas molecule is $\sqrt{3kT/m}$.

• Find the average speed of a gas molecule using the Maxwell–Boltzmann speed distribution function.

• Take the ratio of the two speeds.

14. Compare the root mean square speed of a gas molecule with the most probable speed by finding their ratio.

How to Solve It

• The most probable speed is the speed for which the Maxwell–Boltzmann speed distribution function is a maximum. Using differentiation, find an expression for this speed.

• The rms speed of a gas molecule is $\sqrt{3kT/m}$.

• Take the ratio of the two speeds.

Solutions

1. The gas in a constant-volume gas thermometer is at low density, so the ideal gas law ($PV = nRT$) is a suitable equation of state. Because the number of moles and the volume of the gas do not change, the ratio of pressure to temperature is a constant. That is,

$$\frac{P}{T} = \frac{P_3}{T_3}$$

where P_3 and T_3 are the temperature and pressure of the gas at the triple point of water. We know that T_3 is 273 K and the problem states that P_3 is 88 torr.

(*a*) When $T = 310$ K,

$$P = \frac{P_3 T}{T_3} = \frac{(88\text{ torr})(310\text{ K})}{273\text{ K}} = \underline{99.9\text{ torr}}$$

(*b*) When $P = 70$ torr,

$$T = \frac{T_3 P}{P_3} = \frac{(273\text{ K})(70\text{ torr})}{88\text{ torr}} = \underline{217\text{ K}}$$

2. $-38.2°\text{F}$, $\underline{234\text{ K}}$

3. The pressure of the air in the tire is only a few atmospheres so the ideal gas law ($PV = nRT$) is a suitable equation of state. Because the number of moles and the volume of the confined air do not change, the ratio of pressure to temperature is a constant. That is,

$$\frac{P_2}{T_2} = \frac{P_1}{T_1}$$

In this equation, both the temperatures and the pressures must be absolute. In order to convert the temperatures to kelvins we will first convert them to degrees Celsius. When the Fahrenheit temperature is 52°,

$$t_C = \tfrac{5}{9}(t_F - 32°) = \tfrac{5}{9}(52° - 32°) = 11°\text{C}$$
$$T_1 = 11 + 273 = 284\text{ K}$$

In just the same way, 125°F corresponds to $T_2 = 325$ K. The gauge pressure of 75 lb/in² corresponds to an absolute pressure of

$$P_1 = 75\text{ lb/in}^2 + 14.7\text{ lb/in}^2 = 89.7\text{ lb/in}^2$$

Therefore

$$P_2 = \frac{P_1 T_2}{T_1} = \frac{(89.7\text{ lb/in}^2)(325\text{ K})}{284\text{ K}}$$
$$= 103\text{ lb/in}^2\text{ absolute} = \underline{88\text{ lb/in}^2\text{ gauge}}$$

You can't read a tire gauge to a precision of ± 0.1 lb/in².

4. (*a*) $\underline{162\text{ K}}$ (*b*) $\underline{67.4\text{ L}}$

5. From the ideal-gas law

$$PV = nRT$$

we obtain

$$\frac{n}{V} = \frac{P}{RT}$$

$$= \frac{10^{-21} \text{ atm}}{(0.0821 \text{ atm·L/mol·K})(3 \text{ K})} \approx 4.1 \times 10^{-21} \text{ mol/L}$$

Since there are Avogadro's number of atoms in 1 mol of a gas, the number of atoms per cubic centimeter is

$$N = nN_A$$

$$= (4.1 \times 10^{-21} \text{ mol/L})$$
$$\times (6.02 \times 10^{23} \text{ atoms/mol})$$

$$= (2500 \text{ atoms/L})(10^{-3} \text{ L/cm}^3)$$

$$= 2.5 \text{ atoms/cm}^3$$

6. $\underline{3.3 \times 10^8 \text{ molecules/cm}^3}$

7. From the ideal-gas law

$$PV = nRT$$

so

$$n = \frac{PV}{RT}$$

The molar mass

$$M = \frac{m}{n} = \frac{mRT}{PV}$$

$$= \frac{(55 \text{ g})(0.0821 \text{ atm·L/mol·K})(293 \text{ K})}{(10 \text{ atm})(3 \text{ L})}$$

$$= 44.0 \text{ g/mol}$$

This may be any gas with a molar mass of 44.0. I had <u>carbon dioxide</u>, CO_2 $(12 + 16 + 16 = 44)$, in mind when I wrote the problem, but it could just as well be <u>nitrous oxide</u>, N_2O $(14 + 14 + 16 = 44)$, or <u>propane</u>, C_3H_8.

8. $\underline{252 \text{ atm (or } 2.55 \times 10^7 \text{ Pa)}}$

9. From Appendix C we have that the atomic masses of hydrogen, oxygen, and argon are 1, 16, and 39.9, respectively. Thus, the molecular mass of water is 18 and the molecular mass of argon is 39.9.

Equating the average translational kinetic energies of the two species gives

$$\tfrac{1}{2} m_{H_2O} v_{rms,H_2O}^2 = \tfrac{1}{2} m_{Ar} v_{rms,Ar}^2$$

or

$$\frac{v_{rms,H_2O}}{v_{rms,Ar}} = \left(\frac{m_{Ar}}{m_{H_2O}}\right)^{1/2} = \sqrt{\frac{39.9}{18}} = 1.49$$

10. $\underline{0.996}$

11. The escape speed at the surface of the sun is given by

$$v_e = \sqrt{\frac{2GM_s}{R_s}}$$

The rms speed of monatomic hydrogen is related to the temperature T by the formula

$$v_{rms} = \sqrt{\frac{3RT}{M}}$$

where M is the molar mass of monatomic hydrogen and R is the universal gas constant. Then

$$\left(\frac{v_{rms}}{v_e}\right)^2 = \frac{3RTR_s}{2MGM_s}$$

$$= \frac{3(8.314 \text{ J/mol·K})(6 \times 10^3 \text{ K})(6.96 \times 10^8 \text{ m})}{2(0.001 \text{ kg/mol})(6.67 \times 10^{-11} \text{ N·m}^2/\text{kg}^2)(1.99 \times 10^{30} \text{ kg})}$$

$$= 3.92 \times 10^{-4}$$

so

$$\frac{v_{rms}}{v_e} \approx \underline{0.02}$$

Our result is that the escape speed is about 50 times the rms speed. This is in agreement with the fact that the sun's gravity is able to hold the sun together.

12. (a) $K = \tfrac{3}{2} PV = \underline{152 \text{ J}}$ (b) $\underline{\text{It does not change.}}$

13. The root mean square speed is given by $v_{rms} = (3kT/m)^{1/2}$. To find the v_{av} we use the Maxwell–Boltzmann speed distribution function

$$f(v) = \frac{4}{\sqrt{\pi}}\left(\frac{m}{2kT}\right)^{3/2} v^2 e^{-mv^2/2kT}$$

Using this distribution function to find the average we get

$$v_{av} = \int_0^\infty v f(v) \, dv$$

$$= \frac{4}{\sqrt{\pi}}\left(\frac{m}{2kT}\right)^{3/2} \int_0^\infty v^3 e^{-mv^2/2kT} \, dv$$

We are given that

$$\int_0^\infty x^3 e^{-a^2 x^2}\, dx = \frac{1}{2a^4}$$

so with $x = v$ and with $a^2 = m/2kT$ we have

$$v_{av} = \frac{4}{\sqrt{\pi}}\left(\frac{m}{2kT}\right)^{3/2} \frac{1}{2\left(\frac{m}{2kT}\right)^2}$$

$$= \frac{2}{\sqrt{\pi}}\left(\frac{m}{2kT}\right)^{-1/2} = \left(\frac{8kT}{\pi m}\right)^{1/2}$$

Taking the ratio of the two speeds gives

$$\frac{v_{rms}}{v_{av}} = \frac{\left(\dfrac{3kT}{m}\right)^{1/2}}{\left(\dfrac{8kT}{\pi m}\right)^{1/2}} = \sqrt{\frac{3\pi}{8}} = \underline{1.09}$$

Remark: The rms speed is about 9 percent larger than the mean speed.

14. $\dfrac{v_{rms}}{v_{most\ probable}} = \sqrt{3/2} = \underline{1.22}$

Remark: The rms speed is about 22 percent larger than the most probable speed.

Chapter 19

Heat and the First Law of Thermodynamics

I. Key Ideas

Heat is energy in transit from a warmer object to a colder object because of the temperature difference between them. When energy is transferred to or from an object as heat, a change in the internal energy (thermal energy) of the object occurs. Neither thermal energy nor mechanical energy is necessarily conserved. What is always conserved in an isolated system is the sum of the mechanical energy and the thermal energy.

19-1 Heat Capacity and Specific Heat The transfer of energy via heat is often accompanied by changes in temperature. The amount of heat Q needed to raise the temperature of an object by one degree is called its **heat capacity** C:

Heat capacity

$$Q = C \, \Delta T$$

The **specific heat** c is the heat capacity per unit mass of a substance. The **molar specific heat** c' of a substance is the heat capacity per mole:

Specific heat and molar heat capacity

$$C = mc = nc'$$

where m is the mass of the substance and n is the number of moles.

The specific heat of water is quite large compared to that of other ordinary materials. The specific heat of liquid water is nearly constant over a wide range of temperature, and the specific heats of other materials are often measured by comparison with water. This is done by heating a sample of a material to some known temperature, placing the sample in a water bath of known mass and temperature, and then measuring the final equilibrium temperature. If this system is isolated from its surroundings, the heat leaving the object equals the heat

entering the water and its container. This process is called *calorimetry* and the insulated water container is called a *calorimeter*.

The traditional unit of heat is the **calorie** (cal). It is approximately the amount of heat required to raise the temperature of 1 gram of water by 1 K. It is defined in terms of the joule:

Definition of calorie

$$1 \text{ cal} = 4.184 \text{ J}$$

The "calorie" used in measuring the nutritional value of food is actually a kilocalorie. The U.S. customary unit of heat is the **Btu** (British thermal unit). It was originally defined as the amount of energy needed to raise the temperature of one pound of water by one Fahrenheit degree. The Btu is now defined as

Definition of Btu

$$1 \text{ Btu} = 1.054 \text{ kJ}$$

19-2 Change of Phase and Latent Heat Under certain conditions, a substance will remain at the same temperature while energy is being transferred to or from it in the form of heat. This happens when the substance undergoes a **phase change** such as **fusion** (a change from the liquid phase to the solid phase). The other common phase changes are **vaporization** (a change from a liquid to a vapor or gas) and **sublimation** (a change from a solid directly into a gas). The heat required to melt a substance of mass m with no change in its temperature is

Latent heat of fusion

$$Q = mL_f$$

where L_f is the **latent heat of fusion** of the substance. For the melting of ice to water at a pressure of 1 atm, the latent heat of fusion is 333.5 kJ/kg = 79.7 kcal/kg. When the phase change is from liquid to gas, the heat required is

Latent heat of vaporization

$$Q = mL_v$$

where L_v is the **latent heat of vaporization**. The latent heat of vaporization for water at 1 atm is 2.26 MJ/kg = 540 kcal/kg.

19-3 Joule's Experiment and the First Law of Thermodynamics

When Joule measured the amount of mechanical work required to produce a given amount of thermal energy, he found it took 4.18 joules of work (or mechanical energy) to raise the temperature of 1 g of water 1 C°. This quantity (≈ 4.18 J/cal) is known as the **mechanical equivalent of heat**. This led to the first law of thermodynamics, which states that energy is conserved. Conservation of energy means that the sum of all types of energy in a system at an instant in time equals the sum at an earlier time, but that the distribution of the different types of energy, such as potential, kinetic, and thermal, may vary. If the sum varies, then energy has been added to or taken away from the system.

First law of thermodynamics

$$Q = \Delta U + W$$

The heat added to a system over a time interval equals the change in the internal energy of the system plus the work done by the system during the same time interval.

(For very small amounts of heat added, work done, or changes in internal energy, it is customary to express this as $dQ = dU + dW$.) If the work W done by the system on its surroundings is positive, this work represents energy transferred from the system to the surroundings; negative W represents energy transferred from the surroundings to the system. The heat Q is taken to be positive if it represents energy transferred from its surroundings to the system; and negative if it represents energy transferred from the system to its surroundings.

19-4 The Internal Energy of an Ideal Gas

At low densities, the total volume of the molecules of a gas, compared with the volume occupied by the gas, is negligible, and the forces the molecules exert on each other during the intervals between collisions are negligible. Gases at densities low enough to fulfill these conditions are called ideal gases. The average translational kinetic energy of the molecules of an ideal gas is proportional to the absolute temperature of the gas. If the internal energy U of a gas is only this translational kinetic energy, then

Internal energy of a monatomic ideal gas

$$U = \tfrac{3}{2}nRT$$

where n is the number of moles, R is the universal gas constant, and T is the absolute temperature. If the molecules have other types of energy in addition to translational energy, such as rotational energy, the internal energy will be greater than $\frac{3}{2}nRT$. According to the equipartition theorem, the average energy associated with any degree of freedom will be $\frac{1}{2}kT$ per molecule ($\frac{1}{2}RT$ per mole). *Note:* The internal energy of an ideal gas depends only on the temperature and not on the volume or pressure. (In a real gas the forces between molecules cannot always be neglected. The internal energy of such a gas includes the potential energy associated with these forces. This potential energy depends upon the intermolecular separation, which does depend on the pressure and volume of such a gas.)

To test whether or not the particles of real gases are noninteracting, Joule performed an experiment in which a gas confined in a rigid, thermally insulated container was allowed to expand into a second rigid, thermally insulated container that, prior to the expansion, was evacuated. This process is called a **free expansion**. In a free expansion the gas does no work on its surroundings and no heat is transferred to or from the gas, so the internal energy of the gas does not change. Joule found that if a gas initially at low density (with the molecules well separated) underwent a free expansion, any changes in its temperature were too small to observe. However, when a gas initially at high density (with the molecules in close proximity) underwent a free expansion, he observed a slight decrease in its temperature. These results demonstrate that for a gas at low density, the interactions between molecules are negligible; however, for a gas that is at a higher density, the attractive forces between molecules have a small effect. When a gas expands, the potential energy associated with the attractive forces between the molecules must increase. Since the internal energy of the gas is constant, this increase in potential energy is accompanied by a corresponding decrease in translational kinetic energy—a decrease that is evidenced by the observed slight temperature drop.

19-5 Work and the PV Diagram for a Gas

In the following discussion, the first law of thermodynamics is applied to gases. For simplicity this discussion will be restricted to gases at low densities that satisfy the ideal-gas law.

Processes such as expansions and compressions may occur so slowly that during the process the gas is never far from an equilibrium state. In this kind of process, called a **quasi-static process**, the gas moves through a series of equilibrium states. In practice it is possible to execute processes slowly enough to approximate quasi-static processes fairly well. If a confined gas is quasi-statically compressed, during the compression the pressure is the same throughout the gas. However, if the compression occurs very rapidly (violently), there will be pressure shock waves in the gas and the pressure of the confined gas will vary with location in the gas at any moment in time.

Consider a gas confined to a cylinder with a movable piston. When the piston slowly (quasi-statically) recedes a small distance dx, the gas expands. The work dW done by the gas on the piston is

$$dW = F\,dx = PA\,dx = P\,dV$$

where A is the area of the piston and $dV = A\,dx$ is the increase in volume of the gas. When the piston moves through a finite distance, from x_1 to x_2, the volume of the gas increases from V_1 to V_2.

The equation of state of a given sample of gas relates the three variables: pressure P, temperature T, and volume V. Thus, the state of the gas can be specified by any two of these variables, say P and V. Therefore, each point on a PV diagram specifies a particular state of the gas. Because specifying P and V specifies the temperature, a temperature T is associated with each point on the graph. The variables P, V, and T are called functions of state (or state functions) because they only depend upon the state of the gas. The internal energy U is also a state function. During an expansion or compression, the work done by the gas is

Work done by a gas

$$W = \int_{V_1}^{V_2} P\,dV = \text{area under } P\text{-versus-}V \text{ curve}$$

It is particularly easy to calculate the work done by a gas during an **isobaric expansion**, which is one that takes place at constant pressure. It is simply

Isobaric work

$$W_{\text{isobaric}} = P\,\Delta V$$

For an **isothermal expansion**—one that takes place at constant temperature—the work is

Isothermal work

$$W_{\text{isothermal}} = \int_{V_1}^{V_2} P\,dV = \int_{V_1}^{V_2} \frac{nRT}{V}\,dV$$
$$= nRT\int_{V_1}^{V_2} \frac{1}{V}\,dV = nRT \ln \frac{V_2}{V_1}$$

Because pressures are often given in atmospheres and volumes are often given in liters, it is convenient to have a conversion factor between liter-atmospheres and joules:

Liter-atmospheres

$$1 \text{ L·atm} = (10^{-3}\text{ m}^3)(1.013 \times 10^5\text{ Pa}) = 101.3\text{ J}$$

If a gas undergoes a **cyclic process**, by definition the final state of the gas is the same as its initial state. Thus, for a cyclic process the path on a PV curve is closed. The total work done by the gas during a cycle equals the area enclosed by the PV curve during the cycle.

19-6 Heat Capacities of Gases If a sample of a material is allowed to expand as it is heated at a constant pressure (as nearly all materials tend to do), more heat input is required for a given temperature increase than would be required for the same temperature increase if the volume of the material were held constant. This is because in expanding the sample does work by pushing back on its surroundings. This loss of energy to the surroundings is made up for by an increased input of heat.

For solids and liquids, which expand only slightly, the additional heat required for a given temperature increase is very small. Thus, the specific heat at constant pressure is approximately equal to the specific heat at constant volume. The specific heat at constant pressure is normally specified because solids and liquids generate enormous pressure if they are not allowed to expand, making it difficult to measure the specific heat at constant volume.

Unlike solids and liquids, the heat capacity at constant pressure for gases is substantially greater than the heat capacity at constant volume. At constant pressure, gases undergo significant fractional increases in volume, so an appreciable increase of heat is required for a given temperature change.

When heat Q is added to a gas confined to a constant volume, no work is done by the gas so all of the heat goes into increasing the internal energy U of the gas. That is,

$$dQ = dU = C_v\,dT$$

so

Heat capacity at constant volume

$$C_v = \frac{dU}{dT}$$

For an ideal gas U depends only on T. Thus $dU = C_v\, dT$ even if V and/or P are not constant. For an ideal gas $V = nRT/P$. Thus, for an expansion at constant pressure, $dV = (nR/P)\, dT$, so

$$dQ = dU + dW = dU + P\, dV = C_v\, dT + nR\, dT$$
$$= C_p\, dT$$

Hence,

Heat capacity at constant pressure

$$C_p = C_v + nR$$

The heat capacity at constant pressure is greater than that at constant volume because of the work done by the expanding gas. The heat capacity per mole c_p' of a gas at constant pressure exceeds the heat capacity per mole c_v' at constant volume by R.

If the translational kinetic energy of the molecules of a gas is the only form of internal energy U, then

$$U = \tfrac{3}{2}RT$$

and the heat capacity at constant volume C_v would be just $\tfrac{3}{2}nR$ (and the heat capacity per mole c_v' at constant volume is $\tfrac{3}{2}R$). Measurements show that this holds for monatomic gases but not for more complex molecules.

Heat Capacities and the Equipartition Theorem
The average translational kinetic energy per molecule of a monatomic gas is $\tfrac{3}{2}kT$ (recall that $R = N_A k$). There are three equivalent, independent directions of motion, so a kinetic energy $\tfrac{1}{2}kT$ associated with each direction contributes to this total. This division of the kinetic energy into three terms in the translational kinetic energy equation ($K = \tfrac{1}{2}mv_x^2 + \tfrac{1}{2}mv_y^2 + \tfrac{1}{2}mv_z^2$) is a special case of the **equipartition theorem**. Each coordinate, velocity component, angular velocity component, and so forth that appears in the expression for the energy of a molecule is called a **degree of freedom**. The equipartition theorem states that there is an average energy of $\tfrac{1}{2}kT$ per molecule associated with each degree of freedom. For gases with complex molecules, the rotational and vibration degrees of freedom must also be considered. Diatomic molecules are found to have molar heat capacities at constant volume equal to $\tfrac{5}{2}R$. This is understood by recognizing that in addition to the three degrees associated with its translational kinetic energy, a diatomic molecule has two degrees of freedom associated with its rotational kinetic energy. It

follows that the energy per molecule, energy per mole, and heat capacity at constant volume are, respectively, $\tfrac{5}{2}kT$, $\tfrac{5}{2}RT$, and $\tfrac{5}{2}nR$.

19-7 Heat Capacities of Solids Experimental measurements show the molar specific heats of most solids to be approximately $3R$. This result is called the **Dulong–Petit law**. The molecules in a solid are fixed in place, but each can vibrate in three dimensions. Solids thus have a total of six degrees of freedom (three for vibrational potential energy and three for vibrational kinetic energy), and the equipartition theorem predicts a heat capacity per mole of $3R$.

19-8 Failure of the Equipartition Theorem Although the equipartition theorem has been highly successful, it also has shortcomings. Ultimately it fails because at the molecular level Newtonian mechanics fails. To predict observed results successfully, the theory of quantum mechanics must be used.

19-9 The Quasi-static Adiabatic Expansion of a Gas An **adiabatic process** is one in which there is no net flow of heat into or out of the system. A **quasi-static process** is one that happens in infinitesimal steps with delays between steps allowing the system to reach equilibrium. If a gas undergoes a quasi-static adiabatic expansion (or compression), it can be shown that

Quasi-static adiabatic process

$$PV^\gamma = \text{constant} \quad \text{and} \quad TV^{\gamma-1} = \text{constant}$$

where $\gamma = C_p/C_v$ is the ratio of two heat capacities. The work done by a gas undergoing a quasi-static adiabatic expansion is

Quasi-static adiabatic work

$$W_{\text{adiabatic}} = -C_v\,\Delta T = \frac{P_1 V_1 - P_2 V_2}{\gamma - 1}$$

The speed of sound is given by $v = \sqrt{B/\rho}$ where ρ is the mass density and B the bulk modulus. For an isothermal process the bulk modulus equals the pressure P, but for an adiabatic process it equals γP. Using the adiabatic bulk modulus it can be shown that

Speed of sound

$$v = \sqrt{\frac{B_{\text{adiab}}}{\rho}} = \sqrt{\frac{\gamma RT}{M}}$$

where M is the molar mass. This result is in close agreement with measured values for the speed of sound.

II. Numbers and Key Equations

Numbers

$1 \text{ cal} = 4.184 \text{ J}$

$1 \text{ Btu} = 1.054 \text{ kJ}$

$1 \text{ L·atm} = (10^{-3} \text{ m}^3)(1.013 \times 10^5 \text{ Pa}) = 101.3 \text{ J}$

Latent heat of fusion for water

$L_f = 333.5 \text{ kJ/kg} = 79.7 \text{ kcal/kg}$

Latent heat of vaporization for water

$L_v = 2.26 \text{ MJ/kg} = 540 \text{ kcal/kg}$

Universal gas constant

$R = 8.31 \text{ J/mol·K} = 0.0821 \text{ L·atm/mol·K}$

Avogadro's number

$N_A = 6.02 \times 10^{23} \text{ molecules/mole}$

Boltzmann's constant

$$k = \frac{R}{N_A} = 1.38 \times 10^{-23} \text{ J/K}$$

Key Equations

Heat capacity

$$Q = C \, \Delta T$$

Specific heat and molar heat capacity

$$C = mc = nc'$$

Latent heat of fusion

$$Q = mL_f$$

Latent heat of vaporization

$$Q = mL_v$$

First law of thermodynamics

$$Q = \Delta U + W$$

Work done by a gas

$$W = \int_{V_1}^{V_2} P \, dV = \text{area under } P\text{-versus-}V \text{ curve}$$

Heat capacity

$$C = \frac{dQ}{dT}$$

Heat capacity at constant volume

$$C_v = \frac{dU}{dT}$$

Except for the Dulong–Petit law, the following formulas are restricted to ideal gases:

Translational kinetic energy of a gas

$$K = \tfrac{3}{2} nRT$$

Internal energy of a monotomic gas

$$U = \tfrac{3}{2} nRT$$

Internal energy of a diatomic gas

$$U = \tfrac{5}{2} nRT$$

Heat capacity per mole for a monatomic gas

$$c_v' = \tfrac{3}{2} R$$

Heat capacity per mole for a diatomic gas

$$c_v' = \tfrac{5}{2} R$$

Heat capacity per mole at constant pressure

$$c_p' = c_v' + R$$

Ratio of specific heats

$$\gamma = \frac{C_P}{C_v} = \frac{c_p'}{c_v'}$$

Quasi-static adiabatic (Q = 0) process

$$PV^\gamma = \text{constant} \quad \text{and} \quad TV^{\gamma-1} = \text{constant}$$

Isobaric (P = constant) work

$$W_{\text{isobaric}} = P \, \Delta V$$

Isothermal (T = constant) work

$$W_{\text{isothermal}} = \int_{V_1}^{V_2} P \, dV = nRT \ln \frac{V_2}{V_1}$$

Quasi-static adiabatic (Q = 0) work

$$W_{\text{adiabatic}} = C_v \, \Delta T = \frac{P_1 V_1 - P_2 V_2}{\gamma - 1}$$

Dulong–Petit law (for solids)

$$c' = 3R$$

III. Potential Pitfalls

Be very careful with signs in first-law problems, heat exchange problems, and such. It's easy to get them mixed up. The convention is that Q is positive when heat flows into the system and W is positive when work is done by the system on its surroundings.

In a free expansion the gas does no work. For any other type of expansion the gas does positive work. As a gas is compressed it does negative work.

For a solid or a liquid, the heat capacity specified in a table is almost always the heat capacity at a pressure of one atmosphere. The heat capacity at constant volume is slightly less than this, but the difference is so small it is usually negligible. This is because, for a given temperature increase, the work done by the solid or liquid in pushing back the atmosphere is small compared to the increase in its internal energy.

In doing first-law-of-thermodynamics problems, don't assume that a heat transfer and a temperature change always go together. A quasi-static adiabatic expansion is an example of a temperature change without heat transfer. An isothermal expansion is an example of a heat transfer without a temperature change. There can be a temperature change without any heat transfer, and there can be a heat transfer without any temperature change.

Remember that the heat capacities C_p and C_v refer to a specific sample of gas and are proportional to the quantity of gas. By contrast the molar heat capacities (heat capacities per mole) c_p' and c_v' are independent of the total amount of gas.

The internal energy of an ideal gas depends only on its temperature. For other systems the internal energy depends on additional conditions, such as the density of the material.

When an ideal gas undergoes some process and thus changes state, the ideal-gas law may be used to relate initial and final states. However, this equation does not specify whether heat flowed into or from the gas in the process. The amount of heat exchanged by the gas depends on how the process was carried out.

IV. True and False and Responses

True and False

_____ **1.** The only way in which the internal energy of a system can increase is for heat to be added to the system.

_____ **2.** The heat capacity of a material is its specific heat per unit mass.

_____ **3.** A quasi-static process is one in which no work is done.

_____ **4.** The specific heat commonly given for solids and liquids is the specific heat at constant pressure.

_____ **5.** The internal energy of an ideal gas depends only on the temperature of the gas.

_____ **6.** In an isothermal process, the temperature of the system remains constant.

_____ **7.** In an adiabatic process, the internal energy of the system remains constant.

Responses

1. False. Doing work on a system can also cause an increase in its internal energy.

2. False. It is the other way around. The specific heat is the heat capacity per unit mass.

3. False. A quasi-static process is one that occurs sufficiently slowly for the system to progress through a sequence of equilibrium states.

4. True. For solids and liquids the difference between the specific heat at constant pressure and the specific heat at constant volume is so small that it is usually considered negligible.

5. True

6. True

7. False. In an adiabatic process, no heat is added to or removed from the system.

V. Questions and Answers

Questions

1. Can a system absorb heat without its temperature increasing? Can a system absorb heat without its internal energy increasing?

2. Can the temperature of 2 L of water be increased from 20 to 30°C without any work being done by the water? Explain.

3. For solids and liquids, we usually don't distinguish between specific heats at constant volume and specific heats at constant pressure. Why not? If we did want to make this distinction, which one would be larger? Which specific heat are you most likely to find tabulated?

4. An ideal gas expands at constant temperature and does work on a piston. Does the internal energy of the gas decrease? If not, where does the energy to do the work come from?

5. Give an example of a process in which the temperature of a system is increased without any heat input or output.

6. An ideal gas expands slowly to twice its initial volume (*a*) at constant pressure and (*b*) at constant temperature. In which case does it do more work on its surroundings? Why?

7. Consider nitrogen (N_2) and helium (He) gases. For which gas is the internal energy per mole greater at a given temperature? For which gas is the internal energy per gram greater at a given temperature?

8. The ratio of molar heat capacities ($\gamma = c_p'/c_v'$) for a monatomic ideal gas is $5/3 = 1.67$; for a diatomic ideal gas it is $7/5 = 1.4$. Would you expect the ratio to be higher or lower than 1.4 for an ideal gas having three or more atoms per molecule?

Answers

1. Consider a process in which heat is added to a system and the system does work on its surroundings. According to the first law of thermodynamics, the internal energy of the system will increase if the heat added to the system is greater than the work done by the system, the internal energy will remain the same if the heat added to the system is equal to the work done by the system, and the internal energy will decrease if the heat added to the system is less than the work done by the system. Thus, it is possible for a system to absorb heat without its temperature and its internal energy increasing, if the work done by the system equals or is greater than the heat it absorbs.

   Even without doing work, a system can absorb energy without its temperature increasing—for example, when it is melting or vaporizing.

2. If the volume of the water is kept the same during the temperature increase (which is very difficult to do, by the way), then the water does no work.

3. For any material that expands when its temperature is increased, c_p is greater than c_v, just as for a gas. That is because, in an expansion at constant pressure, the material does work in pushing back its surroundings. We normally pay no attention to the distinction because the change in volume is very small. The value ordinarily measured and tabulated is c_p.

4. The internal energy of an ideal gas depends only upon its temperature. Thus in any isothermal process the internal energy of the gas remains constant. When a gas expands, it does work on its surroundings. According to the first law of thermodynamics, when the gas does work on its surroundings, its internal energy will decrease by an amount equal to the work done, unless energy is transferred to the gas via

heat. To keep the temperature—and thus the internal energy—constant, the heat added to the gas during expansion must equal the work done by the gas.

5. Consider an ideal gas, confined in a cylinder with insulated walls, being compressed by a piston. The piston does work on the gas, which results in an increase in the internal energy of the gas. The internal energy of the gas is proportional to its temperature, so the increase in internal energy will result in an increase in temperature.

6. An ideal gas is initially at pressure P_0 and volume V_0. If the gas expands to a volume of $2V_0$ at constant pressure, the work it does on its surroundings is $P_0 \Delta V = P_0 V_0$. When a gas expands at constant temperature, the pressure varies inversely with the volume, so the pressure steadily drops as it expands. Thus, during the isothermal expansion the average pressure is always less than P_0, so the work done by the gas is less than $P_0 V_0$.

7. The internal energies per mole of all monatomic and diatomic gases are $1.5RT$ and $2.5RT$, respectively. Helium is a monatomic gas and nitrogen is diatomic. Thus at a given temperature the internal energy per mole for nitrogen is greater than it is for helium. The internal energy per gram of a substance equals the internal energy per mole divided by the molar mass, and the molar masses are 28 g/mol for nitrogen and 4 g/mol for helium. Thus, the internal energy per gram is $(2.5/28)RT = 0.089RT$ for nitrogen and $(1.5/4)RT = 0.375RT$ for helium. The internal energy per unit mass for helium is greater than it is for nitrogen.

8. The ratio of molar heat capacities is

$$\gamma = \frac{c_p'}{c_v'} = \frac{c_v' + R}{c_v'} = 1 + \frac{R}{c_v'}$$

It is smaller when the number of degrees of freedom is larger because c_v' is proportional to the number of degrees of freedom. A gas whose molecules have three or more atoms can be expected to have more degrees of freedom than a diatomic gas. Thus, for a gas with three or more atoms per molecule, the ratio of molar heat capacities is expected to be lower than it is for either monatomic or diatomic gases. Measurements show that many polyatomic gases have a ratio around $\gamma = 1.35$.

VI. Problems and Solutions

Problems

1. A piece of iron (specific heat 0.431 kJ/kg·K) of mass 80 g at a temperature of 98°C is dropped into an insulated vessel containing 120 g of water at 20°C. At what final temperature does the system come to equilibrium?

How to Solve It
• If the heat losses to the container and the surroundings are negligible, the heat lost by the iron is equal to that gained by the water.

• The heat lost (or gained) is equal to the mass times the specific heat times the temperature change.

• Solve for the equilibrium temperature.

2. The specific heat of copper is 0.386 kJ/kg·K. If 180 g of copper at 200°C is dropped into an insulated container containing 280 g of water at 20°C, what is the final equilibrium temperature of the copper and water?

How to Solve It
• Follow the approach outlined for Problem 1.

• This approach neglects any boiling of the water. When this experiment is actually done some water boils when the 200°C surface of the copper comes in contact with it. Boiling continues until the surface temperature of the copper falls below 100°C. Under these circumstances the equilibrium temperature of the copper and the remaining water would be somewhat lower than the temperature calculated without considering the boiling.

3. A 160-g mass of a certain metal at an initial temperature of 88°C is dropped into 140 g of water in an insulated container. The water is initially at 10°C. The system finally comes to equilibrium at 18.4°C. (*a*) If heat losses to the container and the surroundings are negligible, what is the specific heat of the metal? (*b*) If the Dulong–Petit law holds, what is the molar mass of the metal?

How to Solve It
• If the heat losses to the container and the surroundings are negligible, the heat lost by the metal object is equal to that gained by the water.

• The heat lost (or gained) is equal to the mass times the specific heat times the temperature change.

• The specific heat times the molar mass equals the heat capacity per mole.

• Equate the heat capacity per mole with $3R$.

• Knowing the molar mass, use Appendix C on page AP-6 of the text to find out what the metal is.

4. Imagine that you want to take a warm bath, but there's no hot water. You draw 40 kg of tap water at 18°C in the bathtub and heat water on your stove to

warm the bath water up. If you heat the water to 100°C in a 2-L saucepan, how many panfuls must you add to the bath to raise its temperature to 40°C?

How to Solve It
• If the heat losses to the bathtub and the surroundings are negligible, the heat lost by the heated water is equal to that gained by the cold bath water.

• The heat lost (or gained) is equal to the mass times the specific heat times the temperature change.

• Knowing the equilibrium temperature, solve for the mass of the heated water and use the known density of water to determine how many liters of heated water is required. How many panfuls is this?

5. A copper bar of mass 2.5 kg at an initial temperature of 66°C is dropped into an insulated vessel containing 400 g of water and 70 g of ice at 0°C. The specific heat of copper can be found in Table 19-1 on page 567 of the text. At what final temperature does the system come to thermal equilibrium?

How to Solve It
• Before the temperature of the ice–water mixture will change, all the ice must melt. How much heat does this require?

• By how much does the loss of the heat needed to melt the ice reduce the temperature of the copper? What would it mean if this result meant that the final temperature of the copper were below 0°C?

• Once the ice is melted, if the temperature of the copper is still above 0°C, the copper and 470 g of water will come to thermal equilibrium.

6. A cup contains 240 g of fresh-brewed coffee at a temperature of 97°C. If heat losses to the surroundings are negligible, to what final temperature is the coffee cooled if you drop an 18-g ice cube into it?

How to Solve It
• The specific heat of coffee is the same as that of water.

• How much heat is required to melt 18 g of ice?

• This much heat must be extracted from the hot coffee. Then, when the ice is melted, there is 18 g of water at 0°C to come to thermal equilibrium with the coffee.

7. A piece of aluminum (specific heat 0.90 kJ/kg·K) of mass 135 g initially at 20°C is placed in a large container of liquid nitrogen at 77 K (its normal boiling point). If the latent heat of vaporization of nitrogen is 199 kJ/kg, what mass of nitrogen is vaporized in cooling the aluminum to 77 K?

How to Solve It

• Since the amount of liquid nitrogen available is large, the aluminum will be cooled all the way to 77 K.

• The heat extracted from the aluminum is absorbed by the liquid nitrogen. Equate an expression for the heat absorbed by the nitrogen with an expression for the heat loss by the aluminum. Solve for the mass of nitrogen vaporized.

8. An abundance of steam at 100°C is passed into an insulated flask containing 200 g of ice at −25°C. Neglecting any heat loss from the flask, what mass of liquid water at 100°C will finally be present in the flask? The specific heats of ice and liquid water can be found in Table 19-1 on page 567 of the text. The latent heats of fusion and vaporization of water can be found in Table 19-2 on page 570 of the text.

How to Solve It

• The temperature of the 200 g of ice will first be raised to the melting point, then the ice will be melted, and then the temperature will be raised to 100°C.

• The heat absorbed by the ice equals the heat extracted from the steam that condenses. Obtain expressions for each of these heats and equate the heat gained by the ice to the heat lost by the condensed steam.

• Solve for the mass of the condensed steam.

9. Consider a 1.0-mol block of ice as it slowly warms from a temperature of −10°C to −3°C. As it warms it expands 21 mm³. The molar mass of ice is 18 g/mole. (*a*) Calculate the work done by the ice as it expands the 21.0 mm³ at atmospheric pressure. What does it do this work on? (*b*) How much heat is transferred to the block during this process? (*c*) What is the increase in the block's internal energy?

How to Solve It

• The work at constant pressure is the product of the pressure (1 atm) and the change in volume.

• The values for the specific heats and molar heat capacities (molar specific heats) of solids and liquids in Table 19-1 of the text are for processes that take place at a pressure of 1 atm.

• Use the first law of thermodynamics to find the change in internal energy.

10. A sample of gas initially occupies a volume of 15 L at 20°C and a pressure of 240 kPa. It is compressed at constant pressure to a volume of 6 L. How much work is done by the gas in the process?

How to Solve It

• In an isobaric (constant pressure) process, the work that the gas does is the product of the pressure and the change in volume.

11. In the *PV* diagram for a monatomic ideal gas, shown in Figure 19-1, path A is an isothermal expansion and path B is an expansion at constant pressure followed by cooling at constant volume. For each process, calculate the heat input, the work done, and the change in internal energy for 1 mol of the gas.

Figure 19-1

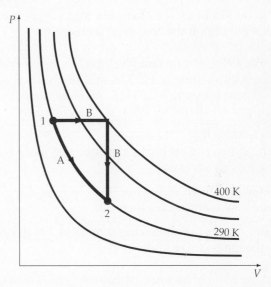

How to Solve It

• Use the formula for the work done in an isothermal expansion of an ideal gas to calculate the work done on path A. The initial and final volumes that appear in this formula are not given, but their ratio can be determined from the temperatures given on the *PV* diagram.

• The internal energy of an ideal gas is a function only of temperature. Thus, on path A the internal energy of the gas does not change. In accordance with the first law, the heat input equals the work done.

• On the constant-pressure leg of path B the work equals the product of the pressure and the change in volume. Use the ideal-gas law to obtain an expression for this work in terms of temperatures given on the *PV* diagram.

• On the constant-volume leg of path B the work is zero. The initial and final states for path B are the same as they are for path A; thus, the changes in internal energies are the same for both paths.

• The heat input for path B can be determined using the first law of thermodynamics.

12. One-half mole of nitrogen gas is heated from room temperature (20°C) and a pressure of 1 atm to a final temperature of 120°C. (*a*) How much heat must be supplied if the volume is kept constant while the gas is heated? (*b*) How much heat must be supplied if the heating is at constant pressure? (*c*) By how much is the internal energy changed?

How to Solve It
• Remember that nitrogen is a diatomic gas.

• You can find the molar heat capacities of nitrogen at constant volume and constant pressure in Table 19-3 on page 580 of the text.

• For an ideal gas, the change in internal energy depends only upon the change in temperature.

13. For 195 g of a certain ideal gas, the heat capacity at constant volume is 145 J/K and the heat capacity at constant pressure is 203 J/K. (*a*) How many moles of the gas are there? (*b*) What is its molar mass?

How to Solve It
• The difference in heat capacities per mole is the same for any ideal gas.

• From the given difference, therefore, you can calculate how many moles there are.

• That number of moles has a mass of 195 g; calculate the mass of one mole.

14. One mole of an ideal monatomic gas is heated at constant volume from 273 K to 500 K. (*a*) Find the heat added to the gas, the work done by it, and the change in its internal energy. (*b*) Repeat the calculations if the gas is heated at constant pressure.

How to Solve It
• For an ideal monotomic gas $c'_v = \frac{3}{2}R$ and $c'_p = c'_v + R$. Use these to calculate the heat inputs.

15. Two moles of nitrogen, initially at a temperature of 293 K, undergo a quasi-static, adiabatic expansion from a pressure of 5 atm to a pressure of 1 atm. Find the work done by the gas.

How to Solve It
• For any process involving an ideal gas, the change in internal energy depends only upon the change in temperature of the gas. For an adiabatic process, the change in internal energy plus the work done by the gas equals zero. Obtain an expression relating the work done with the change in temperature.

• For a quasi-static abiabatic expansion of an ideal gas, PV^γ is constant. Use this together with the ideal-gas law to find the final temperature of the gas.

• Solve for the work.

16. Ten grams of argon at a pressure of 1 atm is placed in a thermally insulated, flexible 4-L container. The container and the argon are slowly lowered into the ocean until the volume occupied by the argon is 1 L. (*a*) What is the pressure of the gas after being compressed? (*b*) How much work is done by the gas as it is compressed from 4 L to 1 L? (*c*) What is the change in internal energy of the gas?

How to Solve It
• Argon is a monatomic gas, so $C_v = \frac{3}{2}nR$, $C_p = C_v + nR$, and $\gamma = C_p/C_v$.

• The number of moles of argon can be obtained from the given mass and the information in Appendix C on page AP-6 of the text.

• For a quasi-static adiabatic compression of an ideal gas, PV^γ is constant. Use this to find the final pressure of the gas.

• There is a formula for the work done by the gas in a quasi-static adiabatic expansion (or compression).

• For any adiabatic process, the change in the gas's internal energy plus the work done by the gas equals zero.

Solutions

1. The heat Q_{out} that leaves the iron object is equal to the heat Q_{in} received by the water. Thus, using the subscript $()_o$ to refer to the object, we have

$$Q_{in} = Q_{out}$$
$$m_w c_w \, \Delta T_w = m_o c_o \, |\Delta T_o|$$
$$m_w c_w (T_f - T_{iw}) = m_o c_o (T_{io} - T_f)$$

Solving for the final temperature T_f we obtain

$$T_f = \frac{m_o c_o T_{io} + m_w c_w T_{iw}}{m_o c_o + m_w c_w}$$

$$= \frac{(0.0345 \text{ kJ/K})(371 \text{ K}) + (0.502 \text{ kJ/K})(293 \text{ K})}{0.034 \text{ kJ/K} + 0.502 \text{ kJ/K}}$$

$$= 298 \text{ K} = \underline{25°C}$$

2. $\underline{30.1°C}$

3. (*a*) The heat lost by the metal object is equal to the heat gained by the water. Thus,

$$Q_{gained} = Q_{lost}$$
$$m_w c_w \, \Delta T_w = m_o c_o \, |\Delta T_o|$$

or

$$c_o = \frac{m_w c_w \, \Delta T_w}{m_o \, |\Delta T_o|}$$

$$= \frac{(0.14 \text{ kg})(4.18 \text{ kJ/kg·K})(18.4°C - 10°C)}{(0.16 \text{ kg})(88°C - 18.4°C)}$$

$$= 0.441 \text{ kJ/kg·K}$$

(b) The heat capacity per mole equals the product of the specific heat and the molar mass M. The Dulong–Petit law states that the heat capacity per mole C_{mo} for a solid is approximately equal to $3R$. Thus,

$$C_{mo} = c_o M$$

or

$$M = \frac{C_{mo}}{c_o} = \frac{3R}{c_o} = \frac{3(8.31 \text{ J/mol·K})}{0.441 \text{ kJ/kg·K}} = \underline{56.5 \text{ g/mol}}$$

The unknown metal is probably iron.

4. $\underline{14.7 \text{ kg}}$ of hot water or $\underline{7.33 \text{ panfuls}}$

5. The heat required to melt 70 g of ice is

$$Q = mL_f = (0.07 \text{ kg})(333.5 \text{ kJ/kg}) = 23.3 \text{ kJ}$$

Next, we determine how much the temperature of the 2.5-kg copper bar will decrease if it gives up this much heat:

$$Q = mc \, \Delta T$$

or

$$\Delta T = \frac{Q}{mc} = \frac{-23.3 \text{ kJ}}{(2.5 \text{ kg})(0.386 \text{ kJ/kg·K})}$$

$$= -24.1 \text{ K} = -24.1 \text{ C°}$$

Thus, when all the ice melts, the temperature of the copper bar will be 66°C $-$ 24.1 C° $= \underline{41.9°C}$. (If there had been more ice or less copper, this result might have come out negative. This would mean that there was not enough heat available from the copper to melt all the ice, and the final temperature would have actually been 0°C.) The copper and the water come to thermal equilibrium at the final temperature T_f. To determine this temperature we equate the heat gained by the 470 g of liquid water to the additional heat lost by the copper bar:

$$Q_{gained} = Q_{lost}$$

$$m_w c_w \, \Delta T_w = m_{Cu} c_{Cu} \, |\Delta T_{Cu}|$$

$$m_w c_w (T_f - T_{iw}) = m_{Cu} c_{Cu} (T_{iCu} - T_f)$$

Solving for T_f we have

$$T_f = \frac{m_{Cu} c_{Cu} T_{iCu} + m_w c_w T_{iw}}{m_{Cu} c_{Cu} + m_w c_w}$$

$$= \frac{40.4 \text{ kJ·°C/K} + 0 \text{ kJ·°C/K}}{0.965 \text{ kJ/K} + 1.96 \text{ kJ/K}}$$

$$= \underline{13.8°C}$$

6. $\underline{84.7°C}$

7. The heat gained by the liquid nitrogen is equal to the heat extracted from the piece of aluminum in cooling it to 77 K:

$$Q_{gained} = Q_{lost}$$

$$mL_v = m_{Al} c_{Al} \, |\Delta T_{Al}|$$

so

$$m = \frac{m_{Al} c_{Al} \, |\Delta T_{Al}|}{L_v}$$

$$= \frac{(0.135 \text{ kg})(0.900 \text{ kJ/kg·K})(293 \text{ K} - 77 \text{ K})}{199 \text{ kJ/kg}}$$

$$= \underline{0.132 \text{ kg}}$$

8. $\underline{271 \text{ g of water}}$

9. (a) As it expands, the ice pushes back the atmosphere and does positive work on it. This work is done at constant pressure so

$$W = \int P \, dV = P \int dV$$

$$= P \, \Delta V = (1 \text{ atm})(21 \text{ mm}^3)$$

$$= (21 \text{ atm·mm}^3) \frac{1.01 \times 10^5 \text{ Pa/atm}}{(10^3 \text{ mm/m})^3}$$

$$= \underline{2.12 \times 10^{-3} \text{ J}}$$

(b) The process occurs at constant pressure so

$$Q = nc_p' \, \Delta T = (1.0 \text{ mol})(36.9 \text{ J/K·mol})(+7 \text{ C°})$$

$$= \underline{258 \text{ J}}$$

(c) The first law of thermodynamics is $Q = \Delta U + W$ so

$$\Delta U = Q - W = (258 \text{ J}) - (2.12 \times 10^{-3} \text{ J}) = \underline{258 \text{ J}}$$

10. $\underline{-2160 \text{ J}}$

11. Path A is a quasi-static isothermal expansion. The work done for this process is

$$W_A = W_{isothermal} = nRT \ln \frac{V_2}{V_1}$$

In order to evaluate this expression we must obtain a value for the ratio V_2/V_1. This can be done by realizing that the ratio of the final volume to the initial volume on the constant pressure segment of path B is also V_2/V_1, and for an expansion at constant pressure the volume is directly proportional to the temperature. Thus

$$\frac{V_2}{V_1} = \frac{T_{high}}{T_{low}}$$

so the work on path A is

$$W_A = nRT_{low} \ln \frac{T_{high}}{T_{low}}$$

$$= (1 \text{ mol})(8.31 \text{ J/mol·K})(290 \text{ K}) \ln \frac{400 \text{ K}}{290 \text{ K}}$$

$$= \underline{775 \text{ J}}$$

In accord with the first law of thermodynamics, the heat added to the gas equals the change in the gas's internal energy plus the work done by the gas. The *change in the gas's internal energy is zero* because the internal energy of an ideal gas depends only on temperature and the temperature does not change. Therefore the heat added to the gas is 775 J.

Path B consists of an isobaric (constant pressure) expansion followed by cooling at constant volume. No work is done by the gas during the constant-volume cooling, so the total work done by the gas on path B is done during the isobaric expansion. Thus

$$W_B = W_{isobaric} = P_1 \Delta V = P_1(V_2 - V_1)$$
$$= P_1 V_2 - P_1 V_1$$

Initially the pressure, volume, and temperature are P_1, V_1, and T_{low}, and after the expansion they are P_1, V_2, and T_{high}, so $P_1 V_1 = nRT_{low}$ and $P_1 V_2 = nRT_{high}$. Substituting these expressions into our equation we obtain

$$W_{isobaric} = nR(T_{high} - T_{low})$$

$$= (1 \text{ mol})(8.31 \text{ J/mol·K})(400 \text{ K} - 290 \text{ K})$$

$$= \underline{914 \text{ J}}$$

The *change in the gas's internal energy for path B is zero* for the same reason it is zero on path A. As long as the change in temperature is zero the final internal energy is equal to the initial internal energy. On path B the heat added to the gas equals the change in the gas's internal energy plus the work done by the gas. Therefore the heat added to the gas is 914 J.

12. (a) $\underline{1040 \text{ J}}$ (b) $\underline{1450 \text{ J}}$ (c) $\underline{1040 \text{ J}}$

13. (a) For any ideal gas, the difference in the molar heat capacities at constant pressure and constant volume is $R = 8.31$ J/mol·K, so

$$C_p - C_v = nR$$

or

$$n = \frac{C_p - C_v}{R} = \frac{203 \text{ J/K} - 145 \text{ J/K}}{8.31 \text{ J/mol·K}} = \underline{6.98 \text{ mol}}$$

Since the mass of the sample is 195 g, the molar mass of the gas is

$$M = \frac{m}{n} = \frac{195 \text{ g}}{6.98 \text{ mol}} = \underline{27.9 \text{ g/mol}}$$

The gas might be CO or N_2 or C_2H_4.

14. (a) $W = \underline{0}$, $Q = \Delta U = \underline{2830 \text{ J}}$
(b) $Q = \underline{4720 \text{ J}}$, $W = \underline{1890 \text{ J}}$, $\Delta U = \underline{2830 \text{ J}}$

15. For an ideal gas, the change in internal energy of the gas during *any* process is $C_v \Delta T$, where ΔT is the change in temperature of the gas. Thus, in accordance with the first law of thermodynamics, the work done by a gas during an adiabatic process is

$$W_{adiabatic} = -\Delta U = -C_v \Delta T$$

For a diatomic gas like nitrogen, $C_v = \frac{5}{2}nR$.

The number of moles, the initial temperature and pressure, and the final pressure are given in the statement of the problem. Before we can calculate the work we first must obtain an expression for the final temperature. For a quasi-static, adiabatic expansion, PV^γ is constant. Thus,

$$P_1 V_1^\gamma = P_2 V_2^\gamma$$
$$P_1 \left(\frac{nRT_1}{P_1}\right)^\gamma = P_2 \left(\frac{nRT_2}{P_2}\right)^\gamma$$
$$P_1^{(1-\gamma)/\gamma} T_1 = P_2^{(1-\gamma)/\gamma} T_2$$

so

$$T_2 = T_1 \left(\frac{P_1}{P_2}\right)^{(1-\gamma)/\gamma}$$

The work done by the gas is

$$W_{adiabatic} = -C_v \Delta T = -C_v(T_2 - T_1)$$

$$= -\frac{5}{2}nR \left[T_1 \left(\frac{P_1}{P_2}\right)^{(1-\gamma)/\gamma} - T_1 \right]$$

$$= \frac{5}{2}nRT_1 \left[1 - \left(\frac{P_1}{P_2}\right)^{(1-\gamma)/\gamma} \right]$$

For an ideal gas $C_p = C_v + nR$, so

$$\gamma = \frac{C_p}{C_v} = \frac{C_v + nR}{C_v} = \frac{(7/2)nR}{(5/2)nR} = \frac{7}{5}$$

$$\frac{1-\gamma}{\gamma} = \frac{1-7/5}{7/5} = -\frac{2}{7}$$

and

$$W_{\text{adiabatic}} = \frac{5}{2}(2 \text{ mol})(8.31 \text{ J/mol·K})(293 \text{ K})$$

$$\times \left[1 - \left(\frac{5 \text{ atm}}{1 \text{ atm}}\right)^{-2/7}\right]$$

$$= \underline{4490 \text{ J}}$$

16. (a) $\underline{10.1 \text{ atm}}$ (b) $\underline{-9.12 \text{ L·atm}}$
(c) $\underline{-9.12 \text{ L·atm}}$

Chapter 20

The Second Law of Thermodynamics

I. Key Ideas

Recall from Chapter 19 that energy is transferred as either heat or work. Heat is energy in transit due to a temperature difference whereas work is energy in transit due to something other than a temperature difference (the displacement of a piston, for example).

By the first law of thermodynamics, total energy is conserved. In a given case, however, not all the energy is available for use. Consider a block sliding to rest along a horizontal tabletop where friction is not negligible. In this process, the block and table become warmer as mechanical energy, the initial kinetic energy of the block, is transformed into internal energy of the block and table. This is an example of an irreversible process; that is, the reverse process never occurs. Internal energy of the block and table is never spontaneously converted into kinetic energy to send the block sliding along the table while the table and block cool. Another example of an irreversible process occurs when a hot body is placed in thermal contact with a cold body. Heat will flow from the hot body to the cold body until they are at the same temperature. However, the reverse never occurs. Two bodies at the same temperature that are in thermal contact with each other remain at the same temperature. That is, heat does not flow from one to the other, making one colder and the other warmer.

The second law of thermodynamics states that processes of this type do not occur. One statement of the **second law of thermodynamics** is

It is not possible to remove a given amount of energy from a system as heat and use it all to do work without some other change.

The conversion of work and mechanical energy into heat is an irreversible process, and the second law is fundamentally a statement about the direction of irreversible processes.

20-1   Heat Engines and the Second Law of Thermodynamics   Historically, the second law of thermodynamics was first formulated in terms of the efficiency of heat engines. A **heat engine** is a device that operates in a cyclic process. It extracts heat from a high-temperature heat reservoir, converts some of the heat into mechanical energy, and transfers the mechanical energy, as work, to some external agent. In accordance with the first law of thermodynamics, the heat that is not converted into work is rejected to a low-temperature reservoir. A **heat reservoir** is a system with a heat capacity so large that it can absorb or give off heat with no appreciable change in its temperature. The surrounding atmosphere, lakes, or the ocean often act as practical heat reservoirs. A working substance in the engine absorbs heat Q_h from a high-temperature reservoir, converts some of the heat into work W done by the engine on an external agent, and rejects the remaining heat $|Q_c|$ to a low-temperature reservoir. *Note:* Our **sign convention** is that when the system (heat engine) absorbs heat, the sign of the heat term is positive, and when the system rejects heat, the sign of the heat term is negative. Thus, for a heat engine, Q_h is positive and Q_c is negative. To keep track of signs, when a heat term Q is negative it is expressed as $-|Q|$.

The **efficiency** ε of the engine is the ratio of the work done by the heat engine to the heat absorbed by the engine from the high-temperature reservoir:

Efficiency of a heat engine

$$\varepsilon = \frac{W}{Q_h} = \frac{Q_h - |Q_c|}{Q_h} = 1 - \frac{|Q_c|}{Q_h}$$

In practical engines, the efficiency may be as high as 50% or so.

As a matter of convenience, several alternative statements of the second law of thermodynamics are in common use. These statements are completely equivalent. The Kelvin–Planck or **heat-engine statement of the second law of thermodynamics** is

> It is impossible for a heat engine working in a cycle to produce no other effect than that of extracting energy as heat from a reservoir and performing an equivalent amount of work.

An engine that is 100% efficient would absorb heat and convert it all to work with no heat being rejected. This is impossible according to the second law.

20-2   Refrigerators and the Second Law of Thermodynamics   A **refrigerator** is simply a heat engine run backwards. It operates in a cyclic process, in which work W is done by an external agent on the refrigerator, heat Q_c is absorbed from a low-temperature reservoir by the working substance, and heat $|Q_h|$ is rejected to a high-temperature reservoir. For an integral number of cycles, $W + Q_c = |Q_h|$. The coefficient of performance COP of a refrigerator is the ratio of the heat extracted by the refrigerator from the low-temperature reservoir to the work done on the refrigerator by the external agent. That is,

Coefficient of performance for a refrigerator

$$\text{COP} = \frac{Q_c}{W}$$

where both Q_c and W, the work done by the external agent on the refrigerator, are positive quantities. Practical refrigerators have COPs as high as 5 or 6 or more.

The Clausius or **refrigerator statement of the second law of thermodynamics** is

> It is impossible for a refrigerator working in a cycle to produce no other effect than the transfer of energy as heat from a cold object to a hot object.

An ideal refrigerator would require zero work input and would transfer heat from a lower- to a higher-temperature reservoir without any other change to the surroundings. This is impossible according to the second law.

20-3   Equivalence of the Heat-Engine and Refrigerator Statements   Although appearing to be very different, the heat-engine and refrigerator statements of the second law are equivalent in that, if we could make a device that violated one, it would also violate the other. This can be seen from simple heat flow diagrams and is independent of any specific engine design features.

20-4   The Carnot Engine   We have seen that, according to the second law of thermodynamics, it is impossible for a heat engine working between two heat reservoirs to be 100% efficient. What, then, is the maximum possible efficiency for such an engine? This question was answered by Sadi Carnot. Carnot found that all reversible engines working between two heat reservoirs have the same efficiency, and that no engine could have a greater efficiency than that of a reversible engine. This is known as **Carnot's theorem**:

> No engine working between two given heat reservoirs can be more efficient than a reversible engine working between those reservoirs.

Any reversible engine working between two heat reservoirs is called a **Carnot engine**.

The conditions necessary for a process to be reversible are as follows:

1. No mechanical energy can be transformed into thermal energy by friction, viscous forces, or other dissipative forces.
2. There can be no heat conduction due to a finite temperature difference.
3. The process must be quasi-static so that the system is always in an equilibrium state (or infinitesimally near an equilibrium state).

Any process that violates any of the above conditions is irreversible. Most processes in nature are irreversible. Nevertheless, one can come very close to a reversible process, and the concept is very important.

A Carnot cycle consists of a sequence of four stages:

(1) a reversible isothermal expansion,
(2) a reversible adiabatic expansion,
(3) a reversible isothermal compression, and
(4) a reversible adiabatic compression.

Prior to stage (1) the working substance is brought to thermal equilibrium with a high-temperature reservoir of temperature T_h. During stage (1) the working substance, in thermal contact with a high-temperature reservoir, expands isothermally (at constant temperature). During this expansion it does positive work on an external agent and heat is transferred to it from the high-temperature reservoir. Stage (1) ends and stage (2) begins when the working substance is thermally isolated from the high-temperature reservoir. During stage (2) it continues to expand, this time adiabatically; that is, no heat is transferred to or from the material. During this expansion it does positive work on the external agent and the material cools as its internal energy is converted to work. Its temperature continues to drop until it reaches the temperature of the low-temperature reservoir T_c. At this point the working substance is placed in thermal contact with the low-temperature reservoir; stage (2) ends, and stage (3) begins. During stage (3) the working substance, now in thermal contact with the low-temperature reservoir, is isothermally compressed. During the compression the external agent does positive work on it and the working substance rejects heat to the low-temperature reservoir. When the working substance is thermally isolated from the cold-temperature reservoir, stage (3) ends. During stage (4) the compression continues, this time adiabatically. The external agent does additional positive work on the working substance, which results in its internal energy increasing along with its temperature. The compression continues until its temperature again equals T_h, at which point the engine has completed stage (4), the final stage of the cycle. At this moment the working substance is in the exact state it was in when the cycle began.

The efficiency of a reversible engine (a Carnot engine) can be calculated if an ideal gas is used for the working substance. By Carnot's theorem, this must be the efficiency of every reversible engine and the upper limit possible for any real heat engine operating between the same two temperatures. The efficiency for a reversible heat engine using an ideal gas is calculated to be

Efficiency of a Carnot engine

$$\varepsilon_C = 1 - \frac{T_c}{T_h}$$

where T_c and T_h are absolute temperatures.

Since the efficiency is the same for all reversible heat engines operating in a cycle, it can be used to define the temperature of the two reservoirs. This allows the definition of a temperature scale that is independent of any particular material or thermometer. To use this definition to construct a temperature scale we choose one fixed point, say the triple point of water, and assign it a value. Then the temperature of a reservoir is measured by means of a reversible heat engine operating between the reservoir and a second reservoir maintained at the triple point of water. By measuring the efficiency of this engine we determine the temperature of the reservoir. If the temperature of the triple point of water is given the value 273.16 K, then the absolute temperature and the ideal-gas temperature will agree over the range for which gas thermometers are able to be used.

20-5 Heat Pumps

A heat pump, which is essentially a refrigerator, is used to pump heat from a colder region (for example, outdoors) to a warmer region (for example, the interior of a building). The coefficient of performance COP of a heat pump is Q_c/W, the same as the COP of a refrigerator. The maximum COP theoretically attainable (COP_{max}) is that of a reversible cycle. It can be shown that

Maximum coefficient of performance

$$COP_{max} = \frac{T_c}{\Delta T}$$

where ΔT is the difference $T_h - T_c$. However, because perfectly adiabatic and isothermal processes cannot be carried out and because of frictional losses and the like, COP_{max} cannot be attained in practice.

For a heat pump we are interested in the heat $|Q_h|$ delivered into the high-temperature reservoir, the house. Consequently the parameter of interest is usually not the COP (Q_c/W) but the ratio $|Q_h|/W$, where the work W is the amount of energy that will appear on our electric bill. The parameter of interest for a heat pump can be shown to equal one plus the coefficient of performance:

$$\frac{|Q_h|}{W} = 1 + COP$$

20-6 Irreversibility and Disorder

The second law is related to the fact that physical processes go only in one direction. In all irreversible processes, the direction is such that the system and its surroundings, taken together, tend to a less ordered state. **Entropy** S is a thermodynamic quantity that measures the "disorder" of a system. It is defined as

Entropy change defined

$$\Delta S = \int \frac{dQ_{rev}}{T}$$

where dQ_{rev} is the increment of heat absorbed by the system during a reversible process. To calculate the change in entropy, the integral must be evaluated for any reversible process that brings the system from its initial state to its final state. Changes in entropy, like changes in internal energy, depend only on the initial and final states of the system, and not on the process taking the system from the initial state to the final state. Thus entropy, like internal energy, is a state function.

The total entropy of the universe never decreases. In an irreversible process it always increases, and in a reversible process it remains constant.

In an irreversible process, some energy, an amount equal to the product of the entropy change of the universe ΔS and the temperature T of the coldest available reservoir, becomes unavailable for doing work. This energy, or the "work lost," is

Work lost in an irreversible process

$$W_{\text{lost}} = T \Delta S_{\text{u}}$$

20-7 Entropy   If two or more states of a system have the same energy but different amounts of order, there is a greater probability that the system will be found in the state with the greater disorder. If for some reason the system is in one of the more ordered states, it will tend to spontaneously change states to a state with greater disorder. In an irreversible process the entropy of the universe must increase.

II. Numbers and Key Equations

Numbers

There are no new numbers for this chapter.

Key Equations

Efficiency of a heat engine

$$\varepsilon = \frac{W}{Q_{\text{h}}} = 1 - \frac{|Q_{\text{c}}|}{Q_{\text{h}}}$$

Efficiency of a Carnot cycle

$$\varepsilon_{\text{C}} = 1 - \frac{T_{\text{c}}}{T_{\text{h}}}$$

Coefficient of performance

$$\text{COP} = \frac{Q_{\text{c}}}{W}$$

Maximum coefficient of performance

$$\text{COP}_{\text{max}} = \frac{T_{\text{c}}}{\Delta T}$$

Entropy change defined

$$\Delta S = \int \frac{dQ_{\text{rev}}}{T}$$

Work lost in an irreversible process

$$W_{\text{lost}} = T \Delta S_{\text{u}}$$

III. Potential Pitfalls

The work done by a heat engine during one cycle must be calculated as the work done by the working substance during its expansion plus the work done by it during its compression, not just the work during the expansion; it is the total for the entire cycle. (The work done by the engine during the expansion is positive, whereas during the compression the work is negative.)

Changes in entropy may be calculated only for reversible processes. Entropy is a function of state and a given entropy change is associated with a given change in state, but not with any specific process bringing the system from the initial state to the final state. To calculate the entropy change, we must calculate the entropy change for any reversible process bringing the system from the initial state to the final state.

Note that $\int dQ_{\text{rev}}/T$ does not equal the entropy of the initial or final state in a process but only the difference in the entropy between the two states. It is rather like potential energy in that only differences, or changes, are defined.

The second law does not say that the entropy of some object or system cannot decrease; in fact, it can decrease. But the decrease will always be made up (or more than made up) by an increase elsewhere. The second law states that the total entropy of the universe may not decrease.

In calculating the change in the entropy of a system, you must always use the absolute temperature scale.

IV. True and False and Responses

True and False

_____ 1. Work can never be converted completely into internal energy.

_____ 2. When a wooden block is pushed across a wooden desktop at constant speed, energy is transferred as heat to the block–desktop system.

_____ 3. Work can never be converted completely into heat.

_____ 4. Heat can never be converted completely into work.

_____ 5. The transfer of heat across a finite temperature difference is an irreversible process.

_____ 6. According to the second law of thermodynamics, a heat engine cannot have an efficiency of 100%.

_____ 7. According to the second law of thermodynamics, a refrigerator cannot have a COP of 1 or greater.

_____ 8. All heat engines operating between the same two heat reservoirs have the same efficiency.

_____ 9. All quasi-static processes are reversible.

_____ **10.** All reversible processes are quasi-static.
_____ **11.** In a reversible heat engine, heat must be absorbed or rejected isothermally.
_____ **12.** The entropy of a system depends only upon the state of the system.
_____ **13.** In any adiabatic process, the entropy change of the system is zero.
_____ **14.** The entropy of an isolated system cannot decrease.
_____ **15.** When an irreversible process takes place, the universe becomes more disordered.
_____ **16.** Disorder is more probable than order.

Responses

1. False. In an adiabatic compression no heat is transferred into or out of the system so the change in internal energy of the system is equal to the total work done on the system.

2. False. Energy is transferred as *work* to the block–desktop system. The work done by the pushing agent equals the increase in internal energy of the block–desktop system plus any energy that is transferred to the surroundings (such as acoustic energy). Heat is the energy transferred due to a temperature difference; however, there is no temperature difference involved in this transfer.

3. True. In a reversible, isothermal (constant temperature) compression of an ideal gas, the internal energy of the gas remains constant and the work done on the gas equals the heat transferred out of the gas. To be truly reversible the heat transfer must occur with only an infinitesimal temperature difference. Because no real process is 100% reversible this does not occur.

4. False. In a reversible, isothermal (constant temperature) expansion of an ideal gas, the internal energy of the gas remains constant and the heat transferred to the gas is equal to the total work done by the gas. The second law states that heat can never be completely transformed into work without some other change. In our case, we are not contradicting the second law by saying that the statement "Heat can never be converted completely into work" is false because some other change does occur. That "other change" is the increase in the volume of the gas. For a cyclic process, there is no "other change" and the heat absorbed can not be converted completely into work.

5. True. Heat flows from a region of high temperature to a region of low temperature, and never the other way around.

6. True. For an engine operating at 100% efficiency, no heat is rejected to the lower-temperature reservoir. Thus after an integral number of cycles, there are no other effects than that of extracting heat from the high-temperature reservoir and performing an equivalent amount of work. This contradicts the heat-engine statement of the second law of thermodynamics.

7. False. The COP (COP = Q_c / W) of a refrigerator is typically greater than 1.

8. False. All *reversible* heat engines operating between the same two heat reservoirs have the same efficiency. Actual heat engines are not reversible and have lower efficiencies.

9. False. For example, the quasi-static free expansion of a gas is not a reversible process.

10. True

11. True. Any heat flow through a finite temperature difference is irreversible. Heat never flows from a lower temperature to a higher temperature.

12. True

13. False. $\Delta S = \int dQ_{rev}/T$ for a reversible process, but not all adiabatic processes are reversible. For example, an adiabatic free (unrestrained) expansion of a gas is an irreversible process which results in an increase in entropy.

14. True

15. True

16. True. If all states are equally probable, there are many more disordered states than ordered states. Just look at your desk. Most of the time, unless you continuously straighten it up, it will become more disordered as time goes on.

V. Questions and Answers

Questions

1. When we say "engine," we think of something mechanical with moving parts. In such an engine, friction always reduces the engine's efficiency. Why is this?

2. There are people who try to keep cool on a hot summer day by leaving their refrigerator doors open, but you can't cool your kitchen this way! Why not?

3. Why do engineers designing a steam–electric generating plant always try to design for as high a feed-steam temperature as possible?

4. The conduction of heat across a temperature difference is an irreversible process, but the object that lost heat can always be rewarmed, and the one that gained it can be recooled. The dissipation of mechanical energy, as in the case of an object sliding across a rough table and slowing down, is irreversible, but the object can be cooled and set moving again at its original speed. So in just what sense are these processes "irreversible"?

5. In a slow, steady isothermal expansion of an ideal gas against a piston, the work done is equal to the heat input. Is this consistent with the first law?

6. If a gas expands freely into a larger volume in an insulated container so that no heat is added to the gas, its entropy increases. In view of the definition of ΔS, how can this be?

7. Heat flows from a hotter to a colder body. By how much is the entropy of the universe changed? In what sense does this correspond to energy becoming unavailable for doing work?

8. In discussing the Carnot cycle, we say that extracting heat from a reservoir isothermally does not change the entropy of the universe. In a real process, this is a limiting situation that can never quite be reached. Why not? What is the effect on the entropy of the universe?

Answers

1. The force of kinetic friction always transforms mechanical energy into thermal energy. The engine is rated by how well it transfers mechanical energy to an external agent as work, so if some of the mechanical energy is dissipated via kinetic friction, the engine has less mechanical energy to transfer to the external agent. Thus the efficiency of the engine is reduced.

2. Energy is transferred to the refrigerator's working substance both as heat from the things inside the box and as work done on it by the compressor. The refrigerator then transfers all of this energy as heat to the room air. With the refrigerator door kept open, the compressor has to work even harder. This extra energy is transferred by the refrigerator to the room air as heat, causing the room to grow even hotter.

3. High efficiency is almost always a primary objective for a steam–electric generating plant. In such a plant, the steam is the working fluid for the heat engine that drives the electric generator. The temperature of the steam is the temperature of the high-temperature heat source. The temperature of the low-temperature reservoir is usually fixed by circumstances, such as the temperature of a nearby lake. Thus, increasing the feed-steam temperature is the only way to increase the Carnot efficiency limit for the generator.

4. To say that a process is irreversible means, essen-

tially, that the universe as a whole won't go in the other direction. In the examples given, we can put each "system" back where it started, but not without making a permanent change in its "surroundings." That is, the agents used to return each "system" back to where it started are changed in the process, so the *universe* is not restored to its original state.

5. This is consistent with the first law, which states that the heat absorbed by the gas equals the change in the internal energy of the gas plus the work done by the gas. The internal energy of an ideal gas depends only upon its temperature. Thus, in an isothermal expansion the change in internal energy is zero and the heat absorbed equals the work done.

6. For the adiabatic, free expansion of an ideal gas no work is done by the gas. The first law tells us that the internal energy, and thus the temperature, does not change. Since the expansion is adiabatic, no heat is exchanged by the gas, which might tempt you to think that the change in entropy ($\Delta S = \int dQ_{\text{rev}}/T$) is zero. However, this is incorrect. To calculate the change in entropy, we must consider a *reversible* process that brings the gas from its initial to its final state. Thus, we consider a reversible process that consists of two segments: an reversible adiabatic expansion to the final volume (during which work is done by the gas and so its temperature decreases), followed by a reversible, constant-volume warming where heat is absorbed by the gas until it reaches its final temperature. No heat is absorbed (or released) during the reversible adiabatic expansion. Therefore, during that segment the entropy of the gas does not change. During the reversible, constant-volume warming segment, heat is absorbed by the gas and its entropy increases. Therefore, an adiabatic, free expansion, the entropy of an ideal gas must increase.

7. Let Q be the heat transferred and let T_1 be the higher and T_2 be the lower temperature. The change in entropy of the high-temperature object is $-Q/T_1$ and the change in entropy of the low-temperature object is Q/T_2. Therefore the change in entropy of the two bodies, taken together, is

$$\Delta S = \frac{Q}{T_2} - \frac{Q}{T_1}$$

If this is all that happens, then this is the total change in entropy of the universe. If, instead of just letting the heat flow, we had run a Carnot engine between these two bodies as temperature "reservoirs," the work we could have gotten from it is

$$W = \varepsilon_C Q = (1 - T_2/T_1)Q = T_2 \, \Delta S$$

If, instead of running a Carnot engine, we just let the

heat flow, this work would not have been done and the energy that could have done this work would no longer be available.

8. In any real process, the entropy of the universe will increase. In reality, if everything is at exactly the same temperature, it's all at thermal equilibrium and there can be no heat transfer; there must be some small temperature difference to have a heat transfer. However, if there is a temperature difference, the entropy loss Q/T_h of the warmer object is less than the entropy gain Q/T_c of the cooler object. The net entropy change of everything-put-together is always an increase.

VI. Problems and Solutions

Problems

1. A certain engine absorbs 150 J of heat and rejects 88 J in each cycle. (*a*) What is its efficiency? (*b*) If it runs at 200 cycles/min, what is its power output?

How to Solve It
• In any complete cycle, the heat the engine absorbs equals the work it does plus the heat it rejects.

• The efficiency is the ratio of the work done to the heat absorbed.

• The power equals the work done per unit time.

2. An electric refrigerator removes 13 MJ of heat from its interior for each kilowatt-hour of electric energy used. What is its coefficient of performance?

How to Solve It
• Convert 1 kW·h to SI units (joules). This is the work done on the refrigerator.

• The coefficient of performance is the ratio of the heat removed to the work done.

3. A certain refrigerator has a power rating of 88 W. Consider it to be a reversible refrigerator. If the temperature of the room is 26°C, how long will the refrigerator take to freeze 2.5 kg of water that is put into it at 0°C?

How to Solve It
• The power rating is the work input per unit time.

• Heat is extracted from the interior of the refrigerator at 0°C since it is extracted from freezing water.

• Since the cycle is reversible, the refrigerator is ideal and the heats exhausted to and extracted from the hot and cold reservoirs are proportional to their absolute temperatures.

4. A certain refrigerator requires 35 J of work to re-

move 45 cal of heat from its interior. (*a*) What is its coefficient of performance? (*b*) How much heat is rejected to the surroundings at 22°C? (*c*) If the refrigerator cycle is reversible, what is the temperature inside the refrigerator?

How to Solve It
• The coefficient of performance is the ratio of the heat removed to the work done.

• The heat rejected equals the work done plus the heat extracted from the interior.

• Since the cycle is thermodynamically reversible, the heats exhausted to and extracted from the hot and cold reservoirs are proportional to their absolute temperatures.

5. A not very clever idea for a ship's engine goes as follows: A Carnot engine extracts heat from seawater at 18°C and exhausts it to evaporating dry ice, which the ship carries with it, at −78°C. If the ship's engines are to run at 8000 horsepower, what is the minimum amount of dry ice it must carry for a day's running?

How to Solve It
• Find the efficiency of the engine from the temperatures involved.

• Convert 8000 hp to SI units (watts). How much work do the engines do in one day?

• The heat rejected to the low-temperature reservoir in the process of doing this much work determines the amount of dry ice that must be sublimated. The latent heat of sublimation of dry ice (carbon dioxide) can be found in Table 19-2 on page 570 of the text.

6. A certain electric generating plant produces electrical energy by using steam that enters its turbine at a temperature of 320°C and leaves it at 40°C. Over the course of a year, the plant consumes 4.4×10^{16} J of heat and produces an average electric power output of 600 MW. What is its second-law efficiency? (The second-law efficiency is the ratio of the actual efficiency to the Carnot efficiency.)

How to Solve It
• Calculate the Carnot efficiency from the temperatures involved.

• The actual efficiency of the plant is the ratio of the work done (electrical energy produced) to the heat absorbed. Calculate the second-law efficiency.

7. A certain engine has a second-law efficiency of 85%. In each cycle it absorbs 480 J of heat from a reservoir at 300°C and rejects 300 J of heat to a cold-temperature reservoir. (*a*) What is the temperature of the cold reservoir? (*b*) How much more work could be done by a Carnot engine working between the same two reservoirs and extracting the same 480 J of heat in each cycle?

How to Solve It
- The energy data given allow you to obtain an expression for the actual efficiency. You can then calculate the Carnot efficiency.

- Use the Carnot efficiency to determine the temperature of the low-temperature reservoir.

- Calculate the work that would be done by the Carnot engine. The work done by the actual engine is 85% of this.

8. A Carnot engine removes 1200 J of heat from a high-temperature reservoir and rejects 600 J to the atmosphere at a temperature of 20°C. (*a*) What is the efficiency of this engine? (*b*) What is the temperature of the high-temperature reservoir?

How to Solve It
- Determine the efficiency from the given values of the heat input and the work done.

- This is a Carnot engine. Relate its efficiency to the operating temperatures.

9. When 1 kg of steam condenses at a pressure of 1 atm and a temperature of 100°C, by how much has its entropy increased?

How to Solve It
- The change in the entropy of the water equals the heat it absorbs divided by its temperature.

10. When 1 kg of water is frozen under standard conditions, by how much does its entropy change?

How to Solve It
- "Standard conditions" means that the heat is extracted from the water at 0°C.

- The entropy of the water decreases as heat is extracted from it.

11. A heat engine works in a cycle between reservoirs at 273 K and 490 K. In each cycle the engine absorbs 1250 J of heat from the high-temperature reservoir and does 475 J of work. (*a*) What is its efficiency? (*b*) What is the change in entropy of the universe when the engine goes through one complete cycle? (*c*) How much energy becomes unavailable for doing work when this engine goes through one complete cycle?

How to Solve It
- Calculate the efficiency from the heat absorbed and the work done.

- The entropy change in the universe is the difference between the entropy increase of the low-temperature reservoir and the entropy decrease of the high-temperature reservoir.

- The "lost" work is the additional work that a Carnot engine operating between hot and cold reservoirs of the same temperatures can perform when it absorbs the same amount of heat from the high-temperature reservoir.

12. The interior of a refrigerator's freezing compartment is at 10°F, the kitchen is at 78°F. Suppose that heat leaks through the walls into the freezing compartment at a rate of 70 cal/min. (*a*) In one hour, how much has the entropy of the universe been increased by this heat leakage? (*b*) How much energy becomes unavailable for doing work when this heat leaks into the freezer compartment?

How to Solve It
- Convert the temperatures to the Kelvin scale.

- The entropy change of the freezing compartment increases by an amount equal to the heat delivered to it divided by its absolute temperature.

- In the same way, obtain an expression for the entropy decrease of the kitchen outside the refrigerator.

- The entropy change of the universe is that of the freezing compartment plus that of the outside world.

- Relate the "lost" work to the entropy change and the temperature of the low-temperature reservoir.

13. In a vacuum bottle, 350 g of water and 150 g of ice are initially in equilibrium at 0°C. The bottle is not a perfect insulator. Over time, its contents come to thermal equilibrium with the outside air at 25°C. How much does the entropy of universe increase in this process?

How to Solve It
- The universe consists of the ice and water in the bottle and the outside air that supplies the heat.

- As the ice melts, its entropy increases and the entropy of the air decreases.

- Once the ice has melted, there is $m = 500$ g of water warming from 0°C to 25°C. The increase in the entropy of the water as it warms is $\int dQ/T$, where $dQ = mc\, dT$.

- As the water warms, the entropy of the air decreases further.

14. I have a cup containing 220 g of hot water at 75°C. To cool it I pour 60 g of tap water at 26°C into it. How much does the entropy of the universe change during this cooling?

How to Solve It
- Calculate the final temperature of the 280 g of water.

- The entropy of the 60 g of tap water increases as it warms up. This increase in entropy is $\int dQ/T$, where $dQ = mc\,dT$.

- In the same manner calculate the change in entropy of the hot water as it cools.

- The net entropy change of the universe is the change in entropy of the tap water plus the change in entropy of the hot water.

15. Consider an engine in which the working substance is 1.23 mol of an ideal gas for which γ is 1.41. The engine runs reversibly in the cycle shown on the *PV* diagram in Figure 20-1. The cycle consists of an isobaric (constant pressure) expansion *a* at a pressure of 15 atm, during which the temperature of the gas increases from 300 K to 600 K, followed by an isothermal expansion *b* until its pressure becomes 3 atm. Next is an isobaric compression *c* at a pressure of 3 atm, during which the temperature decreases from 600 K to 300 K, followed by an isothermal compression *d* until its pressure returns to 15 atm. Find the work done by the gas, the heat absorbed by the gas, the internal energy change, and the entropy change of the gas, first for each part of the cycle, and then for the complete cycle.

Figure 20-1

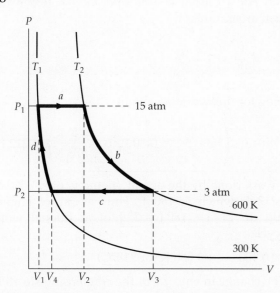

How to Solve It
- The work done by a gas during any quasi-static process is $\int P\,dV$. Thus, the work done by the gas during an isobaric expansion is $P\,\Delta V$.

- The heat absorbed by a gas in an isobaric expansion is $nc'_p\Delta T$. To determine the heat absorbed we must first determine c'_p. For an ideal gas, $c'_p = c'_v + R$. Using the relation $\gamma = c'_p/c'_v$, obtain an expression for c'_p in terms of γ and R.

- For any process, the change in internal energy equals the heat absorbed minus the work done by the gas.

- The change in entropy of the gas during any reversible process is $\int dQ/T$. During an isobaric expansion $dQ = C_p\,dT$.

- To determine the work done by the gas during an isothermal expansion, first we use the ideal-gas law to obtain an expression for the pressure. Thus, the work done is $nRT\int dV/V$.

- During an isothermal expansion of an ideal gas the heat absorbed equals the work done by the gas.

- During the isothermal expansion, the change in entropy of the gas equals the heat absorbed divided by the temperature.

- The same relations hold for the parts of the cycle where the gas is quasi-statically compressed.

Solutions

1. (*a*) The efficiency of the engine is

$$\varepsilon = \frac{W}{Q_h} = \frac{Q_h - |Q_c|}{Q_h} = 1 - \frac{|Q_c|}{Q_h}$$

$$= 1 - \frac{88\,\text{J}}{150\,\text{J}} = \underline{0.413}$$

(*b*) The power output of the engine is the rate at which the engine does work. In 200 cycles the engine performs 200 times the work it does in one cycle:

$$P = \frac{200W}{t} = \frac{200(Q_h - |Q_c|)}{t}$$

$$= \frac{200(150\,\text{J} - 88\,\text{J})}{60\,\text{s}} = \underline{207\,\text{W}}$$

2. $\underline{3.61}$

3. Because the heat is extracted from freezing water, the temperature inside the refrigerator is 0°C. The COP of an ideal refrigerator is

$$\text{COP} = \frac{Q_c}{W} = \frac{T_c}{\Delta T}$$

where the work W equals the power P times the

time t, and the heat extracted Q_c equals the mass m of water times the latent heat of fusion L_f. Thus,

$$\frac{mL_f}{Pt} = \frac{T_c}{\Delta T}$$

so

$$t = \frac{mL_f \, \Delta T}{PT_c} = \frac{(2.5 \text{ kg})(334 \text{ kJ/kg})(26 \text{ K})}{(88 \text{ W})(273 \text{ K})} = \underline{904 \text{ s}}$$

4. (a) $\underline{5.37}$ (b) $\underline{223 \text{ J}}$ (c) $\underline{249 \text{ K}}$ (or $\underline{-24°C}$)

5. First we will convert to SI units. Thus,

$$8000 \text{ hp} = (8000 \text{ hp})(746 \text{ W/hp}) = 5.97 \text{ MW}$$
$$1 \text{ d} = (1 \text{ d})(24 \text{ h/d})(3600 \text{ s/h}) = 8.64 \times 10^4 \text{ s}$$
$$+18°C = 273 \text{ K} + 18 \text{ K} = 291 \text{ K}$$
$$-78°C = 273 \text{ K} - 78 \text{ K} = 195 \text{ K}$$

The efficiency of an Carnot engine is

$$\varepsilon_C = \frac{W}{Q_h} = 1 - \frac{T_c}{T_h}$$

where $W = Pt$ (work equals power times time). Since $Q_h = W + |Q_c|$, we have

$$\frac{W}{W + |Q_c|} = 1 - \frac{T_c}{T_h}$$

The heat absorbed by the dry ice equals the product of the mass m of sublimated dry ice and the latent heat of sublimation L_s, so $|Q_c| = mL_s$. Also, $W = Pt$ (work equals power times time). Substituting these expressions for $|Q_c|$ and W we obtain

$$\frac{Pt}{Pt + mL_s} = 1 - \frac{T_c}{T_h}$$

Solving for the mass we have

$$m = \frac{Pt}{L_s}\left(\frac{T_h}{T_c} - 1\right)^{-1}$$

$$= \frac{(5.97 \times 10^6 \text{ W})(8.64 \times 10^4 \text{ s})}{573 \text{ kJ/kg}}\left(\frac{291 \text{ K}}{195 \text{ K}} - 1\right)^{-1}$$

$$= \underline{1.83 \times 10^6 \text{ kg}}$$

That is about 2000 tons of dry ice for a day's sailing. Well, I said it wasn't a very good scheme.

6. $\underline{0.911}$

7. (a) The actual efficiency ε of the engine is

$$\varepsilon = \frac{W}{Q_h} = \frac{Q_h - |Q_c|}{Q_h} = 1 - \frac{|Q_c|}{Q_h}$$

The efficiency ε_C of a Carnot engine operating between the same temperatures is

$$\varepsilon_C = 1 - \frac{T_c}{T_h}$$

The second law efficiency ε_{sl} is

$$\varepsilon_{sl} = \frac{\varepsilon}{\varepsilon_C}$$

so

$$\varepsilon = \varepsilon_{sl} \, \varepsilon_C$$

$$1 - \frac{|Q_c|}{Q_h} = \varepsilon_{sl}\left(1 - \frac{T_c}{T_h}\right)$$

Solving for T_c we obtain

$$T_c = T_h\left[1 - \frac{1}{\varepsilon_{sl}}\left(1 - \frac{|Q_c|}{Q_h}\right)\right]$$

$$= 573 \text{ K}\left[1 - \frac{1}{0.85}\left(1 - \frac{300 \text{ J}}{480 \text{ J}}\right)\right]$$

$$= \underline{320 \text{ K}} \text{ (or } \underline{47°C})$$

(b) For a Carnot engine operating between the same temperatures

$$\varepsilon_C = \frac{W}{Q_h} = 1 - \frac{T_c}{T_h}$$

Solving for this work we obtain

$$W = Q_h\left(1 - \frac{T_c}{T_h}\right) = (480 \text{ J})\left(1 - \frac{320 \text{ K}}{573 \text{ K}}\right) = 212 \text{ J}$$

The work done by the actual engine is only 85% of the 212 J, which is 180 J. Thus a Carnot engine would do 212 J − 180 J = $\underline{32 \text{ J}}$ of additional work per cycle.

8. (a) $\underline{0.500}$ (b) $\underline{586 \text{ K}}$ (or $\underline{313°C}$)

9. The change in entropy of the condensing steam is

$$\Delta s = \frac{Q}{T} = \frac{mL_v}{T}$$

$$= \frac{(1 \text{ kg})(2257 \text{ kJ/kg})}{373 \text{ K}} = \underline{6.05 \text{ kJ/K}}$$

10. $- \underline{1221 \text{ J/K}}$

11. (a) The actual efficiency of the engine is

$$\varepsilon = \frac{W}{Q_h} = \frac{475 \text{ J}}{1250 \text{ J}} = \underline{0.380}$$

(b) The change in entropy ΔS_u of the universe equals the change in entropy of the high-temperature reservoir plus the change in entropy of the low-temperature reservoir, plus the change in entropy of the engine. The change in entropy of the working substance of the engine is zero because the engine has completed an entire cycle and is thus unchanged. The change in entropy ΔS_h of the high-temperature reservoir is negative because heat is extracted from it. The change in entropy ΔS_c of the low-temperature reservoir is positive.

$$\Delta S_u = \Delta S_h + \Delta S_c$$

$$= -\frac{|Q_h|}{T_h} + \frac{|Q_c|}{T_c} = -\frac{|Q_h|}{T_h} + \frac{|Q_h - W|}{T_c}$$

$$= -\frac{1250 \text{ J}}{490 \text{ K}} + \frac{1250 \text{ J} - 475 \text{ J}}{273 \text{ K}} = \underline{0.288 \text{ J/K}}$$

(c) The "lost work" W_{lost} is the additional work that a Carnot engine operating between the same reservoirs would do after extracting the same amount of heat from the high-temperature reservoir:

$$W_{lost} = W_C - W = \varepsilon_C Q_h - \varepsilon Q_h = (\varepsilon_C - \varepsilon) Q_h$$

$$= \left[\left(1 - \frac{T_c}{T_h} \right) - \varepsilon \right] Q_h$$

$$= \left[\left(1 - \frac{273 \text{ K}}{490 \text{ K}} \right) - 0.380 \right] 1250 \text{ J} = \underline{78.6 \text{ J}}$$

where W_C is the work that the Carnot engine would do and W is the actual work done by the engine.

An alternative method for calculating the "lost work" is to use the result

$$W_{lost} = T_c \Delta S_u = (273 \text{ K})(0.288 \text{ J/K}) = \underline{78.6 \text{ J}}$$

12. (a) 8.6 J/K   (b) 2.24 kJ

13. The total increase in entropy ΔS_{iw} of the ice and water equals its increase ΔS_m during melting at $T_1 = 273$ K plus its increase ΔS_w as it warms to $T_2 = 298$ K.

$$\Delta S_{iw} = \Delta S_m + \Delta S_w = \frac{Q_m}{T_1} + \int \frac{dQ_w}{T}$$

$$= \frac{m_i L_f}{T_1} + \int_{T_1}^{T_2} \frac{mc \, dT}{T} = \frac{m_i L_f}{T_1} + mc \int_{T_1}^{T_2} \frac{dT}{T}$$

$$= \frac{m_i L_f}{T_1} + mc \ln \frac{T_2}{T_1}$$

The change in entropy of the air is

$$\Delta S_{air} = -\frac{|Q|}{T_2} = -\frac{m_i L_f + mc \, \Delta T}{T_2}$$

The change in entropy of the universe is

$$\Delta S_u = \Delta S_{iw} + \Delta S_{air}$$

$$= \frac{m_i L_f}{T_1} + mc \ln \frac{T_2}{T_1} - \frac{m_i L_f + mc \, \Delta T}{T_2}$$

$$= m_i L_f \left(\frac{1}{T_1} - \frac{1}{T_2} \right) - mc \left(\frac{\Delta T}{T_2} - \ln \frac{T_2}{T_1} \right)$$

$$= (0.15 \text{ kg})(334 \text{ kJ/kg}) \left(\frac{1}{273 \text{ K}} - \frac{1}{298 \text{ K}} \right)$$

$$\quad - (0.5 \text{ kg})(4.18 \text{ kJ/kg·K}) \left(\frac{25 \text{ K}}{298 \text{ K}} - \ln \frac{298 \text{ K}}{273 \text{ K}} \right)$$

$$= \underline{23.2 \text{ J/K}}$$

14. 2.21 J/K

15. Starting from state (P_1, V_1, T_1) the gas expands isobarically to state (P_1, V_2, T_2). The work done by the gas is

$$W_a = P_1 \Delta V = P_1(V_2 - V_1) = P_1 V_2 - P_1 V_1$$

Using the ideal-gas law ($PV = nRT$) this can be expressed in terms of the temperatures T_1 and T_2:

$$W_a = nRT_2 - nRT_1 = nR \, \Delta T$$

$$= (1.23 \text{ mol})(8.31 \text{ J/mol·K})(300 \text{ K}) = \underline{3070 \text{ J}}$$

During this expansion a the heat absorbed is $Q_a = nc'_p \Delta T$. To determine Q_a we must first determine c'_p. Using the relations $\gamma = c'_p/c'_v$ and $c'_p = c'_v + R$ we have

$$c'_p = c'_v + R = \frac{c'_p}{\gamma} + R$$

so

$$c'_p = \frac{\gamma R}{\gamma - 1} = \frac{1.41(8.31 \text{ J/mol·K})}{0.41}$$

$$= 28.6 \text{ J/mol·K}$$

and

$$Q_a = nc'_p \, \Delta T = (1.23 \text{ mol})(28.6 \text{ J/mol·K})(300 \text{ K})$$
$$= \underline{10{,}550 \text{ J}}$$

The change in internal energy ΔU_a is

$$\Delta U_a = Q_a - W_a = nc'_p \, \Delta T - nR \, \Delta T$$
$$= n(c'_p - R) \, \Delta T = nc'_v \, \Delta T$$

where

$$c'_v = \frac{c'_p}{\gamma} = \frac{28.6 \text{ J/mol·K}}{1.41} = 20.3 \text{ J/mol·K}$$

so

$$\Delta U_a = nc'_v \, \Delta T$$
$$= (1.23 \text{ mol})(20.3 \text{ J/mol·K})(300 \text{ K}) = \underline{7490 \text{ J}}$$

The entropy change is

$$\Delta S_a = \int \frac{dQ}{T} = nc'_p \int_{T_1}^{T_2} \frac{dT}{T} = nc'_p \ln \frac{T_2}{T_1}$$
$$= (1.23 \text{ mol})(28.6 \text{ J/mol·K}) \ln \frac{600 \text{ K}}{300 \text{ K}}$$
$$= \underline{24.4 \text{ J/K}}$$

The gas next expands isothermally along path b to state (P_2, V_3, T_2). The work done is

$$W_b = \int_{V_2}^{V_3} P \, dV$$

Using the ideal-gas law $(PV = nRT)$ the pressure can be expressed in terms of the volume and the temperature. Thus,

$$W_b = nRT_2 \int_{V_2}^{V_3} \frac{dV}{V} = nRT_2 \ln \frac{V_3}{V_2}$$

Because the temperature is constant on path b, the ratio of the final volume V_3 to the initial volume V_2 equals the ratio of the initial pressure P_1 to the final pressure P_2. Thus,

$$W_b = nRT_2 \ln \frac{P_1}{P_2}$$
$$= (1.23 \text{ mol})(8.31 \text{ J/mol·K})(600 \text{ K}) \ln \frac{15 \text{ atm}}{3 \text{ atm}}$$
$$= \underline{9870 \text{ J}}$$

The internal energy of the gas depends only upon its temperature. Therefore during an isothermal process there is no change in the internal energy. Thus,

$$\Delta U_b = \underline{0}$$

Because there is no change in internal energy, the heat absorbed equals the work done:

$$Q_b = W_b = \underline{9870 \text{ J}}$$

The change in entropy is

$$\Delta S_b = \int \frac{dQ}{T} = \frac{1}{T_2} \int dQ = \frac{Q_b}{T_2}$$
$$= \frac{9870 \text{ J}}{600 \text{ K}} = \underline{16.5 \text{ J/K}}$$

Along path c the gas is compressed isobarically to state (P_2, V_4, T_1). Following the same procedures we used for the isobaric expansion, we obtain

$$W_c = nR \, \Delta T$$
$$= (1.23 \text{ mol})(8.31 \text{ J/mol·K})(-300 \text{ K})$$
$$= \underline{-3070 \text{ J}}$$
$$Q_c = nc'_p \, \Delta T$$
$$= (1.23 \text{ mol})(28.6 \text{ J/mol·K})(-300 \text{ K})$$
$$= \underline{-10{,}600 \text{ J}}$$
$$\Delta U_c = nc'_v \, \Delta T$$
$$= (1.23 \text{ mol})(20.3 \text{ J/mol·K})(-300 \text{ K})$$
$$= \underline{-7490 \text{ J}}$$
$$\Delta S_c = nc'_p \ln \frac{T_1}{T_2}$$
$$= (1.23 \text{ mol})(28.6 \text{ J/mol·K}) \ln \frac{300 \text{ K}}{600 \text{ K}}$$
$$= \underline{-24.4 \text{ J/K}}$$

Along path d the gas is compressed isothermally, returning it to its initial state. Following the same procedures we used for the isothermal expansion, we obtain

$$W_d = nRT_1 \ln \frac{P_1}{P_2}$$
$$= (1.23 \text{ mol})(8.31 \text{ J/mol·K})(300 \text{ K}) \ln \frac{3 \text{ atm}}{15 \text{ atm}}$$
$$= \underline{-4940 \text{ J}}$$
$$\Delta U_d = \underline{0}$$
$$Q_d = W_d = \underline{-4940 \text{ J}}$$
$$\Delta S_d = \frac{Q_d}{T_1} = \frac{-4940 \text{ J}}{300 \text{ K}} = \underline{-16.5 \text{ J/K}}$$

To calculate the net work done by the gas, the net heat absorbed by the gas, the net internal energy change of the gas, and the net entropy change of the gas during a complete cycle, we simply sum the results for each segment of the cycle. Therefore,

$$W_{net} = W_a + W_b + W_c + W_d$$
$$= 3070 \text{ J} + 9870 \text{ J} - 3070 \text{ J} - 4940 \text{J}$$
$$= \underline{4930 \text{ J}}$$

$$Q_{net} = Q_a + Q_b + Q_c + Q_d$$
$$= 10{,}600 \text{ J} + 9870 \text{ J} - 10{,}600 \text{ J} - 4940 \text{ J}$$
$$= \underline{4930 \text{ J}}$$

$$\Delta U_{net} = \Delta U_a + \Delta U_b + \Delta U_c + \Delta U_d$$
$$= 7490 \text{ J} + 0 - 7490 \text{ J} + 0 = \underline{0}$$

$$\Delta S_{net} = \Delta S_a + \Delta S_b + \Delta S_c + \Delta S_d$$
$$= 24.4 \text{ J/K} + 16.5 \text{ J/K} - 24.4 \text{ J/K} - 16.5 \text{ J/K}$$
$$= \underline{0}$$

Chapter 21

Thermal Properties and Processes

I. Key Ideas

21-1 Thermal Expansion When the temperature of a solid or liquid increases, it usually expands. When a bar of length L is heated, raising its temperature by ΔT, the length of the bar changes by ΔL. The fractional change in the length is $\Delta L/L$. The ratio of the fractional change in the length of an object to the change in temperature is called the **coefficient of linear expansion** α:

Coefficient of linear expansion

$$\alpha = \frac{\Delta L/L}{\Delta T}$$

Similarly, the **coefficient of volume expansion** β is defined as the ratio of the fractional change in volume to the change in temperature (at constant pressure):

Coefficient of volume expansion

$$\beta = \frac{\Delta V/V}{\Delta T}$$

For a given material, the relation between the coefficient of volume expansion and the coefficient of linear expansion is $\beta = 3\alpha$.

21-2 The van der Waals Equation and Liquid–Vapor Isotherms The van der Waals equation of state describes the behavior of gases over a wide range of pressures more accurately than does the ideal-gas equation of state ($PV = nRT$). The van der Waals equation of state for n moles of a gas is

The van der Waals equation of state

$$\left(P + \frac{an^2}{V^2}\right)(V - bn) = nRT$$

The constants a and b provide for the attractive forces between molecules and the finite size of the molecules, respectively.

The pressure at which a liquid coexists in equilibrium with its own vapor is called the **vapor pressure**. The temperature at which the vapor pressure equals one atmosphere is called the **normal boiling point**.

21-3 Phase Diagrams The equilibrium state of a substance held at constant volume depends on its temperature and pressure. A plot of pressure versus temperature for such a substance is called a **phase diagram** where the state of a substance can be represented by a point. On a phase diagram, one region represents states of solid phase, another region represents states of liquid phase, and yet another region represents states of gaseous phase.

The point on a phase diagram where the liquid and vapor phases have the same density is called the **critical point**, and the temperature at this point is called the critical-point temperature T_c. At this point and above it, there is no distinction between the liquid and gas phases. Every substance has a unique triple point, the point at which the vapor, liquid, and solid phases can coexist in equilibrium. The triple-point temperature for water is 273.16 K = 0.01°C and the triple-point pressure is 4.58 mmHg.

21-4 The Transfer of Thermal Energy Temperature differences result in the transfer of energy from one place to another by three processes: conduction, convection, and radiation. In **conduction**, heat is transferred by interactions among atoms or molecules, though there is no transport of the atoms or molecules themselves. In **convection**, heat is transferred via mass transport. This occurs, for example, in a flowing fluid. In **radiation**, heat is transported via electromagnetic radiation that is emitted and

239

absorbed. In all three processes the net flow of heat (that is, the transfer of energy) is from a region of higher temperature to a region of lower temperature.

When heat flows by conduction along a solid bar, the rate at which heat flows is called the **thermal current** I, measured in units such as joules per second. The thermal current is related to the temperature gradient $\Delta T/\Delta x$ by

Thermal conduction

$$I = \frac{\Delta Q}{\Delta t} = kA\frac{\Delta T}{\Delta x}$$

where ΔQ is the net flow of heat during time ΔT, A is the cross-sectional area of the bar, and the constant k is the **coefficient of thermal conductivity** or just the **thermal conductivity** of the substance. Solving this equation for the temperature difference ΔT we obtain

$$\Delta T = I\frac{\Delta x}{kA} = IR$$

where R is the **thermal resistance** of the material of thickness Δx:

Thermal resistance

$$R = \frac{\Delta x}{kA}$$

The equation $\Delta T = IR$ is analogous to the equation for the viscous flow of a fluid through a pipe, $\Delta P = I_v R$, except that I stands for the flow of heat rather than the volume flow rate of a fluid, R is the thermal resistance, and ΔT has replaced the pressure difference ΔP.

If heat flows from a warmer region to a cooler region through a series of slabs, the equivalent thermal resistance R_{eq} of the series is the sum of the thermal resistances of the individual slabs:

Resistances in series

$$R_{eq} = R_1 + R_2 + \cdots$$

Similarly, if heat is conducted from a warmer region to a cooler region through two or more parallel paths, the reciprocal of the equivalent thermal resistance of the paths is the sum of the reciprocals of the thermal resistances of the individual paths. That is,

Resistances in parallel

$$\frac{1}{R_{eq}} = \frac{1}{R_1} + \frac{1}{R_2} + \cdots$$

The heat transferred to or from a body by convection is approximately proportional to the surface area of the body and to the difference in temperature between the body and the surrounding fluid.

The rate at which a surface radiates heat is given by the **Stefan–Boltzmann law**:

Stefan–Boltzmann law

$$I = e\sigma AT^4$$

where e is the **emissivity** of the surface ($0 \le e \le 1$), A is the area of the surface, T is the surface temperature in kelvins, and σ is Stefan's constant:

Stefan's constant

$$\sigma = 5.6703 \times 10^{-8}\ \text{W/m}^2\cdot\text{K}^4$$

The rate I_a at which a body absorbs radiant heat from its surroundings is

Absorption of radiation

$$I_a = e\sigma AT_0^4$$

where T_0 is the temperature of its surroundings. Therefore the net radiative thermal current I_{net} between a body and its surrounding is

Net radiative thermal current

$$I_{net} = I - I_a = e\sigma A(T^4 - T_0^4)$$

Note that a good emitter (an object having a high emissivity e) is also a good absorber.

An object that absorbs all the radiation incident upon it has an emissivity equal to 1 and is called a **blackbody**. A blackbody is also an ideal radiator. When a body emits thermal radiant energy it does so over a continuum of wavelengths. The wavelength λ_{max} at which a blackbody emits radiant energy at the greatest rate is inversely proportional to the absolute temperature of the body. This result is known as Wien's displacement law:

Wien's displacement law

$$\lambda_{max} = \frac{2.898\ \text{mm}\cdot\text{K}}{T}$$

All three mechanisms of heat flow are driven by a difference in temperature. Independent of which of the mechanisms are at work, the rate of cooling of a warm body is approximately proportional to the temperature difference between the body and its surroundings. This result, called **Newton's law of** cooling, holds whether heat is being transferred by conduction, convection, radiation, or some combination of the three. It is established by applying the differential approximation to the various heat transfer mechanisms and is most accurate when the temperature differences are small.

II. Numbers and Key Equations

Numbers

Stefan's constant

$$\sigma = 5.6703 \times 10^{-8} \text{ W/m}^2 \cdot \text{K}^4$$

Key Equations

Coefficient of linear expansion

$$\alpha = \frac{\Delta L/L}{\Delta T}$$

Coefficient of volume expansion

$$\beta = \frac{\Delta V/V}{\Delta T}$$

The van der Waals equation of state

$$\left(P + \frac{an^2}{V^2}\right)(V - bn) = nRT$$

Thermal conduction

$$I = \frac{\Delta Q}{\Delta t} = kA\frac{\Delta T}{\Delta x}$$

Thermal resistance

$$R = \frac{\Delta x}{kA}$$

so $\quad \Delta T = I\dfrac{\Delta x}{kA} = IR$

Resistances in series

$$R_{eq} = R_1 + R_2 + \cdots$$

Resistances in parallel

$$\frac{1}{R_{eq}} = \frac{1}{R_1} + \frac{1}{R_2} + \cdots$$

Stefan–Boltzmann law

$$I = e\sigma AT^4$$

Absorption of radiation

$$I_a = e\sigma AT_0^4$$

Net radiative thermal current

$$I_{net} = I - I_a = e\sigma A(T^4 - T_0^4)$$

Wien's displacement law

$$\lambda_{max} = \frac{2.898 \text{ mm} \cdot \text{K}}{T}$$

III. Potential Pitfalls

In problems involving thermal expansion, the units of temperature must be consistent with those of α (or β). On the other hand, any units may be used for L and ΔL (or V and ΔV) as long as they are the same.

There are three and only three mechanisms for the transfer of energy via heat. They are conduction, convection, and radiation. Remember, by definition, heat is the transfer of energy due to a difference in temperature.

In conduction and convection problems, only temperature *differences* matter, so Celsius degrees and kelvins may be used interchangeably. In radiation problems this is not true; you must use absolute temperatures. Remember, heat transfers are always from warmer regions to cooler regions.

IV. True and False and Responses

True and False

_____1. The van der Waals equation is a more accurate equation of state for real gases than the ideal-gas equation.
_____2. The constant *a* that appears in the van der Waals equation is related to the volume of the molecules themselves.

Responses

1. True.

2. False. The constant *b* is related to the volume of the molecules. The constant *a* is related to the weak attractive forces between the gas molecules.

V. Questions and Answers

Questions

1. A metal plate with a circular hole drilled through it is uniformly heated. As the plate gets hotter, it expands. Does the hole get bigger or smaller?

2. The ordinary thermometers we see every day are mostly alcohol in glass. How would the use of such a thermometer be affected if alcohol and glass had the same thermal coefficient of volume expansion?

3. If an ordinary (liquid-in-glass) thermometer is placed in something quite hot, the liquid column may actually drop a little before it starts to rise. What is going on here?

4. When you are trying to open a glass jar of food with a stuck lid, it often helps to run hot water over the metal lid for a little while. Why?

5. Under what conditions does the van der Waals equation of state reduce to the ideal-gas law?

6. Vessels like vacuum bottles or dewar flasks, which are designed to keep fluids very cold, are made with double-glass walls. The inner surfaces are silvered and there is a vacuum between the walls. Discuss how this design minimizes heat losses.

7. Materials used commercially for building insulation tend to have relatively little mass and occupy a lot of space: they are foamy, porous materials or masses of compacted fibers or some such. Why?

Answers

1. The hole enlarges with the plate. Every linear dimension of the metal plate undergoes the same fractional increase in length, including the diameter of the hole.

2. This would make them useless, of course. The reason the top of the fluid column rises against the scale marked on the glass is that the volume of the fluid expands more (for a given temperature change) than does the volume of the glass.

3. If this happens, it is because the glass envelope holding the liquid gets hot first and expands a little before the temperature of the liquid inside has had time to rise. Then, as the glass, the liquid, and the surroundings all come to thermal equilibrium, the liquid rises with respect to the glass tube.

4. You may just be dissolving away sticky guck that is gluing the lid to the jar. More relevant to our discussion here is the fact that the metal lid will expand more as you increase its temperature than the glass jar, so the lid loosens.

5. The van der Waals equation of state reduces to the ideal-gas law when the volume per mole is very large and thus the density of the gas is very low.

6. The vacuum between the double-glass walls prevents heat transfer through the walls by conduction and convection. Silvering the inner surfaces of the double walls cuts down on heat loss by radiation by

reducing their emissivity. The double walls must join at the neck, and there will be some heat loss by conduction there. This is reduced by using glass that is a relatively poor conductor.

7. The molecules of a gas spend most of their time between collisions. Thus, gases are good insulators with regard to the transfer of heat by conduction. They are poor insulators with regard to convection, however. By using materials that prevent convection currents from building up—materials that confine the gas in numerous small pockets—convective transfers of heat are all but eliminated. Thus, products used for insulation are light because they consist of numerous pockets of trapped gas. They occupy lots of space in order to minimize the temperature gradient, which in turn minimizes the thermal current.

VI. Problems and Solutions

Problems

1. A brass pin is exactly 5 cm long when it is at a temperature of 140°C. What is its length when it cools to 20°C?

How to Solve It
• Look up the coefficient of thermal expansion for brass in Table 21-1 on page 634 of the text.

• The coefficient of thermal expansion is the fractional change in length per kelvin. Use this relation to determine the new length.

2. A surveyor's steel tape is manufactured to be accurate at 20°C. On a day when the temperature is 38°C it is used to mark off a 100-m-long track. If the length as measured with this tape is 100 m, what is the error in the length of the track due to the thermal expansion of the steel tape?

How to Solve It
• Look up the coefficient of thermal expansion for steel in Table 21-1 on page 634 of the text.

• The coefficient of thermal expansion is the fractional change in length per kelvin. Consider a section of the tape that has a length of 100 m at 20°C. Determine the change in the length of this section when the tape is heated from 20°C to 38°C.

3. A 10-L Pyrex flask is filled to the brim with acetone at 12.4°C. To what temperature must the flask and its contents be heated, or cooled, so that 85 cm^3 of acetone overflow the flask?

How to Solve It
• Look up the coefficients of thermal expansion for Pyrex glass and for acetone in Table 21-1 on page 634 of the text.

- Obtain an expression for the change in the volume of the flask in terms of the temperature change. Remember, for a solid the relation between the volume coefficient of thermal expansion β and the linear coefficient of thermal expansion α is $\beta = 3\alpha$.

- Obtain an expression for the change in the volume of the acetone in terms of the temperature change.

- These two changes in volume must differ by 85 cm^3. Solve for the temperature at which this condition is satisfied.

4. My car has a 40-L gasoline tank. If I fill it completely full of gasoline at a temperature of 12°C and then let the car sit in the sun until the temperatures of the gasoline and the tank reach 30°C, how much gasoline spills out of the tank? The thermal coefficient of volume expansion for gasoline is 9×10^{-4} K^{-1} and the tank is made of steel.

How to Solve It
- See the solution for Problem 3.

5. One mole of steam is confined in a 30-L container at 600°C. (*a*) Calculate the pressure using the ideal-gas equation. (*b*) The van der Waals constants for steam are $a = 5.43$ atm·L^2/mol^2 and $b = 0.03$ L/mol. Calculate the pressure using the van der Waals equation. (*c*) Actual measurements show the pressure to be 2.39 atm. Compare the pressures calculated using the ideal-gas equation and the van der Waals equation with the measured pressure.

How to Solve It
- Calculate the pressure using the ideal-gas equation.

- Repeat the calculations using the van der Waals equation.

- Calculate the percentage difference between both calculated pressures and the measured pressure.

6. Fifty moles of steam are confined in a 30-L container at 600°C. (*a*) Calculate the pressure using the ideal-gas equation. (*b*) The van der Waals constants for steam are $a = 5.43$ atm·L^2/mol^2 and $b = 0.03$ L/mol. Calculate the pressure using the van der Waals equation. (*c*) Actual measurements show the pressure to be 113 atm. Compare the pressures calculated using the ideal-gas equation and the van der Waals equation with the measured pressure.

How to Solve It
- See the solution for Problem 3.

- Use the van der Waals equation to calculate the pressure.

- Repeat the calculation using the ideal-gas law.

- Take the ratio of the calculated pressures and take the ratio of the corresponding pressures from the solution of Problem 5.

7. Figure 21-1 shows a system holding liquid helium at 4.2 K (the boiling point of helium). A cylindrical can 5 cm in diameter and 7 cm high is supported by two stainless steel pins from the walls, which are at 77 K (the boiling point of nitrogen). The space between the can and the walls is evacuated. The steel pins are each 1 mm in diameter and 6 cm long and have a thermal conductivity of 13.4 W/m·K. If the latent heat of vaporization of helium is 21 kJ/kg, at what rate does the helium boil off? (Assume that the emissivity of the can's exterior is 0.25.)

Figure 21-1

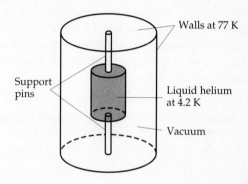

How to Solve It
- Obtain an expression for the thermal currents in the two pins. Also, obtain an expression for the radiative thermal current through the evacuated space.

- The net thermal current from the 77-K walls to the helium equals the rate at which the liquid helium absorbs heat. The absorption of heat results in vaporization of the helium. Equate an expression for the rate at which the helium absorbs heat with the net thermal current. Solve for the rate at which the helium vaporizes.

8. A vacuum bottle with an inside diameter of 6 cm contains 150 g of water and 75 g of ice at 0°C (see Figure 21-2). Heat leakage through the walls of the bottle

Figure 21-2

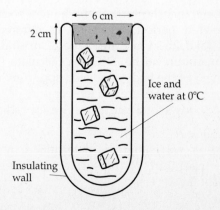

is negligible, but it is closed with a cork stopper 2 cm thick whose thermal conductivity is 0.5 W/m·K. If the surroundings are at 28°C, how long does it take for all the ice to melt?

How to Solve It
• Equate an expression for the product of the time and the thermal current through the cork stopper with an expression for the heat absorbed by the melting ice.

• Solve for the time at which all the ice is melted.

9. A home has a window area of 263 ft² of single-pane glass 0.135 in. thick. (*a*) What is the rate of conductive heat loss through the glass when the temperature inside is 72°F and the temperature outside is 10°F? (*b*) For comparison, calculate the rate of conductive heat loss that would occur if the entire temperature difference of 62 F° were applied across the glass. (Use 5.6 Btu·in/h·ft²·F° as the thermal conductivity of the glass.)

How to Solve It
• Use Table 21-1 on page 644 of the text to find the R factor for glass (note that this is independent of the thickness of the glass). The thermal resistance equals the R factor divided by the area.

• The thermal current is the temperature difference divided by the thermal resistance.

• Calculate the thermal current assuming that the entire temperature difference occurs across the glass.

10. The walls of a certain house consist of 160 m² of brick, 10 cm thick, with a thermal conductivity of 0.8 W/m·K. At what rate is the heat lost by conduction through the walls if the temperature of the inside surface of the wall is 16°C and the outside temperature is 5°C?

How to Solve It
• Calculate the thermal current through the wall. The thermal current equals the thermal conductivity times the product of the temperature gradient and the thickness.

11. Assume that the surface of a human body can be considered a perfect blackbody at infrared wavelengths. Take the surface area to be 2.2 m² and the surface temperature to be 33°C. (*a*) Calculate the peak wavelength of the body's radiated spectrum. (*b*) If the body's surroundings are at a temperature of 22°C, calculate the body's total rate of heat loss by radiation.

How to Solve It
• The peak wavelength is determined by the absolute temperature of the surface. Use Wien's displacement law.

• Obtain an expression for the net radiative thermal current and solve for this current.

• For comparison, the total thermal current of a clothed resting person in an environment at room temperature is of the order of 100 W.

12. In a certain experiment, heat is transferred from a source at 227°C to water at 30°C through a rod 3 cm in diameter. (See Figure 21-3.) The rod consists of a 12-cm-long aluminum rod butted end to end with a 5-cm-long copper rod. What is the temperature of the aluminum–copper junction if the aluminum rod is in thermal contact with the 227°C heat source and the copper rod is in contact with the 30°C water bath? (The thermal conductivities can be found in Figure 21-8 on page 641 of the text.)

Figure 21-3

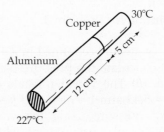

How to Solve It
• Obtain expressions relating the thermal current to the conductivities, cross-sectional areas, and temperature gradients of each rod.

• Equate the two expressions for the thermal current and solve for the temperature at the junction.

Solutions

1. The coefficient of linear expansion α is the ratio of the fractional change in length to the change in temperature. Thus,

$$\alpha = \frac{\Delta L/L}{\Delta T}$$

Therefore

$$L' = L + \Delta L$$
$$= L + \alpha L\,\Delta T = L(1 + \alpha\,\Delta T)$$
$$= (0.05\text{ m})[1 + (19 \times 10^{-6}\text{ K}^{-1})(-120\text{ K})]$$
$$= \underline{0.0499\text{ m}}\text{ (or }\underline{4.99\text{ cm}})$$

2. The track is about <u>2 cm longer than 100 m</u>, a 0.02% error.

3. The coefficient of volume expansion β_P for Pyrex is the ratio of the fractional change in volume of the flask to its change in temperature. Thus,

$$\beta_P = 3\alpha_P = \frac{\Delta V_f/V_0}{\Delta T}$$

so

$$\Delta V_f = 3\alpha_P V_0 \Delta T$$

With the same change in temperature, the volume of the acetone changes by

$$\Delta V_a = \beta_a V_0 \Delta T$$

To determine the change in temperature for which 85 cm³ overflows, we equate the difference between the changes in volume to the overflow volume $V_{of} = 85$ cm³:

$$\beta_a V_0 \Delta T - 3\alpha_P V_0 \Delta T = V_{of}$$

or

$$\Delta T = \frac{V_{of}/V_0}{(\beta_a - 3\alpha_P)}$$

$$= \frac{(85 \text{ cm}^3)(10^{-3} \text{ L/cm}^3)/(10 \text{ L})}{(1.5 \times 10^{-3} \text{ K}^{-1}) - 3(3.2 \times 10^{-6} \text{ K}^{-1})}$$

$$= +5.7 \text{ K}$$

The initial temperature was 12.4°C. Because the Celsius degree and the kelvin are the same size, an increase of 5.7 K is the same as an increase of 5.7 C°. Thus, the final temperature T_{final} is

$$T_{final} = T_0 + \Delta T = 12.4°C + 5.7 \text{ C}° = \underline{18.1°C}$$

4. 0.624 L

5. (*a*) Using the ideal-gas law, we obtain

$$P_1 = \frac{nRT}{V}$$

$$= \frac{(1 \text{ mol})(0.0821 \text{ L·atm/mol·K})(873 \text{ K})}{30 \text{ L}}$$

$$= \underline{2.39 \text{ atm}}$$

(*b*) The van der Waals equation of state is

$$\left(P + \frac{an^2}{V^2}\right)(V - bn) = nRT$$

Solving for the pressure P at 600°C gives

$$P_2 = \frac{nRT}{V - bn} - \frac{an^2}{V^2}$$

$$= \frac{(1 \text{ mol})(0.0821 \text{ L·atm/mol·K})(873 \text{ K})}{30 \text{ L} - (0.03 \text{ L/mol})(1 \text{ mol})}$$
$$- \frac{(5.43 \text{ atm·L}^2/\text{mol}^2)(1 \text{ mol})^2}{(30 \text{ L})^2}$$

$$= \underline{2.39 \text{ atm}}$$

(*c*) The actual pressure is also 2.39 atm. Therefore, both calculated pressures agree with the measured pressure to better than one part in 239. That is, they differ from the measured pressure by <u>less than 0.4 percent</u>.

6. (*a*) <u>119 atm</u>   (*b*) <u>111 atm</u>
(*c*) The ideal-gas pressure is about <u>5%</u> higher than the measured pressure. The van der Waals pressure is about <u>2%</u> lower than the measured pressure.
     The pressures calculated using the van der Waals equation are closer to experimentally measured values than are the pressures calculated using the ideal-gas law. However, the results from both of these equations are in close agreement with the measured pressure when the density of the gas is low. This is evidenced by the results from Problem 5, where both calculated pressures differed from the measured pressure by less than 0.4%.
     The constant a that appears in the van der Waals equation is a measure of the forces exerted by the gas molecules on each other when the molecules are between collisions. Water molecules are electrically polarized. Therefore the forces they exert on each other are large in comparison with the forces exerted by many other molecules. The constant a is least for the inert gases such as helium, neon, and argon.

7. The heat that evaporates the helium is delivered to it through the support pins and by radiation from the 77-K walls. Thus the rate at which the helium must absorb heat is equal to the conductive thermal currents through the pins plus the radiative thermal current:

$$\frac{dQ_{gained}}{dt} = \frac{dQ_{lost}}{dt}$$

$$\frac{d(mL_v)}{dt} = I_{conductive} + I_{radiative}$$

or

$$\frac{dm}{dt} = \frac{I_{conductive} + I_{radiative}}{L_v} \tag{1}$$

where

$$I_{\text{conductive}} = 2kA_{\text{pin}}\frac{\Delta T}{\Delta x}$$

$$= 2(13.4 \text{ W/m·K})(7.85 \times 10^{-7} \text{ m}^2)\frac{77 \text{ K} - 4.2 \text{ K}}{0.06 \text{ m}}$$

$$= 0.0255 \text{ W}$$

The 2 appears in the expression for $I_{\text{conductive}}$ because the heat is conducted through two pins, and $A_{\text{pin}} = \pi r_{\text{pin}}^2 = \pi(0.0005 \text{ m})^2 = 7.85 \times 10^{-7} \text{ m}^2$.

$$I_{\text{radiative}} = e\sigma A_{\text{can}}(T^4 - T_0^4)$$

$$= (0.25)(5.67 \times 10^{-8} \text{ W/m}^2\text{·K})(0.0149 \text{ m}^2)$$
$$\times [(77 \text{ K})^4 - (4.2 \text{ K})^4]$$

$$= 0.00743 \text{ W}$$

In this expression A_{can} denotes the surface area of the can of helium, given by

$$A_{\text{can}} = 2(\pi r_{\text{can}}^2) + 2\pi r_{\text{can}}h_{\text{can}}$$

$$= 2\pi(0.025 \text{ m})^2 + 2\pi(0.025 \text{ m})(0.07 \text{ m})$$
$$= 0.0149 \text{ m}^2$$

where πr_{can}^2 is the area of the can's top (or bottom) and $2\pi r_{\text{can}}h_{\text{can}}$ is the area of its side. Substituting into Equation (1) we obtain

$$\frac{dm}{dt} = \frac{0.0255 \text{ W} + 0.00743 \text{ W}}{2.1 \times 10^4 \text{ J/kg}} = 1.57 \times 10^{-6} \text{ kg/s}$$

8. 3.50 h

9. (a) The thermal resistance of the windows is $R = R_f/A$. Thus, the thermal current through the windows is

$$I = \frac{\Delta T}{R} = \frac{A\,\Delta T}{R_f} = \frac{(263 \text{ ft}^2)(72°\text{F} - 10°\text{F})}{0.9 \text{ h·ft}^2\text{·F°/Btu}}$$

$$= 1.81 \times 10^4 \text{ Btu/h}$$

Note that we used a value for the R factor that is independent of the thickness of glass. This is because the thermal resistance of a pane of glass is domi-

nated by the resistances of the layers of air on either side of the window. The effective resistance of the window is the sum of the resistances of the glass and the two layers of air.

(b) In part (a) we used a value for the R factor that takes into account the temperature gradient in the air near the window. If the entire temperature gradient were across the glass, the thermal current would be

$$I = \frac{kA\,\Delta T}{\Delta x}$$

$$= \frac{(5.6 \text{ Btu·in/h·ft}^2\text{·F°})(263 \text{ ft}^2)(72°\text{F} - 10°\text{F})}{0.135 \text{ in}}$$

$$= 6.76 \times 10^5 \text{ Btu/h}$$

10. 14.1 kW if all the temperature gradient is in the brick. The result of Problem 9 makes us wonder if this is an overestimate, however.

11. (a) For a body with a surface temperature of 33°C = 306 K the peak radiation wavelength is given by Wien's displacement law

$$\lambda_{\text{max}} = \frac{2.898 \text{ mm·K}}{T} = \frac{2.90 \times 10^{-3} \text{ m·K}}{306 \text{ K}}$$

$$= 9.48 \times 10^{-6} \text{ m} = 9.48 \text{ μm}$$

(b) Taking the emissivity of the body's surface to be 1, we have

$$I_{\text{rad}} = e\sigma A(T^4 - T_0^4)$$

$$= (1)(5.67 \times 10^{-8} \text{ W/m}^2\text{·K}^4)(2.2 \text{ m}^2)$$
$$\times [(306 \text{ K})^4 - (295 \text{ K})^4]$$

$$= 149 \text{ W}$$

This result, which estimates only radiative heat-loss rate, is larger than the body's entire resting heat-loss rate, so some of the assumptions made must be incorrect. In fact, the emissivity of the human body is not quite 1, and it is not ordinarily considered socially acceptable to expose our entire surface area directly to the surroundings.

12. 70°C